PRIMERS FOR PROTEOMICS

PRIMERS FOR PROTEOMICS

Editors

Paulina Czaplewska
University of Gdańsk, Poland
Medical University of Gdańsk, Poland

Katarzyna Macur
University of Gdańsk, Poland
Medical University of Gdańsk, Poland

Paweł Ciborowski
University of Nebraska Medical Center, USA

World Scientific

NEW JERSEY • LONDON • SINGAPORE • BEIJING • SHANGHAI • HONG KONG • TAIPEI • CHENNAI • TOKYO

Published by

World Scientific Publishing Co. Pte. Ltd.
5 Toh Tuck Link, Singapore 596224
USA office: 27 Warren Street, Suite 401-402, Hackensack, NJ 07601
UK office: 57 Shelton Street, Covent Garden, London WC2H 9HE

British Library Cataloguing-in-Publication Data
A catalogue record for this book is available from the British Library.

PRIMERS FOR PROTEOMICS

ISBN 978-981-12-8336-9 (hardcover)
ISBN 978-981-12-8337-6 (ebook for institutions)
ISBN 978-981-12-8338-3 (ebook for individuals)

For any available supplementary material, please visit
https://www.worldscientific.com/worldscibooks/10.1142/13595#t=suppl

Typeset by Stallion Press
Email: enquiries@stallionpress.com

https://doi.org/10.1142/9789811283376_fmatter

Preface

This book contains examples written by internationally recognized experts applying proteomics and analysis applications of mass spectrometry in their everyday research. They dedicate this book to a broad audience of all scientists to encourage them to use the omics approach to investigate biological processes. In the first chapter, we offer a short introduction to the basics of analytical chemistry. The following chapters are dedicated to how much high-throughput proteomic studies were applied in various areas of biological and biomedical sciences providing valuable insights into how biological systems work at the molecular level. We mostly keep in mind young scientists, undergraduate, and graduate students. We hope that those who decide to read will find inspiration for their future scientific endeavors.

https://doi.org/10.1142/9789811283376_fmatter

Contents

https://doi.org/10.1142/9789811283376_0001

Chapter 1

Analytical Primer

Katarzyna Pawlak[1,*] and Pawel Ciborowski[2]

[1]*Faculty of Chemistry, Warsaw University of Technology, Warsaw, Poland*
[2]*Department of Pharmacology and Experimental Neuroscience, University of Nebraska Medical Center, Omaha, NE, USA*

Contents

*Corresponding author.
Katarzyna.pawlak@pw.edu.pl

Abstract

This chapter presents the basic parameters of liquid chromatography (efficiency and selectivity) that affect chromatographic resolution. It will be explained how the elements of the chromatographic set affect the width of the chromatographic peaks and why. In the second part of the chapter, we will discuss what interactions are responsible for the ability to separate and distinguish substances in mixtures and how the stationary phase's chemical structure and the mobile phase's composition can influence them.

1. Principles of liquid chromatography

Liquid chromatography (LC) is a fast technique for separating substances present in synthetic or natural origin samples. It allows the separation of the desired group of substances from the matrix components to enrich these compounds so that the detector connected with the system can detect compounds whose amount is from 10^6 to 10^{12} times lower than the main components.

The chromatographic process requires the use of two phases: stationary and mobile. The stationary phase is often a solid adsorbent on the surface of which there are compounds with specific properties, for example, hydrophobic aliphatic chains. The mobile phase is a mixture of solvents with opposite properties: in this case, hydrophilic (e.g., a mixture of water and methanol). The analyte can distribute itself in a certain ratio between two phases, and the ratio represents the similarity of the analyte to both phases ($K = C_{SP}/C_{MP}$, where C_{SP} and C_{MP} is the concentration of a compound in the stationary and mobile phase, respectively).

Suppose the compound shows no affinity to the stationary phase. In that case, it remains in the mobile phase solution moving at a constant velocity — it arrives first in the detector after time-related to the specific volume of connecting tubes, chromatographic pump, and column — referred to as t_0. On the other hand, if a part of the molecule is chemically like the stationary phase components, the compound is slowed down — retained — proportionally to the degree of interactions (retention time, t_r). The retention time and retention

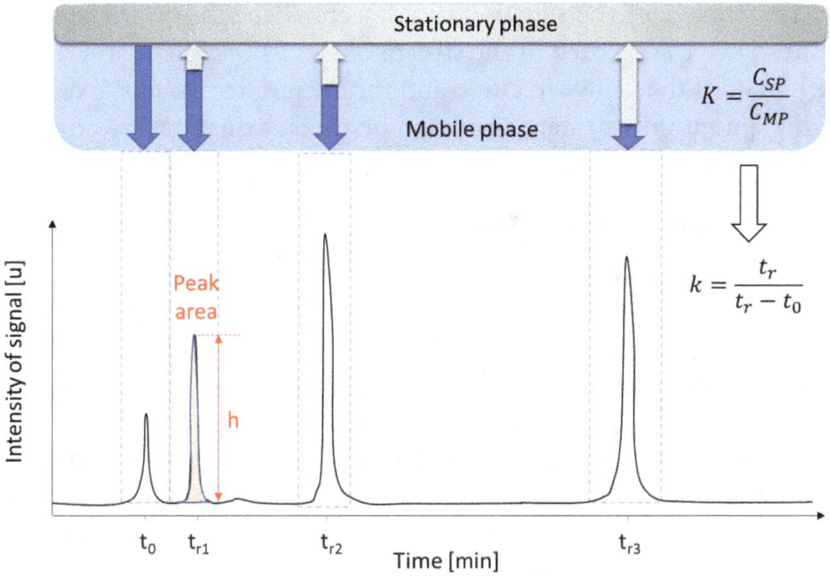

Figure 1. Dependence of the retention time on the degree of affinity of compounds to the stationary and mobile phases. Main chromatographic parameters: t_r — retention time, h — peak's height, and peak area. Arrows reflect the affinity of compounds to stationary (gray) and mobile (blue) phases.

coefficient "k" (equation presented in Figure 1, which should always be higher than 0) is, therefore, a characteristic parameter reflecting the physical or chemical properties of the compound and can be used in qualitative analysis.

The presence of the compound in the detector is responsible for a change in signal intensity with time relative to the mobile phase solution. The signal obtained for the mobile phase composition without the tested compounds is called the baseline. In contrast, the change of the signal due to the flow of the compound through the detector cell recreates a Gaussian curve reflecting the distribution of analyte molecules (profile of laminar flow) in the stream of the mobile phase. The obtained diagram of the change in the detector's signal during the chromatographic process is called the chromatogram (Figure 1).

The change of the signal in the detector depends on the amount of substance transported in the stream of the mobile phase; therefore, the height of the Gaussian curve and the area under the curve depend on the amount of substance and can be used in quantitative analysis.

2. Chromatographic resolution

One of the most critical parameters of a chromatographic method is the resolution describing the yield with which the compounds were purified, that is, the extent to which the chromatographic peaks were separated. From the chromatogram (Figure 2), it can be quickly noticed that despite the same distance between peaks 1–2 and 3–4 referred to as the selectivity factor (α), the width of the peaks decides how much the peaks will overlap. In the first case, compounds are properly purified; peaks overlap by more than 5% in the second. The lowest required resolution for quantitative analysis is 1.5 because the peak overlap is only 0.15% for the Gaussian curve shape.

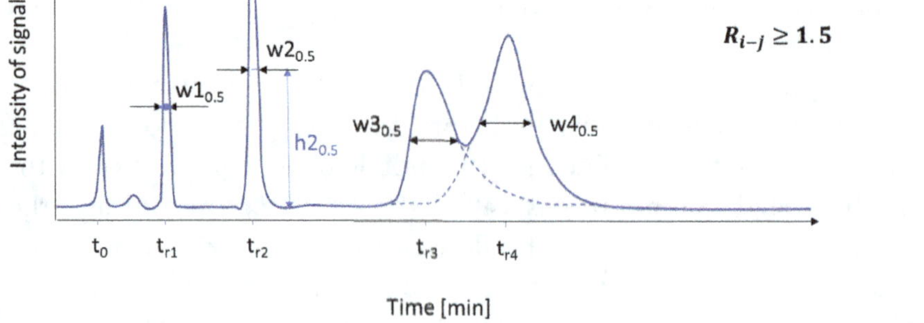

Figure 2. Determination of the chromatographic resolution (R) based on the data read from the chromatogram: w — width of the chromatographic peak in the middle of its height and α — selectivity factor.

A good chromatographic resolution requires the peaks to be narrow and sufficiently spaced apart. The peak's width depends on the chromatographic process efficiency, while the distance depends on the selectivity and retention factor (Equation 1). Many factors affect all three parameters, which will be discussed separately in the following chapters.

$$R = 1/4 \cdot \underbrace{\sqrt{N}}_{\text{Efficiency}} \cdot \underbrace{\left(\frac{\alpha - 1}{\alpha}\right)}_{\text{Selectivity}} \cdot \underbrace{\left(\frac{k}{1+k}\right)}_{\text{Retention}} \quad (1)$$

where N is the number of theoretical plates.

$$N = 2\pi \cdot \left(\frac{t_r - \mathrm{h}}{A}\right)^2 \quad (2)$$

where h is the peak height and A is the peak area.

3. Efficiency of the chromatographic process

Chromatographic efficiency can be described as obtaining the highest number of separated compounds in the shortest possible time. The efficiency of a chromatographic process depends on the performance of its components: column, pump, sample dispensing system, and detector. Each of these system components must be matched to the others taking into account their accuracy and economic factor.

3.1. *Never too much stationary phase?*

How to understand the influence of chromatographic columns on the separation process? The easiest way is to imagine sieves through which we sift ground material (e.g., ground grain). The more sieves with different mesh sizes we stack one on top of the other, the more fractions of grains we get, that is, the greater the separation capacity. In LC, the number of sieves corresponds to the surface area of the stationary phase, that is, the total area of the particles packed in a tube made of steel, plastic, or glass. In short, the greater the surface area of the stationary phase, the greater the number of compounds we can introduce into the column knowing that they will all interact with active groups on the surface of the stationary phase, the greater the number of peaks we will obtain.

The surface area of the stationary phase can be increased by:

- lengthening the chromatographic column, that is, a tube filled with an appropriate particle that serves as the stationary phase, but it will increase retention and significantly prolong analysis time (obvious but not recommended),
- using a homogeneous particle in terms of its diameter and shape. A stationary phase is always most densely packed if it has a spherical shape,
- using ultrasounds or pressure to dance particles in the column with the aim of avoiding gaps in the stationary phase; the presence of such gaps usually leads to a widening front arm of chromatographic peak ("peak fronting" — opposite to "peak tailing")
- reduction of the diameter of the packed particles (Figure 3), and

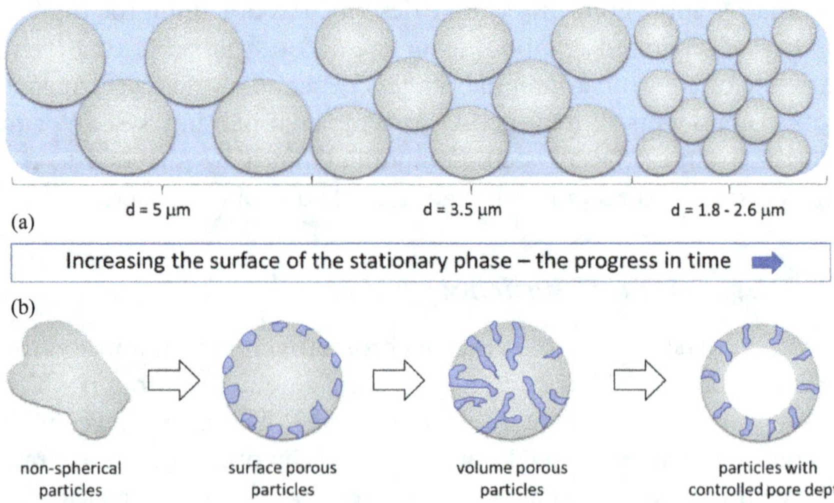

Figure 3. Changes in the stationary phase's particle size, shape, and porosity lead to increased chromatographic efficiency.

- production of pores on the surface or in the volume of the particles (Figure 3).

Increasing the surface area of the stationary phase by increasing the number and depth of pores due to controlled chemical etching has proven to be one of the most critical factors affecting the efficiency of the separation process. Why? Because it is in the pores filled with a liquid solution, the interactions between the solid and liquid phases responsible for the selectivity of the process occur. The diameter of these pores is not large, usually about 100–150 or 300 Å. In the former case, the pore diameter no longer allows penetration by macromolecular compounds such as proteins or polysaccharides, so the pores must be bigger to achieve full separation capability. This also means that the molecules of the solution filling the pores of the stationary phase can interact so strongly with the stationary phase that their exchange does not occur in real time — it is delayed so much that the pore-filling liquid can also be considered immobile. This means that the molecules of the separated compounds to reach the surface of the mobile phase must overcome a barrier in the form of this immobile liquid (other words: "immobile–mobile phase"). Therefore, the rate of this step is controlled by diffusion, which depends on density, the viscosity of the mobile phase solution, and of course, temperature. Therefore, increasing the temperature usually leads to a narrowing of the chromatographic peaks (Figure 4).

Interestingly, the pore size and shape should also be homogeneous so that all molecules of the separated compounds have to, by way of diffusion, overcome an equal path in the volume of the column. This results in a narrower peak. The pore depth is nowadays controlled in the technological process by producing a hybrid silica particle containing a core with higher chemical resistance, that is, not susceptible to chemical etching; in the name of such a column, there is a reference to the term core (Accu-Core, Fused-core, Solid-core, Core-shell, etc.). This type of chromatographic bed ushered in the era of ultra-fast chromatography, in which chromatographic peaks reached a width of around a few seconds, which was reserved for gas chromatography.

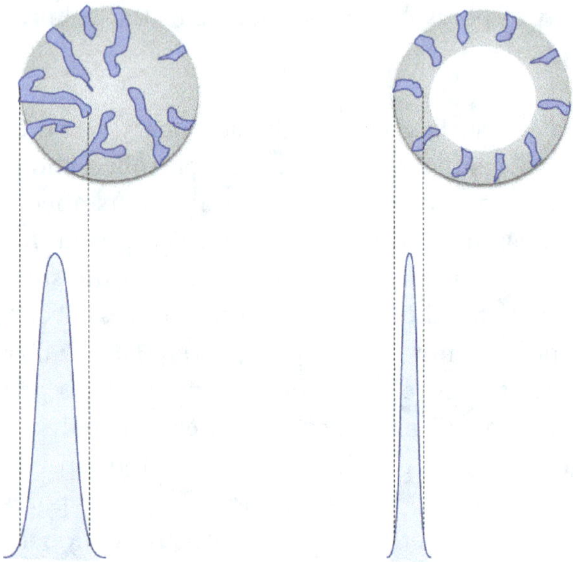

Figure 4. Influence of homogeneity of pore depth in the stationary-phase particles on chromatographic peak width.

3.2. *Does the speed of the mobile phase matter?*

In the early 1990s, more than 50 mL of liquid had to be pumped to completely change the solution composition in the chromatography pump: today, 10 times less. This difference is due to the volume of the pump components, mixer, connecting hoses, and filters that comprise the entire system. The larger the intrinsic volume of the system, the greater the deviation of the mobile phase composition from that expected, especially for smaller flows, which often broadens the chromatographic peak.

It should also be noted that the connecting elements and seals were also very different from those used today. However, they allowed the apparatus to work reproducibly up to 300 bar (high-pressure chromatography). This high pressure was obtained because the liquid must flow through the gaps created by the most densely packed stationary phase providing resistance — back-pressure (observed in the system in front of the column!). If the system is leaky or has a large dead volume, the liquid does not flow evenly but pulses with the rhythm of the pistons, which can be seen as a regular noise in the chromatogram. Increasing the efficiency of chromatographic pumps is work that leads to reducing their volume and increasing the tightness and speed of mixing solutions.

We can represent all pump elements, sample injectors, and columns in a single model tube with a certain internal diameter through which the liquid flows (Figure 5). Liquid molecules close to the tube wall interact with its surface — the closer they are to the tube, the stronger the interactions and the more they are slowed down compared to molecules flowing in the center of the tube. This is a characteristic laminar flow phenomenon when the fluid flows as organized layers. Differences in the velocity of the fluid layers, often referred to as "longitudinal diffusion," can be reduced by uniformly increasing the temperature of the column and the mobile phase solution (since this reduces the viscosity of the liquid) or increasing the ratio of the liquid velocity to the inner diameter of the tubing (i.e., the liquid flow rate can be increased, or the column and tubing can be replaced with ones that have a smaller inner diameter — note, this principle also applies to the detector chamber). Otherwise, increasing the inner

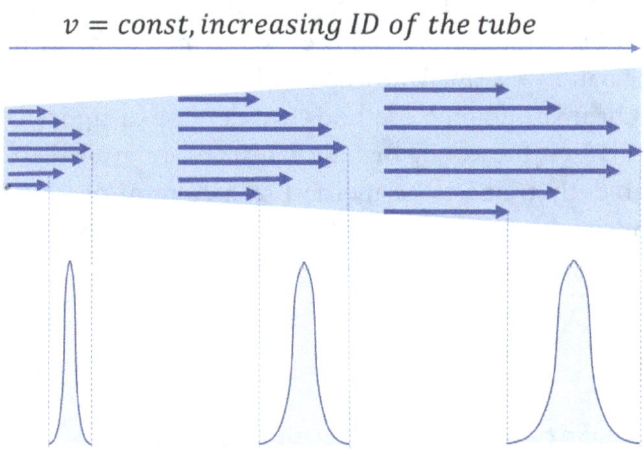

Figure 5. The influence of internal diameter (ID) of LC system elements on peak broadening.

diameter of the LC system components will lead to chromatographic band blurring and chromatographic peak broadening (Figure 5). Not surprisingly, increasing the liquid flow rate in a chromatographic system improves its efficiency and sharpens peaks.

Fortunately, ultra-high performance chromatography pumps are now being developed that can pump liquid without pulsation (i.e., without flow or pressure changes) even when the stationary phase of the chromatography column resists up to 600 bar (ultra-high performance/pressure liquid chromatography, UHPLC) or even over 900 bar (ultra-fast UHPLC).

It should also be mentioned that, for similar reasons, the sample volume entering the chromatography column must not be too large. It should not exceed 1% of the column volume. This can be explained most simply because compounds are separated because of their movement speed through the chromatographic column. It is, therefore, a kind of race. In the case of a 100-m sprint, it does not last longer than 10 s, and the time of the competitors differs by hundredths of a second. They must be placed precisely the same distance from the finish line. In the case of the 5-km run, the differences between the competitors are much more significant, so their positioning at the start does not have to be exact. Likewise, in LC, a shorter column and faster analysis, as well as a smaller-column internal diameter and smaller diameter of particles, requires the sample to be introduced as a narrower band — a smaller sample volume.

Most chromatographic methods are developed under conditions typical of HPLC. However, many calculators are available to quickly and conveniently transfer the method to a column or HPLC system of different sizes and capabilities offered for free by producers of chromatographic systems and columns (Agilent Technologies, Analytical Sciences,[1] ACD Labs,[2] Phenomenex,[3] Supelco,[4] ThermoFisher,[5] and Waters[6]).

3.3. *And what about sorption and desorption?*

Since compounds interact with hydrophobic chains or polar groups present on the surface of the stationary phase, the rate of sorption and desorption also affect the chromatographic process. Suppose the concentration of a component similar to the active groups of the stationary phase is too high in the mobile phase solution. In that case, sorption occurs to a limited extent or not at all. In contrast, a deficiency of the active ingredient in the mobile phase solution causes sorption to begin to dominate, and compounds desorb only partially and very slowly.

It is, therefore, necessary to maintain an appropriate equilibrium between sorption and desorption, which is achieved by a balanced competition of the active components of the mobile phase with the surface of the stationary phase for interaction with the compounds to be separated.

The composition of the mobile phase can be fixed during the separation process (isocratic elution) or changed to increase the amount of the competing component (gradient elution): organic solvent in reversed-phase chromatography, water in hydrophilic interaction chromatography, and salt in ion exchange chromatography. This change in mobile phase composition creates an unusual situation in which adsorption and desorption are perfectly balanced. This is because the molecules of each of the compounds being separated are in their unique equilibrium zones. There is a narrow band in front of them where adsorption dominates and desorption behind them (Figure 6, right panel). This ensures that the chromatographic peaks are narrow regardless of retention time. In contrast, with isocratic chromatography, desorption (short retention times) dominates all the time for some compounds and adsorption (broadening the peak with retention time) for others (Figure 6, left panel).

The optimal mobile phase composition is preferably determined by elution, with the mobile phase solution varying from no active ingredient to its total content (e.g., from 0% to 100% methanol). Compounds are typically eluted at a time when the mobile phase composition provides optimal equilibration of compound interactions with the stationary phase. It is, therefore, possible to determine the initial and final mobile phase composition for gradient elution and the average mobile phase composition for isocratic elution (provided only when the peaks on the chromatogram occur as a single cluster within

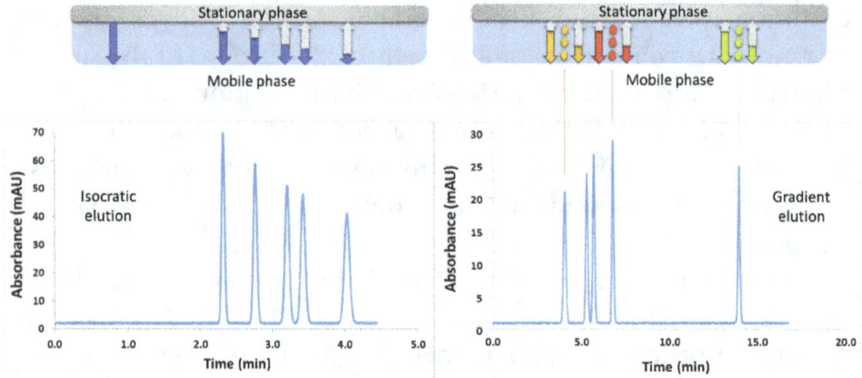

Figure 6. Examples of chromatograms obtained via isocratic elution and gradient elution generated by the HPLC teaching assistance program.[7]

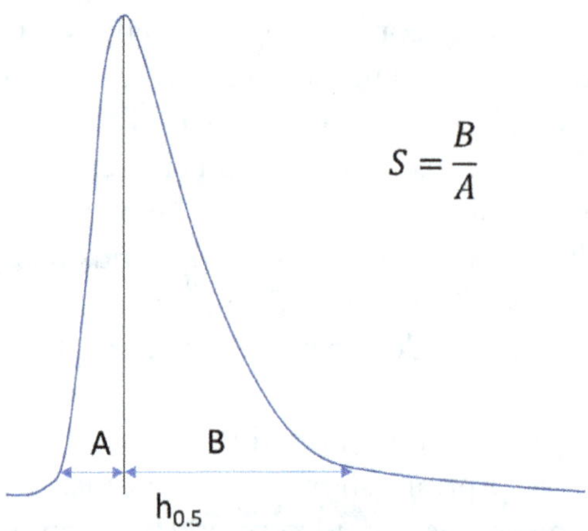

Figure 7. Peak tailing and peak asymmetry (S).

a quarter of the analysis time, that is, in Figure 6 — right panel, the peaks are scattered and their separation by isocratic elution will fail). It is possible to exercise.

It is possible to see how changes in mobile phase composition affect the behavior of chromatographic peaks using the "HPLC teaching assistance" training program, available for free at the Université de Genéve.[8]

Despite the use of gradient elution, the mobile phase composition does not sufficiently balance the adsorption of compounds on the surface of the stationary phase due to an additional type of interaction. Most often, these are ionic interactions between amine groups of the separated compounds and acidic silanol groups -SiOH present on the surface of silica particles, when hydrophobic interactions with aliphatic chains attached to silica should dominate. This phenomenon leads to the broadening of the chromatographic peak and its tailing, which can be described by the asymmetry factor (Figure 7).

The degree of interactions with acidic silica can be reduced by changing the pH of the mobile phase to less than 3.0 (pKa of silica 3.8–4.2) or greater than 8.0 and adding a small-molecule amine to the mobile phase solution to interact with the silanol groups instead of the analytes.

3.4. *How fast does the detector need to be?*

Although the chromatographic process is optimized, the chromatographic resolution can still be lost when the detector conditions are not properly set. Depending on the chromatographic method's efficiency, the chromatographic peak's width varies, about 1 min in LC, 30 s in HPLC, 10 s in UHPLC, and 2 s in ultra-fast UHPLC.

A chromatographic peak observed in a chromatogram consists of points connected by a line. Each point corresponds to a signal intensity measured over a defined period of time and represents a summed value for that signal intensity. This means that the longer a single measurement (called a measurement cycle) lasts, the higher the signal will be. Unfortunately, at the same time, the risk that the peak is not correctly defined increases (Figure 8). As a gold standard for quantitative analyses, a minimum of 20 measurement points (15 for symmetric peaks) is required to adequately describe the chromatographic peak, ensuring good resolution and repeatability of the chromatographic method. This means that the detector cycle may take 3 s for LC, and for ultra-fast UHPLC, it should not exceed 10 ms. During this cycle, the detector can scan the entire wavelength range (spectroscopic techniques) or the mass-to-charge ratio (mass spectrometers)

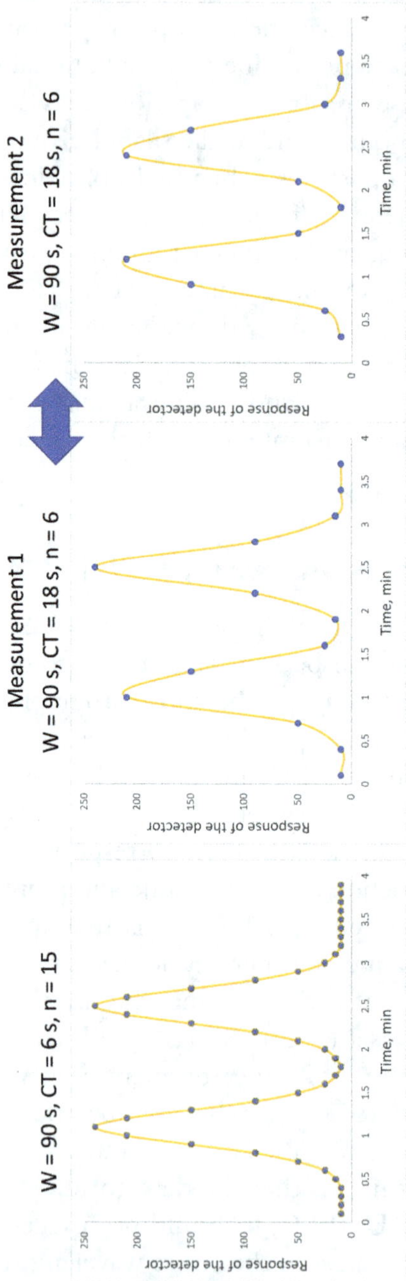

Figure 8. Effect of detector cycle time (CT) on chromatographic peak definition and precision.

and monitor the signal for only selected wavelengths or m/z. The more information we want to obtain about the compound, the more time it takes for the detector to provide it.

Figure 8 shows the chromatographic peaks obtained for a correctly defined detector rate (cycle time, CT = 6 s) for a chromatographic peak with width W = 90 s and how the shape and height of the peaks change for repeated measurements when the cycle time is three times larger (CT = 18 s). The reproducibility of such a method is immediately worse and can also lead to the loss of resolution when retention times change or the distance between peaks is smaller.

4. Selectivity of the chromatographic process

The selectivity of the chromatographic process depends principally on the type of stationary phase, the composition of the mobile phase solution, and the composition of the sample solution.

4.1. *Stationary phase complexity*

Many types of stationary phases are used in LC. However, they can be divided into two main types:

1. polymeric (the most popular are hydrophobic polystyrene cross-linked with divinylbenzene, polyvinyl alcohols, and hydrophilic polyhydroxymethacrylate). They are chemically more resistant and allow to work in a wide pH range of 1–13 and
2. silica [often modified with polyalkylsiloxanes (PMOS, PMODS, PDMS, PMPS, PMTDS, PBD) for better chemical resistance]. The pH range is narrower (2–8).

The most commonly used support in analytical LC is, without a doubt, silica. This material offers good mechanical strength, high surface area, and controlled pore depth and volume. But most of all, its surface can be easily modified through a process called functionalization. This way, a variety of organic stationary phases can be crafted.

On the other hand, the stationary phase itself is liable to degradation. With continued use, the ligands are hydrolyzed and eluted by the mobile phase — the effect is known as "column bleed." The loss of the stationary phase brings a loss of retention and efficiency and induces further hydrolysis and degradation. Also, some detectors can respond to column bleed, gradually distorting the baseline, increasing noise, reducing sensitivity, etc. All these effects combined eventually result in column failure. This phenomenon can be prevented using hybrid silica beads (alkylsiloxanes) or by surface modification of silica with polymers.[9]

Polymer or silica beds can be functionalized by attaching the groups responsible for the selected type of interactions. When hydrophobic interactions are needed [reversed-phase LC (RPLC) or ion-pair LC], the polymeric bed, being sufficiently hydrophobic, is usually not modified. In the case of a hydrophilic silica bed, aliphatic chains of different lengths are attached (the longer chain, the higher hydrophobicity).

However, it should be noted that the silica surface is not modified entirely, and a number of polar/acidic silanol groups remain exposed. These highly reactive groups can interact with amine-containing compounds (majority of low-molecular-weight biologically active compounds) present in the analyzed sample, resulting in distorted chromatographic peaks, low recoveries due to sample adsorption, shifting retention times, and, in some cases, chemical degradation of the sample. These effects can be minimized by additional modification of silanol groups with branched aliphatic chains (Figure 9c and d), which shield the neighboring ones. The best results can be obtained when the amide group is present in the aliphatic chains. It forms hydrogen bonds with silanol groups and efficiently prevents the adsorption of amines on the stationary phase (Figure 9e).

On the other hand, a large number of active silanol groups are desirable for the analysis of polar compounds. Silanol groups interact with water molecules, filling the gaps between hydrophobic chains and preventing them from collapsing (Figure 9a). Hydrophobic aliphatic chains can end with phenyl, nitrile, amine, or hydroxyl groups to introduce different types of interactions leading to changes in the selectivity of the chromatographic method.

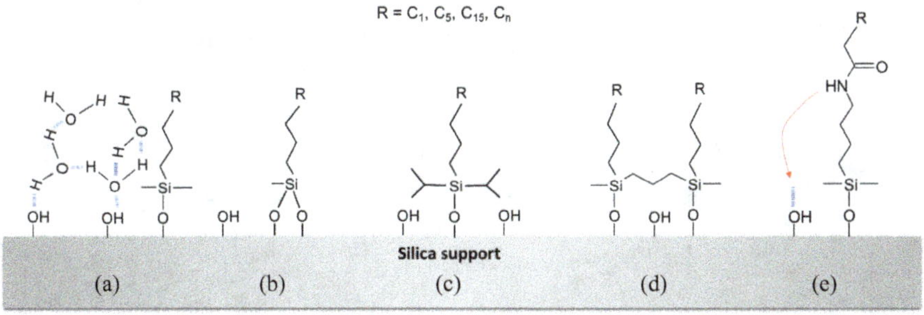

Figure 9. Examples of silica functionalization with aliphatic hydrophobic chains with different properties: (a) resistance to water-rich mobile phase solution (prevention of chain collapsing), (b and c) higher chemical resistance toward acidic hydrolysis (pH 2–8), (d) higher chemical resistance toward acidic hydrolysis and silica dissolution in alkaline solutions (pH 2–11). The activity of silanol groups toward interactions with amines a>b>c,d>>e. Figure adapted from Ref. 10.

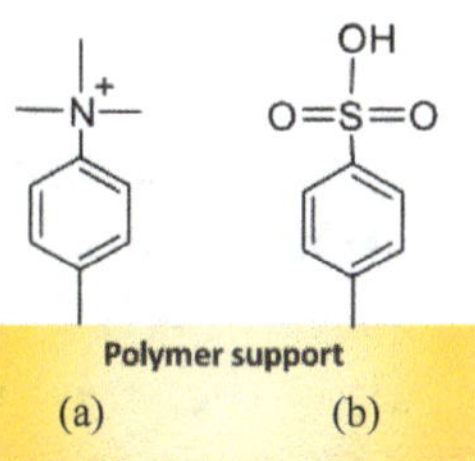

Figure 10. Functionalization of polymer stationary phase toward anion (a) and cation (b) exchange chromatography.

The second type of functionalization leads to a polar stationary phase that is supposed to enforce hydrophilic interactions with the separated compounds. The usual way is to attach short aliphatic chains or phenyl groups terminated with acidic (carboxyl, sulfonate) or alkaline (tertiary, quaternary amines) groups (Figure 10). In the case of an acidic stationary phase, the surface of the stationary phase is endowed with a negative charge due to deprotonation, which will lead to the retention of compounds with a positive charge (cation exchange chromatography). On the other hand, the presence of

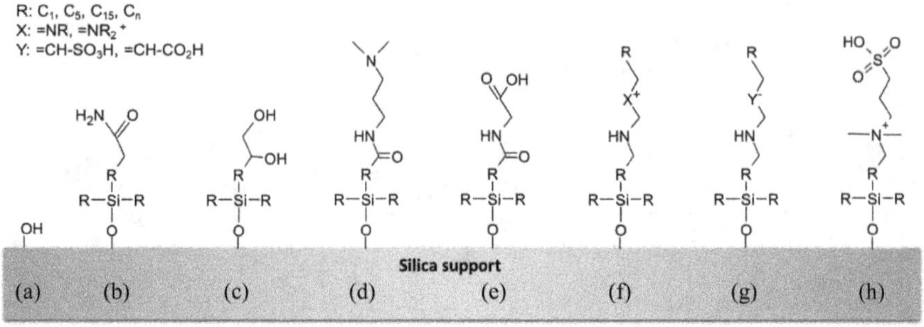

Figure 11. Examples of silica-based stationary phases applied for HILIC with different properties: (a) none modified silica, (b) amide, (c) diol, (d, f) mixed-mode anion exchange, (e, g) mixed-mode cation exchange, and (h) zwitterionic polar. The strength of hydrophobic interactions in mixed mode depends on the aliphatic chain length.

amine groups in the stationary phase leads to the retention of anions (anion exchange chromatography).

Similar functionalization of the stationary phases with different polar groups (amines, nitrile, diols, modified phenyl groups) is applied for silica particles for normal-phase chromatography and hydrophilic interaction liquid chromatography (HILIC). The selectivity of these stationary phases can depend on the strength of the hydrogen bond or ionic interactions. It should be mentioned that hybrid stationary phases are also created, containing both acidic and alkaline groups that occur in the middle or at the end of the hydrophobic chain (Figure 11). This provides a unique opportunity to exploit the full range of hydrophilic and hydrophobic interactions simultaneously. The effect is that molecules can be "attracted" by different functional groups on the surface of the stationary phase but can also be "attracted/repelled" simultaneously. These interactions are also used in chiral chromatography, except that the compounds modifying the silica bed are chiral, that is, the location of the polar groups is spatially defined (R, S) and uniform.

It should be noted that modification of the stationary phase is half the battle. For the compounds to properly interact with the stationary phase, it is necessary to provide the appropriate conditions using a mobile phase solution.

4.2. *Molecular interactions and mobile phase composition*

To fully understand how the composition of the mobile phase can affect the solubility of a compound and the degree of interactions with the stationary phase, we must accurately describe the intermolecular interactions, which until now have been generally classified as hydrophilic and hydrophobic.

First, let us recall that a water molecule is polar; it has positive and negative poles (higher electron density). Such an arrangement is called a dipole. We should expect the presence of such poles when elements in the molecule differ significantly in electronegativity (for O and H is 3.5–2.1 = 1.4). Although for C and H atoms, the difference in electronegativity is small (2.5 – 2.1 = 0.4), due to the presence of molecules in the immediate vicinity, electrons can be rearranged to create electron-rich and electron-poor zones, which is called an induced dipole (Figure 12a). These interactions are quite weak, but their strength increases with the size of the molecule (the number of carbons and multiple bonds that allow electrons to move). These highly flexible interactions cause large molecules to adhere to each other, creating exclusion zones for polar compounds.

Stronger dipole–dipole, hydrogen-bonding interactions are observed when the molecules have electronegative atoms. Nevertheless, the strongest ones occur with dissociated compounds. Figure 12e and f shows the same compounds with polar groups whose nature of interactions with the stationary phase (silica) can vary, as it depends on whether the compound is in the ionic form or not. Strong acids such as HCl will always dissociate to form an anion, just as quaternary amines occur as stable cations. The anion will form in an alkaline or neutral environment for weaker acids because the acidic group (–OH, –COOH) can easily surrender the proton.

In contrast, an acidic environment promotes cation formation by bases. The degree of dissociation of a compound depends on how far the pH of the solution differs from its pK_a (acidity constant). Remember that for a pH–pK_a difference of ±1 and 2, dissociation occurs with efficiencies of about 90% and 99%, respectively. This rule applies equally to compounds in the sample, mobile phase solution, and active groups on the surface of the stationary phase. Therefore,

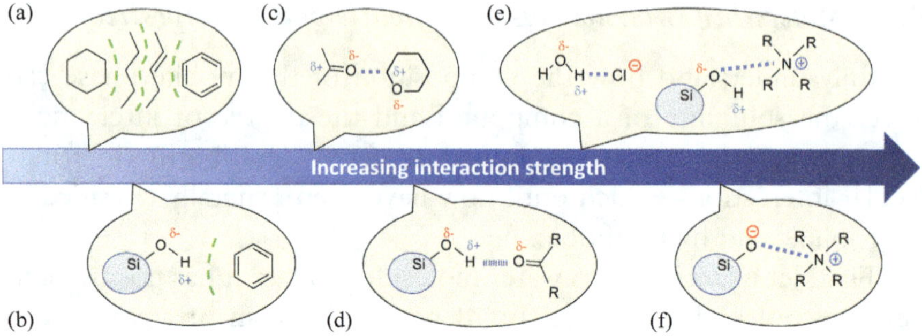

Figure 12. Types of molecular interactions organized by strength: (a) induced dipole–induced dipole (van der Waals or hydrophobic), (b) induced dipole–dipole, (c) dipole–dipole, (d) hydrogen bonds, (e) ion–dipole, and (f) ion–ion (ionic).

the pH of the mobile phase solution will determine which interactions will occur. A neutral compound will interact with aliphatic chains on the surface of the stationary phase. In contrast, an ionized compound will readily interact with water and methanol in the mobile phase solution or with polar groups on the surface of the stationary phase.[11]

For a chromatographic process, the compounds adsorbed on the surface of the stationary phase must be efficiently recovered. For this to occur, the interactions of the compound with the stationary phase must be adequately balanced by the presence of competing interacting molecules in the mobile phase. The strength of the interactions of the mobile phase molecules with the analyte should be similar to or greater than with the stationary phase.

In addition to the pH, the type of interactions can be altered by switching the organic solvent in the mobile phase solution. Methanol and acetonitrile, polar solvents that mix freely with water, are seemingly similar. Methanol (CH_3OH) is a proton-donor compound responsible for several interactions with molecules: dipole–dipole, dipole–ion, hydrogen bonds, and even ion–ion. In contrast, acetonitrile (CH_3CN) is most likely to form hydrogen bonds and promote dipole-based interactions but cannot interact with ions. The differences between these solvents can be exploited in several ways:

- changing the solvent affects the type of interactions and, therefore, the selectivity (order of elution of compounds) in normal-phase and reversed-phase chromatography,
- ion–dipole interactions in the case of methanol are responsible for better solubility of salts. Thus, it is used to prepare mobile phase solutions containing a large amount of salts responsible for pH control, ion–ion interactions, or other effects (e.g., desalting), and
- when one wants to force compounds to ion–ion and ion–dipole interactions with a polar stationary phase, instead of a non-polar solvent (such as hexane or dichloromethane), one can use acetonitrile, which is used in the chromatography of hydrophilic interactions.

Another method of influencing the selectivity of the chromatographic method is the modification of separated compounds. We do not mean a permanent modification as in derivatization, but a reversible modification, which was conducted during the chromatographic process. The suitable modifier is added to the mobile phase solution, and its concentration usually changes gradually to obtain the desired result. This is most often the formation of:

- hydrophobic protein aggregates due to the presence of a high concentration of metal or ammonium sulfate in the mobile phase solution (hydrophobic interaction chromatography),
- complexes–chelates, by adding to the mobile phase metal ions forming complexes with compounds in the sample or a ligand competing with the stationary phase modified by metal ion (affinity and chiral chromatography), and
- neutral ion pairs formed by adding the suitable counterion (ion pair chromatography).

The right combination of the type of stationary phase with the chosen mobile phase composition determines what type of intermolecular interactions will occur and how strong compounds will be retained. It is usually referred to as the separation mechanism (Table 1).

Table 1. Types of chromatographic methods distinguished by different molecular interactions. Active components of the mobile phase (responsible for the recovery of compounds from the stationary phase) are underlined. Letters describing molecular interactions refer to Figure 12.

Type of stationary phase	Interactions in the stationary/ mobile phase	Typical components of the mobile phase	Type of chromatography
Non-polar	A, B, C/D, E, F	Water, methanol, acetonitrile, methyl acetate, pH preventing ionic interactions	Reversed-phase chromatography, RPLC
Polar	B, C, D/A, B, C	Hexan or dichloromethane with methanol or acetonitrile	Normal phase chromatography, NPLC
Polar	C, D, E, F/C, D, E, F	Water, methanol, acetonitrile pH enhancing ionic interactions	Hydrophilic interaction chromatography, HILIC
Non-polar	A, B, C/A, B, C, D, E	Water, methanol, acetonitrile, ethyl acetate, counterion agent pH enhancing ionic interactions	Ion pair liquid chromatography, IPLC
Polar	E, F/E, F	Water, methanol, salts pH enhancing ionic interactions	Ion exchange chromatography, IEC
Polar	Steric interactions, C, D, E/C, D, E, F	Water, methanol, acetonitrile, and salt to compensate for ionic interactions	Size exclusion chromatography, SEC
Non-polar	Steric interactions, A, B, C, D, E/D, E, F	Water, methanol, acetonitrile pH preventing ionic interactions	Gel permission chromatography, GPC
Polar	D, E, F, complex formation/A, B, C, D, E, F complex formation	Water, methanol, metal ion, or ligand (e.g., imidazole) pH enhancing complex formation	Affinity chromatography, AC Chiral chromatography

4.3. *Sample composition*

Remember that the more complex the composition of the mobile phase solution, the longer the components take to exchange in the pores of the chromatographic stationary phase particles, and equilibrium is established. This should also be kept in mind for sample composition. The more matrix components present in the sample, the more likely our resolved compounds in the sample are present in a different form than we assume. The pH, amount and type of solvent, concentration of salts, ion-pair-forming compounds, and complexes all matter. The more the sample composition deviates from the mobile phase composition, the higher the probability of masking the tested compounds. Masking, because in another form, they behave differently than expected, and we can wrongly determine their identity due to the change of the retention time.

The simplest method to eliminate the masking effect is to dilute the sample with a mobile phase solution (for gradient elution, dilute in a solution of the initial composition or a composition corresponding to at most one-third of the elution program run). In this way, the number of matrix molecules is reduced, and some of them will be exchanged in interactions with the desired mobile phase components. The pH in the sample solution will be established, forcing the compounds to interact with the stationary phase appropriately. Retention times and peak areas will be more reproducible. In particular, the appearance of the beginning of the chromatogram will be improved because the time for pH adjustment, ligand exchange, and counterions, will be provided during sample preparation. Otherwise, one should expect fronting and overlayed chromatographic peaks in the initial part of the chromatogram and well separated in the later part (Figure 13).

Suppose dilution is not sufficient because the components to be separated are present in a minimal amount in the sample, and other interfering components are present in excess. In that case, sample purification methods can be used to enrich the tested compounds and remove the interfering ones (provided they differ sufficiently in physicochemical properties). These are liquid–liquid extraction and solid-phase extraction (SPE) (also implemented in micro-scale). Interestingly,

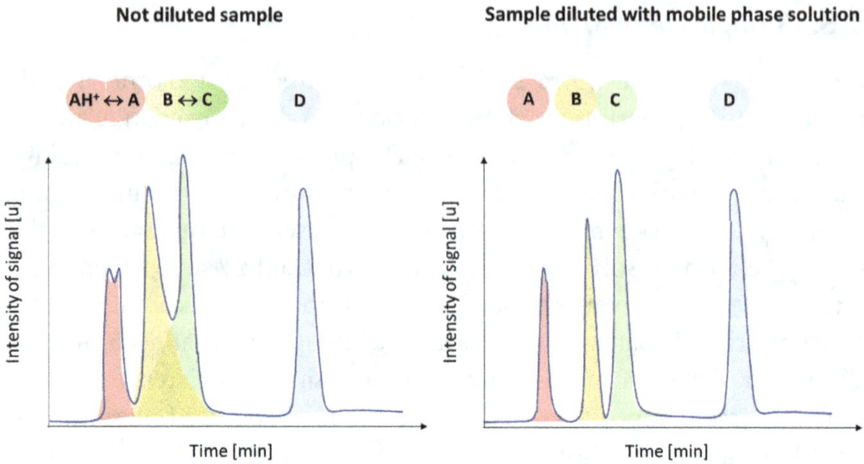

Figure 13. The influence of sample composition on the selectivity in reversed-phase chromatography.

the design of the sample purification process most often duplicates the method of the planned chromatographic separation. For example, peptides that we plan to separate with RPLC will be enriched using SPE column with C8 or C18 sorbent (i.e., a twin to the chromatographic one); organic acids and sugars in wine will be separated and enriched as anions using a bed containing amino groups. The purification step is quite time-consuming, but in the case of analysis of compounds of biological origin — necessary to accomplish.

What if compounds are unstable and dilution/purification increases the likelihood of degradation? Still, dilute and still enrich. Do this before the analysis in the shortest possible time. Chromatographic systems equipped with an additional mobile phase stream divider and an additional injection valve are beneficial for this purpose. It can then automatically enrich the sample and elute sorbed compounds onto the chromatographic column. If we do not have such a system, the enrichment can still be carried out manually on the column during the gradient chromatographic process. The gradient program is started with a long isocratic step. While pumping the mobile phase with a constant initial composition, we introduce the sample several times (e.g., three). The sample components mix with the mobile phase; those interacting with the stationary phase are retained at the head of the column (as one narrow band). Once the dosing sequence is complete and the mobile phase flushes out the interfering agents, we increase the elution force of the mobile phase. Only then do the compounds retained at the head of the column begin to move. This way, we obtain a threefold sample enrichment combined with a prior washout of the interfering compounds. The initial isocratic step combines sequential sample purification with enrichment.

5. Retention in the chromatographic process

The primary advantage of the chromatographic process is the ability to determine/detect components in complex mixtures simultaneously. Thus, its main advantage is time-saving. Even a 90-min analysis can be advantageous because several hundred compounds are separated and compacted into a narrow band observed by the detector. The general rule that the retention factor should be less than 20 is frequently ignored by many chromatographers. Why is this the case? Several reasons contribute to this:

- good selectivity takes time,
- small molecules move more slowly because they have a greater ability to penetrate pores, and diffusion significantly affects process efficiency,
- the high rate of change in mobile phase composition requires a chromatographic pump with a small dead volume and an efficient liquid mixing system, and
- complementary qualitative data takes more time.

Why does selectivity take time? Let us go back to an old cue when we wonder what solvent will dissolve the studied compound: "similar likes similar." How long does it take us to evaluate who we like and don't or do not like among people we have just met? This process takes time. Likewise, during the chromatographic process, a compound selects its preferred interactions with the active groups of either the stationary or mobile phases. One must also consider the time for diffusion of compounds in the pore space and the equilibration due to changes in the composition of the mobile phase. Selectivity will, therefore, increase with time.

Why do we seek to reduce analysis time? There are several factors to consider:

- economic — longer analysis time means increased solvent consumption and performing fewer analyses in 1 day,
- IT — longer analysis time increases the risk of computer-to-instrument communication errors, and the much larger file size poses problems when it comes to data analysis and archiving, and
- analytical — higher retention usually leads to lower substance recovery from the stationary phase, which leads to poorer method sensitivity; increasing peak tailing with retention time is responsible for non-uniform peak integration, which leads to decreased method reproducibility.

Thus, retention is the interface between selectivity, chromatographic efficiency, and other factors that determine the utility of a developed chromatographic method.

6. Summary

Chromatography is a highly dynamic technique that requires constant self-education. Significant changes can be expected in the case of selectivity achieved by using new functionalized stationary phases.

It should be noted, however, that increasing chromatographic efficiency by decreasing the particle diameter of the stationary phase is limited by Moor's law. When the diameter of the stationary phase particles is reduced twice, separation efficiency doubles and the pressure significantly increases. This requires serious improvement of the chromatographic system, which will be challenging to maintain for the user. It is possible that with nanoparticles used as a stationary phase, a new technique will be required.

References

1. Guillarme D, Nguyen D, Rudaz S, *et al. Eur. J. Pharm. Biopharm.* 2007;66:475–482; Guillarme D, Nguyen D, Rudaz S, *et al. Eur. J. Pharm. Biopharm.* 2008;68:430–440.
2. https://www.acdlabs.com/resources/freeware/translator/.
3. https://www.phenomenex.com/tools/kinetexcalculator.
4. https://www.sigmaaldrich.com/analytical-chromatography/hplc/method-transfer-calculator.html?gclid=CjwKCAiA9bmABhBbEiwASb35V9akDin75YU_B_Nc2mYLyBlAHB8PWzzukMp78rGSZdUKC9kSFsXrmxoCcTIQAvD_BwE.
5. https://appslab.thermofisher.com/lc-method-transfer-calculator.
6. https://www.waters.com/webassets/other/lp/prep_calc/AnalyticaltoPrep/NewPrepcalculator.htm?locale=en_PL.
7. Guillarme D, Veuthey JL. "HPLC Teaching Assistant: A New Tool for Learning and Teaching Liquid Chromatography — Part I." *LC-GC N. Am.* 2016;34:804–811.
8. Guillarme D, Veuthey JL. "HPLC Teaching Assistant: A New Tool for Learning and Teaching Liquid Chromatography — Part II." *LG-GC N. Am.* 2016;34:906–915.
9. Faria AM, Collins CH, Jardim, ICSF. "State-of-the-Art in Immobilized Polymer Stationary Phases for High-Performance Liquid Chromatography." *J. Brazilian Chem. Soc.* 2009;20(8):1385–1398. doi: 10.1590/S0103-50532009000800002.
10. Lopez DA, Green AI, Bell DS. "What Is on Your HPLC Particle? Looking at Stationary Phase Chemistry Synthesis LCGC Europe." *LCGC Europe.* 33(9):460–465.
11. *Life Science Ebook 1a.* Copyright © 2021 The President and Fellows of Harvard College. https://projects.iq.harvard.edu/lifesciences1abookv1.

https://doi.org/10.1142/9789811283376_0002

Chapter 2

Ionization — the First Step of Mass Spectrometry–Based Proteomic Analysis

Katarzyna Macur[1,*]

[1]*Core Facility Laboratories, Intercollegiate Faculty of Biotechnology, University of Gdańsk and Medical University of Gdańsk, Gdańsk, Poland*

Contents

---

[*]Corresponding author.

Abstract

Mass spectrometry (MS) is nowadays a leading analytical technique for high-throughput protein identification and quantification in proteomics. Detection success depends largely on the effective volatilization and ionization of the analyzed molecules, as the mass spectrometer measures ions in the gas phase. Among different existing ionization techniques, two types of soft ionization are mainly applied for the analysis of large biomolecules — proteins and peptides, which are the focus of proteomic studies: electrospray ionization (ESI) and matrix-assisted desorption ionization (MALDI). This chapter discusses the importance of ionization for proteomic analysis by MS and provides a basic description and practical information of ESI and MALDI along with the characteristics of mass spectra of proteins and peptides obtained by both techniques. Finally, a brief description of typical hyphenations of ESI and MALDI ion sources is given, both with separation techniques and mass analyzers predominantly used in proteomics nowadays.

1. Introduction

Mass spectrometry (MS) has become a leading analytical technique for proteomics, largely applied, especially in high-throughput studies that focus on the identification and quantification of proteins and allow for the sensitive profiling of proteomes. The MS-based protein analysis from its early developments has been significantly surpassing the capabilities of methods traditionally used for protein characterization, such as sodium dodecyl sulfate–polyacrylamide gel electrophoresis (SDS-PAGE) or Edman degradation. The crucial role of MS in protein analysis results from its ability to determine the molecular weight of modified and unmodified peptides and proteins up to

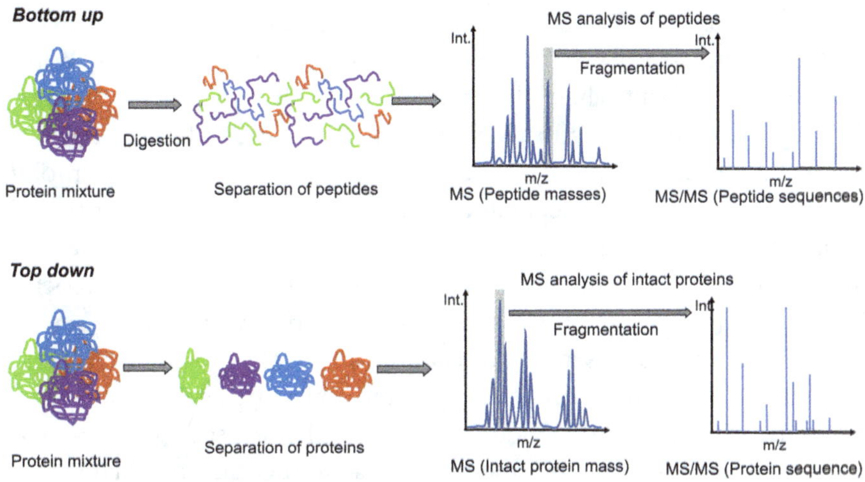

Figure 1. The main MS-based proteomics workflows: *bottom-up* proteomics (upper panel) and *top-down* proteomics (lower panel) (based on Ref. 3).

several hundred kilodaltons with high accuracy and from as low as femtomolar (or less) amounts of even complex samples.[1] There are two main workflows utilized in MS-based proteomics studies called *bottom-up* and *top-down* (Figure 1). In *top-down* approach, the intact proteins are studied by MS. Contrary to that, in *bottom-up* proteomics, the proteins are first proteolytically digested preferably by a protease that cleaves by specific amino acid residues to a mixture of peptides, which is then separated typically by liquid chromatography (LC) and analyzed by MS (Figure 1). The detected peptides have to be later assigned, usually by a database search, to the respective proteins of their origin. Despite this, the *bottom-up* approach is the most commonly chosen for proteomic studies because of a better understanding and predictability of peptides' behavior than proteins on both the experimental and computational levels.[2]

International Union of Pure and Applied Chemistry defines MS as a "study of matter through the formation of gas phase ions that are characterized using mass spectrometers by their mass, charge, structure, and/or physico-chemical properties."[4] From the measured *mass-to-charge* (*m/z*) ratios, one can calculate the accurate mass of the analyzed molecule. Such accurate mass is an intrinsic feature of the molecule and therefore in many cases allows us to identify the compound, particularly when also masses of its characteristic fragments from tandem mass spectra (MS/MS) are available. The MS analysis starts with the formation of the gas phase ions from neutral molecules, which are then separated according to their *m/z* ratio and finally detected based on their *m/z* abundance. This takes place in a mass spectrometer, which is typically built up from three main blocks: (1) ions source, in which the ionization occurs; (2) mass analyzer(s), where the ions are separated according to their *m/z* ratio; and (3) detector, which detects the separated ions; and the instrument is controlled by vendor-dependent software. The ion source, mass analyzer, and detector operate under high vacuum conditions (10^{-8}–10^{-4} Torr), with the exception of, for example, the electrospray ionization (ESI) mass spectrometers, in which the ion source is under atmospheric pressure.[5]

Because of the fact that mass spectrometer measures ions in the gas phase, volatilization and ionization of the analytes are crucial for their effective detection by MS. Different ionization techniques have been developed and they can be divided, for example, according to their relative hardness and softness, that is, the capability to fragment the created ions already in the ion source. The choice of the ionization technique depends on the chemical properties of the investigated molecules, such as their polarity, expected molecular weight, or thermostability. Due to the fact that proteins and peptides — the targets in proteomic analysis — are thermolabile and non-volatile biomolecules, they require soft ionization to prevent their in-source fragmentation during ionization and hence to enable their successful detection by MS.[5]

The expansion of the MS application as a key technique for high-throughput proteomics started with the emergence of two soft ionization techniques: ESI[6] and matrix-assisted laser desorption ionization (MALDI).[7] Their breakthrough role in the analysis of large biomolecules was acknowledged by awarding the Nobel Prize in Chemistry to John B. Fenn and Koichi Tanaka who worked on ESI and MALDI, respectively,[8,9] for the analysis of biological macromolecules. ESI and

MALDI are the main ionization techniques used in proteomics, and they will be described in detail in the following paragraphs.

2. Electrospray ionization

In ESI, the ions are created in solution and transferred into a gas phase. ESI, in the presence of an electric field, produces a fine mist of charged droplets that evaporate releasing ions of the analyzed molecules.

2.1. *Principles of ESI process*

The ESI process takes place under atmospheric pressure. A low-flow (1–10 μL/min) stream of liquid-containing analytes passes from the sample inlet through a metal capillary, where it is introduced into a strong electric field. The electric field is created by applying a high voltage (typically ~2–5 kV) between the tip of the metal capillary containing the solution and a counter electrode of the opposite charge. Depending on the type of compounds to be analyzed, positive or negative voltage (positive or negative ion mode) can be applied, resulting in the electron flow from or to the metal capillary, respectively. The ESI ion source can be considered as an electrolysis cell, because in the absence of redox-active analytes, the redox reactions at the metal/solution contact in the capillary support creation of positive and negative ions: oxidation mainly in positive ion mode and reduction in the negative ion mode, accordingly.[10] The high voltage causes a split of the uniformly charged solution that passes through the capillary, creating portions enriched in positively and negatively charged species. The solution–air interface at the end tip of the capillary from a hemispherical shape becomes conical, as a result of the accumulation of charges in the surface layer of the liquid. When the surface tension and electrostatic force are in perfect balance, a fully conical shape, the so-called Taylor cone, is created. Further increase of the potential destabilizes the Taylor cone. As a result, the surface layer breaks on the apex of the cone, where the electric field is at maximum, and the nebulizing liquid forms a fine spray of charged droplets in the evaporation chamber. The droplets are charged due to the presence of excess ions. In positive ion mode, these excess charge carriers include H^+, NH_4^+, K^+, or Na^+, with protons mainly contributing to the net charge of the droplets.[11] Solvent evaporation causes shrinkage of the droplets. When charges in these droplets get closer together, the charge density at their surface increases, to the point, when the Coulomb repulsion forces between like charges exceed the level of their cohesive forces (Rayleigh instability limit), causing a burst of the droplets. The resulting aerosol undergoes subsequent sequential fission into smaller and smaller charged droplets while the solvent continuously evaporates. In the end, the gaseous ions are released from the few-nanometer-sized droplets. The solvent evaporation is typically aided with the use of a gas, for example, nitrogen and/or heat. This additionally facilitates the separation of the ions from neutral molecules, the solvent and clusters, which are also produced in the ESI process, and it also helps in the introduction of ions from the ESI source to the later stages of the mass spectrometer.[12] However, the exact mechanism of the gas phase ions release is still under investigation and different models have been proposed.[11,13] According to Ion Evaporation Model (IEM), the ions are desorbed from the droplets when the electric field on the droplet's surface becomes sufficiently high. Alternatively, Charge Residue Model (CRM) suggests that the process continues until total evaporation of the solvent results in the formation of the charged molecules. Evidence shows that small ions tend to follow IEM, while entering the gas phase by larger, multiply charged ions (≥1000 Da, e.g., globular proteins) is better explained by CRM.[14,15] In the case of unfolded proteins, the research indicates that the formation of gaseous ions is comprehensively described by the Chain Ejection Model (CEM). There, the non-polar and exposed protein residues are involved in driving the molecule to the droplet surface, from which the stretched-out chains are gradually expelled.[15] Finally, the ESI-formed ions are electrostatically directed to the high vacuum region of the mass spectrometer for their separation in the mass analyzer and detection. One may encounter problems in ESI–MS detection in both the lower and upper concentration ranges. The ion transmission in the mass spectrometer and the presence of interferences, for example, solvent

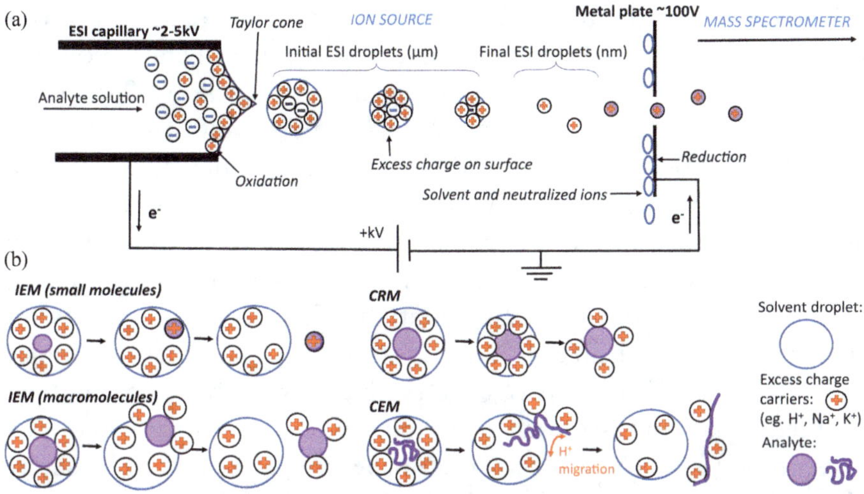

Figure 2. Electrospray ionization (ESI) in a positive ion mode (a). Formation of ions in ESI+ according to Ion Evaporation model (IEM), Charge Residue Model (CRM), and Chain Ejection Model (CEM) (b) (based on Refs. 11, 13, 15).

clusters, contribute mainly to the sensitivity of the ESI–MS measurement and have an impact on the detection limit. On the contrary, saturation of the ESI signal observable at the higher concentration levels may be due to limitations of the number of ions produced in the ESI process.[12]

2.2. *Ions created in the ESI and the ESI mass spectrum*

The ion formation in ESI typically does not (or to a very low extend) induce any changes to the molecular structure of the analytes. Therefore, ESI is classified as a soft ionization technique, which means that the ionized molecules are not fragmented in the source, so we can detect *m/z* of the intact molecule.

Ions may be formed in ESI by protonation, deprotonation, cationization, or a transfer of the charged molecule into the gas phase.[5,12] In the case of proteomics samples, which are typically analyzed in a positive ion mode and introduced into the ion source in a mixture of acidified water with an organic modifier, we mainly observe protonated species, in which a proton is added to an investigated molecule ($[M+H]^+$, where M is the molecule and H is the proton). Sodium or potassium adducts ($[M+Na]^+$ or $[M+K]^+$, respectively) are also often seen. Such cation adducts are produced via cationization, where a positively charged ion (in this example, K^+ or Na^+) is non-covalently added to a neutral molecule. However, adduct ions are not desirable for MS/MS experiments, often applied in proteomics workflows, because they generate low fragmentation information. Depending on the presence of ionizable sites in the structure of the analyzed molecules, the ions obtained in ESI can be singly or, more often, multiply charged with a number of charges typically equal to the number of its sites prone to proton attachment. Smaller peptides tend to be singly charged. However, peptide and protein ions formed in ESI are frequently multiply charged. The abundance of the higher charge state of peptide ions in ESI mass spectra increases with the peptide size, while the occurrence of the lower charge state of ions decreases. At the condition of low pH usually applied in proteomic analysis, peptides will typically accept protons on the free amino termini and side chains of basic amino acids: lysine (Lys), arginine (Arg), and histidine (His). Tryptic peptides, obtained after trypsin digestion of proteomic samples, mainly have two potential protonation sites: the *N*-terminus and basic Lys/Arg residues at the *C*-terminus. Therefore, $[M+2H]^{2+}$ ions are predominantly observed for tryptic peptides (Figure 3b). Because of multiple charging, the ESI mass spectra consist of a series of peaks corresponding to the successive, multiply charged protonated molecule, $[M+nH]^{n+}$, when acquired in positive ion mode (ESI+), or deprotonated molecule, $[M-nH]^{n-}$, when acquired in negative ion mode (ESI−) (M is the molecule, *n* is the number of protons and H is the proton).[16] Detection in ESI+ or ESI− mode is possible, provided the analyte possesses at least one positive or one negative elementary charge, respectively. It is also observed that the ESI signal increases with the increase of the molecule's hydrophobicity. The ESI+ is generally used for the analysis of positively charged ions, such as protonated organic bases or inorganic cations. The ESI− is preferable, when negatively charged ions, such as deprotonated

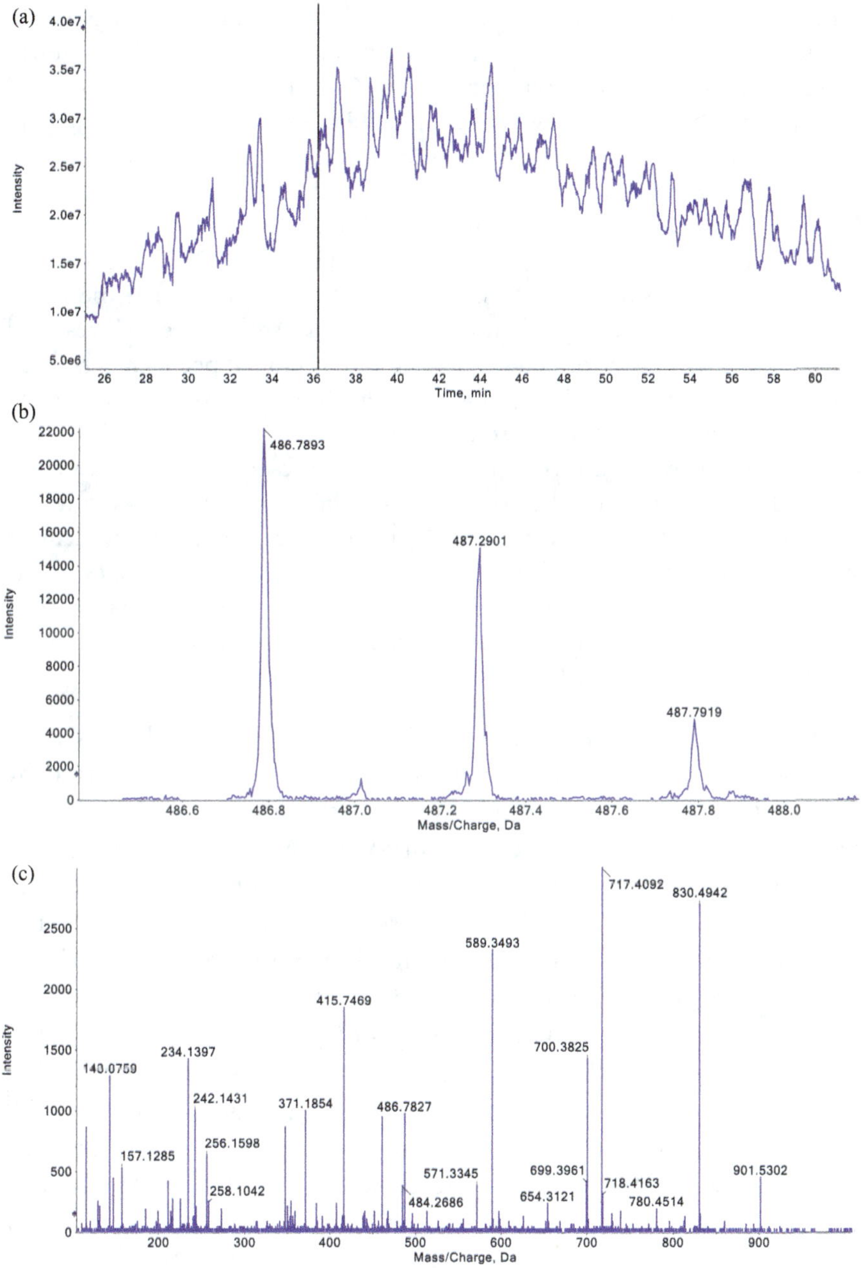

Figure 3. (*Continued*)

Figure 3. (*Continued*) Total ion current (TIC) (a) from the LC–MS/MS (liquid chromatography–tandem mass spectrometry) analysis of the human protein digest standard on the microLC–ESI–Q–TOF (micro-liquid chromatography–electrospray ionization quadrupole–time-of-flight) system. MS (b) and MS/MS (c) spectra were extracted from the total ion current (TIC) at 36.2 min, which correspond to the peptide of the sequence AALQELLSK. The signal of *m/z* 486.7893 in the MS spectrum corresponds to the monoisotopic $[M+2H]^{2+}$ peak of the peptide AALQELLSK (b). The distance between isotopic peaks is 0.5 *m/z* unit, indicating that the ion is +2-charged. The isotopic peak clusters in ESI mass spectra are well-resolved when the analysis is performed on high-resolution mass spectrometers, such as those equipped with Orbitrap (OT) mass analyzers. However, when analyses are performed on lower-resolution mass spectrometers, such as those with Q mass analyzers, isotopic peaks may not be resolved in ESI mass spectra. As a result, it may not be possible to see the peak corresponding to the monoisotopic mass (i.e., the mass of the molecule calculated using accurate masses of the most abundant naturally occurring stable isotopes of its constituent elements, e.g., ^{12}C, and ^{1}H). In such cases, the average mass (i.e., the mass of the molecules calculated using isotope-weighted masses of its elements)[19] should be used.

organic acids or inorganic anions, are analyzed.[12] The ESI mass spectra practically do not contain any peaks of ions that were created by dissociation or fragmentation because of the low internal energy of the ions created in this process.[17] Although such ESI mass spectrum with multiply charged ions present is more complicated, it enables more precise determination of the mass of the analyte. This is because the mass can be calculated independently from each of the multiply charged ions. Moreover, due to the fact that in ESI we often observe the formation of multiply charged ions of proteins (Figure 4a), it is possible to analyze them with the use of mass spectrometers that operate in relatively small nominal mass range, for example, <2000 Da, such as those equipped with quadrupole (Q) mass analyzers. To obtain masses of the analyzed molecules, the multiply charged ESI spectra have to be deconvoluted (Figure 4b).[18] Another advantage of multiple charging is that such precursor ions create more fragment ions on fragmentation in MS/MS experiment, which enables better characterization of the analyzed molecule.[5]

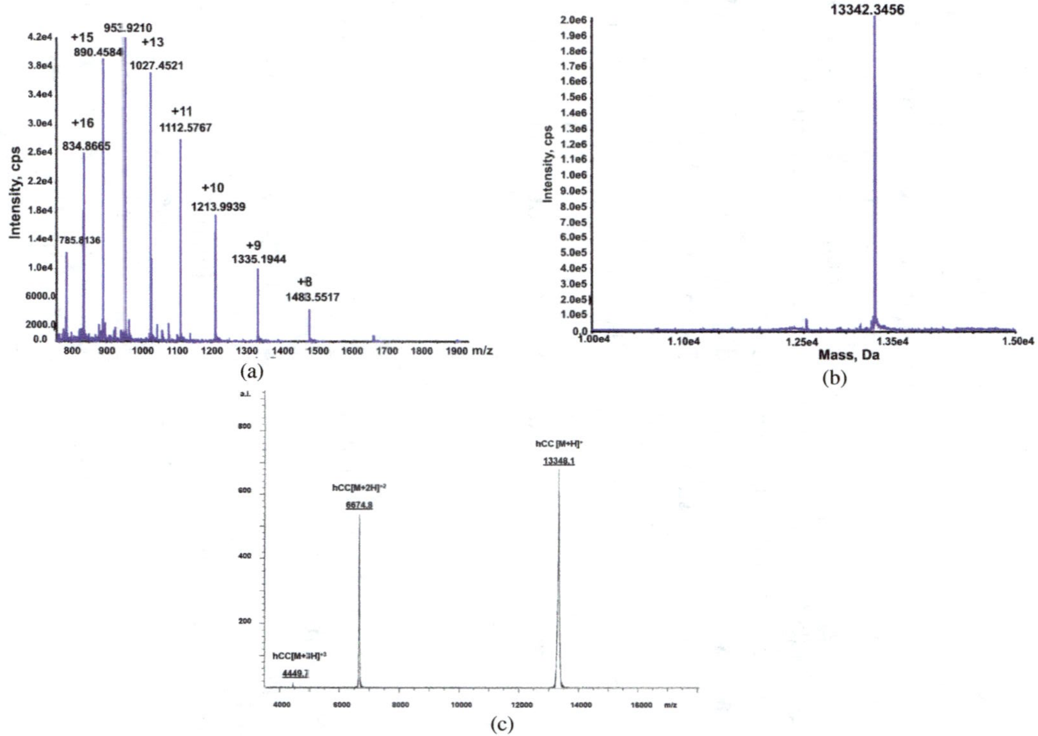

Figure 4. MS spectra of an intact human cystatin C obtained using SCIEX TripleTOF ESI–Q–TOF system [before (a) and after (b) deconvolution], and MALDI–TOF/TOF 5800 system (c). Deconvolution of the ESI–MS spectrum was performed with the aid of PeakView 2.2. software (SCIEX).

The determination of charge state distribution (CSD) along with the indication of a dominant and average charge state of intact protein ions obtained in ESI is more complex than in the case of peptides and has been under investigation by different groups. It has been shown that these features depend mainly on the structure of the protein of interest in solution. The protein CSD is affected by various factors. Some of them include the pH of the solution, the presence of acidic and basic functional groups in the proteins, protein surface area in the solution, the presence of non-volatile salts and electrolyte impurities in the protein solution, the addition of ammonium acetate and/or supercharging substances, physical properties of the solvent, protein concentration, or applied instrument parameters. The denatured proteins have a less compact structure and a bigger spatial size, which exposes additional acidic sites for protonation (or basic sites for deprotonation) and facilitates solvent accessibility leading to the increase in CSD over higher *m/z* range. Hence, the transfer of denatured proteins into the gas phase via ESI generally leads to the production of more charge states than when the proteins are natively folded.[11,20]

2.3. *Nanoelectrospray ionization (nanoESI)*

Nanoelectrospray ionization is an ESI that operates at very low flow rates, typically tens to hundreds of nanoliters per minute. An example is the nanospray, where the replacement of the metal capillary with gold-coated pulled glass capillary with a fine tip of ~1 μm inner diameter and of several microliters of volume enabled ESI at flow rates typically of 20–50 nL/min.[21] NanoESI also requires only a very low sample concentration (of nanomoles per milliliter range) and volume. Application of the ~0.7–1.5 kV spray voltage to the capillary generates the spray, while the aperture is being refilled by the analyzed solution that flows driven by the capillary forces. Initial droplets created by nanoESI are less than 200 nm in diameter, which results in 100–1000 times smaller volume than conventional ESI droplets that are 1–2 μm in diameter. The difference in the initial droplet size between nanoESI and ESI has an important impact on the performance. Due to the high surface-to-volume ratio of small droplets, more molecules are available for desorption in nanoESI than in ESI.[21] Additionally, smaller initial droplets require lower number of fission events to release gas phase ions than bigger droplets, where more analyte remains in a lowly charged fission residue. These features, together with decreased flow rates that assure longer measurement time, result in higher sensitivity of MS measurements performed with nanoESI compared to ESI. Moreover, with each fission event that occurs during ESI, the offspring droplets have higher salts and analyte concentrations than the droplets of their origin because of gradual solvent evaporation. Therefore, nanoESI process will result in lower salt concentration than ESI, owing to smaller initial droplets. Thus, nanoESI will more effectively produce gas phase ions from samples of higher salts and buffer contents than the conventional ESI. This has a significant practical implementation, especially for the analysis of proteins in a native state, which are typically solubilized in buffers to maintain their spatial structure. It has been shown that nanoESI produces spectra comparable to ESI, even when the salt concentration is one order of magnitude higher.[22] Therefore, the advantage of the nanoESI compared to conventional ESI is not only more efficient ionization that leads to increased sensitivity but also smaller sample consumption and its long-lasting capability that enables the performance of multiple experiments. Moreover, nanoESI is also more resistant to salt contamination and higher aqueous solutions.[22] Nonetheless, the presence of impurities in the samples and/or solvents and reagents used is still one of the main sources of problems that hamper obtaining good results with nanoESI, so, as in the case of conventional ESI, some care should be taken to remove the contaminants. The nanoESI maintenance is also more demanding. It requires a proper installment and positioning of the capillary and management of the emitters, as the capillaries tend to plug.[21,23] Moreover, a special attention to stable spray production is necessary for effective ionization.[23]

3. Matrix-assisted desorption ionization (MALDI)

MALDI uses laser beam, usually in the ultraviolet (UV) range, to irradiate a small molecular matrix co-crystallized with the investigated sample. The ionization of the analyte occurs by transferring a proton from previously ionized matrix molecules to the analyte molecule.

Proteomic applications of MALDI employ mainly the UV lasers, for example, nitrogen laser or frequency tripled/quadrupled neodymium-doped yttrium aluminum garnet laser that operates at 337 nm or 355 nm wavelength, respectively.[24] Therefore, the information in this chapter will refer to the MALDI–MS analyses with the use of instrumentation equipped with UV lasers.[25]

3.1. *Principles of the MALDI process*

In MALDI, the sample is co-crystallized with an excess amount of matrix on the MALDI target plate prior to its direct insertion through a vacuum interlock into the ion source. The matrix separates molecules from each other as the solvent evaporates and co-crystallization of the sample with the matrix occurs. This protects from the formation of analytes' complexes, which might be too large for desorption. Then, the small area of the co-crystallized sample/matrix spot (~0.05–2 mm in diameter) is irradiated shortly in nanosecond range pulses by the focused laser beam.[26] The matrix absorbs large amounts of the energy from the laser beam and transfers it to the analyzed molecules. This helps to vaporize and ionize the sample and at the same time prevents the sample decomposition due to the excessive laser beam energy.[24] The MALDI ionization process is still under investigation, but it can be explained by a two-stage mechanism, which includes primary and secondary ionization events.[27] The primary ionization, where the matrix is ionized and desorbed, is a result of photochemical reactions that occur in response to the pulsed laser beam irradiation: photoionization and proton disproportionation. A creation of primary matrix ions during photoionization is a result of pooling of electrons in the photoexcited matrix. According to proton disproportionation, the conversion of absorbed proton energy to thermal energy leads to the generation of matrix ions. The photoabsorption of the laser beam by the matrix initiates desorption, during which the sample co-crystallized with the matrix

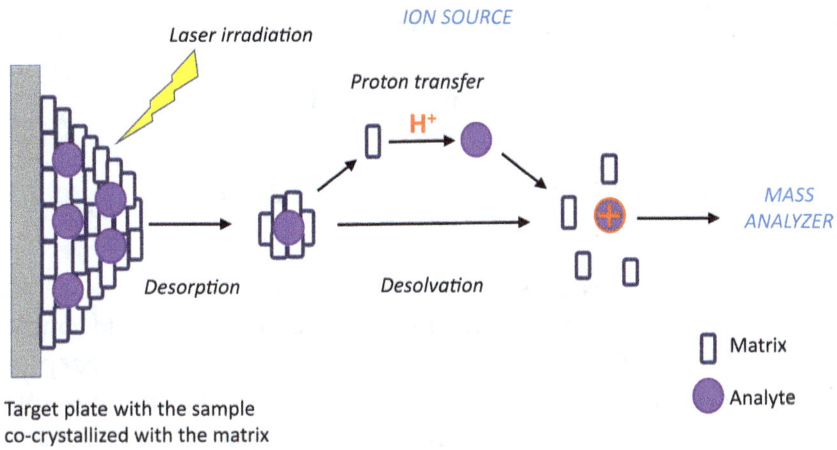

Figure 5. MALDI in a positive ion mode (based on Refs. 24 and 32).

is sublimated into the gas phase in the form of pulsed jet expansion, the plume (Figure 5).[28,29]

In the next step, the analyte molecules are ionized as a result of the ion-molecule reactions occurring in the expanding plume of the desorbed material.[27] These reactions are still the subject of an ongoing debate, but two models are widely accepted: the gas phase protonation and "lucky survivor" model.[30] The occurrence of the processes explained by both the models depends on the analyte, employed matrix, and experimental parameters. In the gas phase protonation, the neutral analytes are released from the sample–matrix cocrystals and collide with matrix ions in the gas phase. The charge transfer from the matrix ions to the analyte leads to the analyte's ionization and uncharging of the matrix molecules (Figure 5). The gas phase protonation is more favorable when increased laser fluency and matrix of lower photon affinities are applied. However, in cases where quantitative analyte protonation in solution and intramolecular ion stabilization are increased, the "lucky survivor" model more comprehensively explains the MALDI ions' formation.[31] The "lucky survivor" model assumes that already preformed analytes' ions are released from charged clusters, which contain the matrix molecules, analyte ions, and counterions. During the desolvation of the neutral matrix molecules from these clusters, singly charged ions are released due to the quantitative charge neutralization (Figure 5).[32]

3.2. *Ions created in the MALDI and the MALDI mass spectrum*

MALDI also belongs to soft ionization techniques. The ionization mechanisms leading to produce ionized protein and peptide molecules in MALDI are as follows: proton transfer (protonation, deprotonation), cationization, and transfer of charged molecules into a gas phase.[5,27] However, in MALDI–MS spectra acquired in the positive ion mode often used for the analysis of proteomic samples we typically observe signals corresponding to protonated and cationized peptide and protein ions. In contrast to ESI (Figure 4a), the characteristic feature of MALDI–MS spectra is the predominant presence of singly charged ions (Figure 4c). The "lucky survivor" model explains this phenomenon as a result of mutual charge neutralization between positive and negative ions that are initially present in the MALDI plume formed in response to the laser beam irradiation. It was observed that for ions of higher initial charge state, this reneutralization process is faster than for singly charged ones. Therefore, finally, singly charged ions remain in the plume and are referred to as the "lucky survivors."[33] Apart from *m/z* of the ions corresponding to the analyte, matrix signals can be observed at the low *m/z* range of the MALDI–MS spectrum, depending on the *m/z* range applied in the analytical method.[24,26]

4. Sample preparation for ESI and MALDI–MS

A proper sample preparation for MS analysis is important for obtaining high-quality data. Usually, the method for the isolation of proteins and peptides, and the preparation for MS analysis have to be established and optimized, depending on the aim of the study and the investigated material.[34–36] In this chapter, the focus is to highlight the key aspects of proteomic sample preparation with respect to ESI and

MALDI ionization. Unfortunately, many buffer components and additives, such as involatile salts, polymers, chaotropic agents, and most of the detergents, which are added for better isolation and solubilization of protein samples, are incompatible with downstream MS analysis and significantly impact ionization. Apart from the poor-quality spectrum, these interfering with MS measurement substances can cause contamination of the MS system, impairing its performance, which often requires extensive cleaning or repairs to retrieve its functionality.[12] Therefore, when developing the proteomic sample preparation protocols, one should keep in mind that the procedure must ensure efficient isolation of the proteins and at the same time include the removal of the components interfering with MS analysis or use the buffers and solutions composed of MS-compatible substances. In general, high-purity combinations of water, volatile polar organic solvent, and small amounts of weak volatile acid or base is the most desirable environment for a sample to be introduced into the mass spectrometer, and such a final MS-sample composition should be a target for proteomic sample preparation procedure.[19]

4.1. *Sample preparation for ESI–MS*

Successful ESI–MS measurements are highly dependent on the sample composition. Particularly, the type, concentration, and purity of the solvents and excipients added mainly during the isolation of proteins of interest have an important impact on the quality of the obtained ESI–MS data. Proteins and peptides that possess basic functional groups are well ionized when they are introduced into ESI+ analysis in a protonated acetonitrile or methanol/water mixtures, acidified with a weak and volatile acid (e.g., acetic or formic acid), which also facilitates protonation of the analytes.[12] As the sample is introduced to ESI in a liquid form and then transferred into the gas phase, all the substances that interfere with the formation of the charged spray and release of the gas phase ions will influence the efficiency of ionization, and thus, contribute to the possibility of MS measurements and their sensitivity. The sample components can compete with the analytes for available charge or access to the droplet surface during gas phase emission. They may also bind to the analytes and co-precipitate or neutralize analytes' ions through acid/base reactions in the gas phase.[37]

Detergents are often added for extraction, solubilization, and denaturation of proteins from cellular structures, for example, sodium dodecyl sulfate (SDS), NP-40, Triton, octyl glucoside, Tween, CHAPS, or sodium deoxycholate. However, detergents concentrate at the surface of the liquid and alter surface tension, interfering with the ions' evaporation during the ESI process, which leads to ion suppression. Therefore, detergents should not be present in the final sample for MS analysis and they preferably ought to be omitted during the protein isolation and solubilization, if possible. The ion suppression is especially profound in the case of SDS, which was additionally shown to bind to the proteins, making it difficult to remove from the proteomic samples.[38] At the same time SDS, belonging to ionic detergents, is particularly useful in proteomics as it denatures proteins and completely disrupts cellular structures. This ensures very efficient isolation of a broad range of different proteins, including membrane proteins.[39] Filter-aided sample preparation (FASP) is an efficient workflow for *bottom-up* proteomic analysis that uses a protein lysis buffer containing 1%–4% of SDS.[40] At the same time, due to the fact that the FASP process is performed on the ultrafiltration units with molecular weight cut-off (MWCO) membrane, the SDS together with other low-molecular-weight sample components are removed from the protein extract prior to proteolytic digestion. This is achieved through extensive washes with 8M urea (facilitates protein–SDS dissociation) and buffer exchange via ultrafiltration, while the proteins remain on the membrane. On contrary to protein–SDS binding interactions, the SDS interactions on the peptide level can be overcome, and the resulting peptides are essentially free of the detergent.[39] One can also exchange MS-incompatible surfactants for the ones that do not interfere with MS analysis. An example is RapiGestSF (chemical name: sodium 3-[(2-methyl-2-undecyl-1,3-dioxolan-4-yl)methoxyl]-1-propanesulfonate, produced by Waters), an acid-labile anionic surfactant, which degrades rapidly in low pH. Its decomposition products are either water immiscible and can be removed by centrifugation or soluble in water, but not interfering with reversed-phase liquid chromatography (RP-LC) or MS analysis.[41] Another example of the MS-compatible surfactant is ProteaseMAX [chemical name: sodium 3-((1-(furan-2-yl) undecyloxy) carbonylamino) propane-1-sulfonate, produced by Promega], which also undergoes degradation in acidic conditions, and its degradation products can be removed by C18 solid phase extraction [solid phase extraction (SPE), see below] and hence do not interfere with the MS analysis. Moreover, this detergent in concentrations used in proteomics degrades also in the neutral conditions over the course of the standard protein digestion procedure, which additionally simplifies the sample preparation.[42] Both ProteaseMAX and RapiGestSF surfactants were found to improve solubility, and they accelerate the digestion of proteins in terms of speed and higher sequence coverage. At the same time, their acid-labile property facilitates sample clean-up, while the degradation products do not interfere with subsequent MS analysis.

Inorganic, ionic salts (e.g., NaCl, KCl) and buffering salts (e.g., Tris-HCl, phosphate-buffered saline, HEPES, MOPS, MES, phosphate buffer) are also common components that stabilize biomolecules by maintaining osmolarity and pH of the solutions. The presence of these salts and buffers in proteomic samples to be analyzed by ESI–MS is undesirable. These substances cause a significant decrease in ESI–MS signal, because they reduce the volatility of the droplets via lowering the vapor pressure and they create clusters with the analytes by non-specific ion-pairing.[5,12] Instead, volatile buffers, for example, ammonium carbonate, ammonium acetate, or ammonium formate, should be used.[43]

Chaotropes (e.g., urea, guanidinium chloride) constitute another group of substances often present in protein preparations. They disrupt hydrophobic interactions in the structure of proteins and denature them.[3] At the same time, however, chaotropic compounds increase the surface tension of the droplets, which leads to decreased vapor pressure and results in the suppression of the ESI–MS signal.[44] Thus, they should be exchanged in the final ESI–MS sample for the solution that does not interfere with the analysis.[45]

In cases when the substances potentially interfering with ESI–MS analysis cannot be excluded or exchanged for ESI-compatible ones, there have been different approaches developed for proteomic sample clean-up. Dilution of the sample is the easiest and most straightforward option to decrease the concentration of its ESI-incompatible components. However, in this case, also the analytes — peptides or proteins — are diluted, which may lead to the inability of their detection because of too low concentration for the applied MS system and method.[37] Ultrafiltration with the use of MWCO membranes may also help remove some small-molecule contaminants from proteomic samples. Protein precipitation from the original sample is another example of its clean-up. Organic solvents precipitate the proteins, while salts and detergents remain solubilized.[45] Yet, still some substances can be absorbed on the proteins and entrapped in the precipitate so the complete dissolution of such pellet may be sometimes problematic.[37] Among different approaches that can be employed for sample clean-up prior to MS analysis, a SPE is an effective and popular way to remove many contaminants from the sample.[46,47] SPE is based on the principles of LC for selective adsorption and elution of the sample components. A common practice in proteomics is to use SPE with reversed phase (RP) sorbent, for example, C18, C8, or C4 silica sorbents. There, the proteins or peptides are retained in the RP-stationary phase, while the interfering substances are washed out.[46] An advantage of the use of RP-LC technique is that the analytes can be finally eluted from the stationary phase to the solution that is compatible with further ESI–MS analysis. The analytes can be eluted with the mixture of water with RP organic solvent, for example, acetonitrile or methanol acidified with acetic or formic acid. This will also support ionization in positive ion mode.[19] Different formats of RP-LC purification can be used for this purpose, including RP-SPE columns, spin well plates, or sorbent loaded into the pipette tips. In certain applications, other types of LC techniques are employed. Apart from the removal of the substances interfering with MS analysis, SPE can also be used for sample concentration or enrichment in the analytes of interest. The enrichment of the sample in the analytes increases the possibility of their MS detection. This also improves their ionization efficiency by removing the competing sample components. Then, other solid phase materials and different types of LC techniques may be used. For example, when the focus of the study is on mapping phosphoproteome, the samples are typically enriched in phosphopeptides. The phosphopeptides can be enriched with the use of TiO_2, ZrO_2 beads, immobilized metal affinity chromatography, or combinations of different techniques.[48] In any case, the final sample composition should be compatible with the downstream MS analysis, which often leads to its clean-up and concentration after the enrichment procedure using RP-SPE technique.[49,50] Sample clean-up with RP-SPE technique is also often employed as a final step of removing the ESI-incompatible substances present in the proteomic samples by means of gel electrophoresis. The use of one-dimensional (1D) SDS-PAGE, apart from the capability of separation and fractionation of the initial sample components, may also be one of the options for the removal of detergents, salts, buffers, and chaotropic agents from proteomic samples. The spots corresponding to the gel-resolved fractions of the protein sample are in this approach subjected to gel-assisted proteolysis. Subsequently, the resulting peptides extracted from the gel are analyzed by MS (Gel LC–MS/MS).[51] Although sample purification methodologies, including the mentioned above, can remove many ESI-incompatible contaminants, it is good to keep the concentrations of buffers, detergents, and other sample additives at possibly low levels, as the efficiency of their removal may vary depending on the analyzed sample and used methodology.

4.2. *Sample preparation for MALDI–MS*

Compared to ESI, MALDI is more resistant to salts and other sample additives.[26,52] However, usually at higher concentrations than in ESI, these substances also negatively affect the quality of MALDI–MS spectra. Hence, it is still a common practice to remove contaminants in a similar manner as it is performed in the case of the samples to be introduced to ESI–MS, and the same principles can be applied.[53,54] Cleaner protein and peptide mixtures, obtained, for example, by desalting with RP-SPE technique, provide MALDI–MS spectra of better quality in terms of intensity and absence of additional peaks that correspond to background ions of sample additives and adducts.[54,55] Later, the liquid sample has to be co-crystallized with a proper matrix on the MALDI target prior to the MS analysis. The MALDI target is often a stainless steel plate, with predefined spot positions (e.g., 16 × 24 spot array, giving 384 positions on one MALDI target). On its surface, between 0.5 and 2 μL of liquid sample is typically deposited with an excess of saturated matrix solution (matrix to sample molar ratio should range between 100:1 and 50,000:1). Co-crystallization occurs while the solution evaporates.[24] Protein and peptide sample preparation procedures and the composition of the final solution applied on the MALDI target, as well as its surface significantly affect the appearance and quality of the spectrum obtained in the MALDI–MS analysis. For example, when the spotted surface is hydrophobic, the evaporating solution sticks to its area and creates a more compact spot, which leads to lower the limits of detection.[25,55] The process of crystallization is also critical for obtaining proper MALDI mass spectra. Small and rectangular crystals should entirely and evenly cover the surface of the spot with only a slightly thickened rim in a sample desirably co-crystallized with the matrix. It ensures more efficient desorption and better mass accuracy.[24,32,53,56] It was observed that different methods of matrix/sample application on MALDI target yield different co-crystallization outcomes. Conventional *dried droplet preparation*, where the mixture of matrix and sample was spotted on the MALDI target, resulted in the formation of relatively large crystals. On the other hand, the *thin layer technique*, where the sample solution is applied on the top of the matrix layer without redissolving it, enabled the formation of a uniform microcrystalline layer.[56] Application of the matrix/sample on the MALDI target is very important when MALDI imaging analysis is performed. The MALDI imaging technique combines sensitive and selective MS detection of multiple analytes along with their spatial distribution in the investigated sample, for example, tissue section.[57] A non-uniform matrix layer may cause errors in MALDI imaging analyses, as the signal intensity will depend rather on matrix coating quality than on the analyte concentration.[58]

The compounds used as UV-MALDI matrices are typically low-molecular-weight crystalline organic substances that strongly absorb the light at the wavelengths appropriate for the laser type employed in the experiment. An aromatic core in the chemical structure of the matrix is a chromophore responsible for light absorption during laser irradiation. This causes electronic excitation of the matrix, which then leads to vaporization of the part of analyte deposited with the matrix and it facilitates ionization.[26] Apart from the ability for strong light absorption, MALDI matrices should protect the analyte from laser beam damages and promote energy transfer from the matrix to the analyte during the matrix vaporization, which enables efficient sample ionization. To be stable in the vacuum, MALDI matrix should have a low sublimation rate. Concurrently, the matrices should be soluble in the solvents, compatible with the investigated sample, but they should not exhibit chemical reactivity with the solvents and sample. Additionally, MALDI matrices are often acidic substances, which support the ionization of the analytes. They also contain polar functional groups in their structure ensuring water solubility of the matrix, which also matches solubilization conditions for proteins and peptides in aqueous solutions.[52] Among various matrices used in MALDI–MS, more often applied in proteomic studies are as follows : α-cyano-4-hydroxy-cinnamic acid (HCCA) for the analysis of peptides and proteins; and 2,5-dihydroxybenzoic acid (DHB) for peptides and proteins below 10 kDa and glycopeptides[59]; but also sinapinic acid for proteins; 2,6-dihydroxyacetophenone for glycopeptides[60] and phosphopeptides[61,62]; or 2-(-4-hydroxyphenazylo)benzoic acid when peptides, proteins, and glycoproteins are to be analyzed.[26] The HCCA is a particularly suitable matrix for *bottom-up* proteomic sample analysis by MALDI–MS. It was shown to produce a good quality spectra even in the presence of typical protein purification and lysis buffer components, including salts, chaotropes, or detergents, provided the is pH below 2. The low pH, achieved by the addition of trifluoroacetic acid (TFA), is also essential for proper HCCA crystal growth, which also contributes to a better quality of MALDI–MS spectrum and more accurate mass determination.[53] Further improvement in the mass spectrum quality is obtained, when the dried sample/matrix spot on the target plate is washed with, for example, (alkali-free) aqueous solution of formic or TFA. This reduces the content of alkali metals (K, Na), which results in higher intensity of a protonated molecule, while the alkali adduct ions are less abundant. This also enables the removal of some contaminants, such as denaturing agents (e.g., urea), which otherwise may form films on the matrix crystals.[53,63] Although matrices like HCCA or DHB are commonly used in proteomics and yield good results, to obtain the best quality of MALDI–MS spectra, it is critical to choose a proper matrix and optimize sample preparation depending on the investigated compounds. A good starting point is to begin optimization with one of the typically used matrices of polarity similar to the investigated molecules. Additionally, information about the acidity of the matrix may also help to evaluate its protonating effect.[25] For example, the application of a mixture of two common matrices, namely HCCA and DHB, improved reproducibility, resistance to impurities, and sequence coverage in the MALDI–MS analysis of standard proteins' tryptic digests.[64] Furthermore, in some cases, the application of liquid MALDI matrices, maintaining a liquid state in the vacuum, proved to perform better than the traditional solid ones. A liquid 3-aminoquinoline/*p*-coumaric acid matrix improved the sensitivity of MALDI–MS analyses of glycopeptides, phosphopeptides, and carbohydrates and was more effective in preventing the dissociation of labile sialic acids and phospho groups in comparison to DHB.[65]

5. Typical hyphenations of the ESI and MALDI ion sources with separation techniques used in proteomics

Due to the capability of generating gas phase ions straight from liquid solutions, ESI is commonly applied in the MS systems hyphenated with modern, robust, and high-throughput separation techniques: mainly LC, or increasingly, capillary electrophoresis (CE). Separation prior to MS detection splits the analyzed sample into smaller portions that need to be analyzed by MS in the time unit. This way more analytes can be efficiently detected, making the MS analysis more sensitive. Moreover, the use of online sample separation prior to MS detection can preconcentrate the sample, which leads to lowering the limit of detection (LOD) and helps to remove impurities, when part of the separated sample is directed to waste instead of MS. Yet, traditional direct infusion as a sample introduction method is also used. Electrospray is affected by a number of factors, which finally have an impact on the observed MS response. These factors include: (1) applied ESI source parameters (e.g., spray position relative to MS entrance, spray potential, flow rates of the gases, such as curtain gas) and (2) composition of the sample and the solvent (e.g., type and content of organic solvent and its ionic strength, nature of the analyte, presence of other analytes and contaminants, such as salts or detergents). The balance between the ESI source voltage, flow rate, and surface tension of the applied solution is necessary to obtain stable spray. Depending on the experimental set-up, the charge distribution in the spectra of multiply charged ions can change or the ESI signal can be significantly suppressed.[12,13] Peptides and proteins are typically introduced to the ESI in a solvent consisting of water with acetonitrile (or other volatile organic modifier, e.g., methanol) and up to a few percentage of volatile acids (e.g., acetic or formic acid). A reaction with a weak acid creates protonated solvent clusters, which promotes protonation of the containing basic functional groups of peptides and proteins, and thus, also enhances their ionization.[12,17] Organic solvents also enhance spray formation by decreasing surface tension in the droplets.[45] These features of the preferable ESI+ solvent match the optimal conditions for protein and peptide separation in the RP-LC. Therefore, RP-LC is the most often applied separation method for the analysis of peptide mixtures obtained in the proteolytic digestion of proteomic samples. However, TFA (or other strong acids), an ion-pairing additive often used in RP-LC separations of peptides,[66] should be avoided, as it affects the decrease of MS signal intensity.[12]

There have been different LC techniques and methodologies developed and applied to broaden the coverage of proteomes, to increase the sensitivity of MS detection, or to target specific groups of peptides. These include, for example, high pH LC separation[67] or multidimensional separations[68] that consist of combined orthogonal chromatographic and/or electrophoretic techniques, such as strong cation exchange–RP-LC,[69] high pH–low pH RP-LC,[70] nano-RP-LC–capillary zone electrophoresis,[71] and size exclusion chromatography–RP-LC–capillary zone electrophoresis.[72] In certain cases, for example, when structural information is required, proteins in their native state are analyzed by ESI–MS to preserve non-covalent interactions in the complexes. The non-denaturing conditions necessary for native MS analysis are achieved by the use of aqueous volatile buffers (mainly ammonium acetate, but also ammonium formate or ammonium bicarbonate up to 0.1 M at pH between 6 and 8) without organic solvents and harsh acidic conditions. However, in 100% aqueous conditions, the vapor pressure is low, which results in the drop in sensitivity. At the same time, the use of volatile salts as buffering additives helps to minimize the structural changes in protein complexes during their transfer from the liquid into the gas phase by reducing perturbations in the concentration of H^+, which occur during the ESI process.[73] Similarly, as in the proteomics sample preparation procedures, the non-volatile salts and buffers, such as sodium and potassium salts or phosphate buffer, often used in biochemical studies of proteins should be avoided in the mobile phase preparation. Apart from decreasing the ESI–MS signal, due to decremented volatility of the droplets they can also create deposits and contaminate the ESI source.[5,12]

Although RP-LC is the most compatible separation technique for efficient ESI and subsequent MS analysis and a large portion of the proteomic workflows use this separation technique, it may not be appropriate for some compounds to be investigated, due to their physicochemical properties or the research question posed. An option may be then to use CE prior to MS, which uses an applied electric field for the separation of compounds charged in solution based on their charge-to-size ratio.[74] Due to a different separation mechanism, CE provides orthogonal separation to LC, which ensures higher proteome coverage, when both techniques are applied for a particular sample analysis.[75] Furthermore, CE overcomes RP-LC limitations in the analysis of small peptides, which are typically unretained on the column stationary phase, or larger peptides that bound irreversibly to it. CE is also useful for the analysis of acidic and hydrophilic peptides, which may undergo ion suppression in RP-LC. Another advantage of applying CE for proteomic sample separation is that the technique requires only small amounts of samples and reagents. At the same time, CE provides highly efficient separation that combined with high-resolution MS detection ensures sensitive and selective measurements of peptides, proteins with their proteoforms, and even complexes. However, when using CE one may also encounter some problems, for example, with migration time reproducibility.[76,77]

The mass spectra are recorded in a continuous fashion in the LC– or CE–ESI–MS experiment, as the separated portions of the analyzed sample elute from the column (LC) or migrate from the capillary (CE). The mass spectra are generated in each scan of the MS analysis in the number depending on the method settings. The mass spectrometer generates a TIC chromatogram (LC)/electropherogram (CE), which is the plot of all ions detected in each scan versus time (Figure 3a). Post analysis, one can extract either a full mass spectrum for a certain time frame (Figure 3b) or a certain *m/z* of the compound(s) of interest over the entire chromatographic or electrophoretic run in the form of extracted ion chromatogram or extracted ion electropherogram, respectively.[19]

In contrast to ESI, the MALDI ionization process requires prior co-crystallization of matrix with the analyte and evaporation of the solvent. Therefore, this type of ionization is well suited for hyphenation with the offline separation techniques. In the case of proteomic samples, the proteins are often separated with the use of PAGE prior to enzymatic cleavage to peptides. Proteins can be separated according to their molecular weight using 1D SDS-PAGE. The second dimension of the separation can be added, where prior to SDS-PAGE, the proteins are separated based on their isoelectric point (two-dimensional electrophoresis, 2-DE).[78] It is also possible to perform relative quantification of the proteins in a few samples simultaneously using two-dimensional differential gel electrophoresis (2D-DIGE) technique. In 2D-DIGE, two or more samples can be labeled with cyanine fluorescent dyes (CyDyes: Cy2, Cy3, and Cy5) and pooled together before 2-DE separation, which simplifies further spot matching between the samples and thus ensures more accurate analysis. Next, the gels are visualized and the proteins in the spots can be quantified, as the particular CyDyes are spectrally resolvable. For protein identification, the spots of interest are excised, in-gel digested, and subjected to MS/MS analysis,[79] typically with the use of MALDI–TOF/TOF,[80] but LC–ESI–MS/MS analyses of such in-gel protein digests are also performed.[81]

6. Typical hyphenations of the ESI and MALDI ion sources with mass analyzers used in proteomics

The mass analyzer is a central part of the mass spectrometer. Mass analyzers commonly applied in proteomics studies include the OT, time-of-flight (TOF), Q, ion traps (IT) — three-dimensional quadrupole ion trap (QIT) and 2D linear ion trap (LIT/LTQ), and the Fourier transform ion cyclotron resonance (FT-ICR). The applicability of various mass analyzers for different types of proteomic analyses depends on their performance characteristics, of which the most important determinants are resolution, accuracy of mass measurements, and sensitivity.[2] The resolution is an ability to distinguish two peaks of a slight difference in m/z. For two adjacent peaks in the mass spectrum of similar intensity and masses (in daltons) of m and $m + \Delta m$, the resolution can be expressed as $m/\Delta m$.[82] The mass accuracy is an ability to accurately measure the mass of a molecule, which is calculated as the ratio of m/z measurement error to the actual m/z and is usually described in ppm (parts per million). Resolution has a crucial impact on the accuracy of mass measurements: the higher the resolution, the better mass accuracy can be obtained. Thanks to more accurately determined masses of proteins and peptides, the analytes can be identified more reliably.[82] Finally, sensitivity refers to the smallest amount of analyte that can be accurately measured. It can be expressed as a signal-to-noise ratio (S/N) but also as a LOD and limit of quantification (LOQ) that refer to the smallest concentration of the studied compound detected and quantified, respectively. The lower LOD and LOQ, the higher the sensitivity of the mass spectrometer.[83]

The construction, physical operation principles, and performance characteristics of different mass analyzers vary, but their common role is to separate the molecules ionized in the ion source according to their *m/z* ratio.[84] The Q is a scanning mass analyzer that works as a mass filter. Only selected ions of the desired *m/z* values have stable trajectories and oscillate in the center of the Q moving toward the detector, while the remaining ions with different *m/z* are scattered. The Q is an inexpensive mass analyzer; however, it has a low resolution. Therefore, it is used in proteomics mainly in combination with other mass analyzers in multistage mass spectrometers, such as another Q, LIT, OT, or TOF.[83] The *m/z* determination in the TOF is achieved by measuring the time necessary for the ions to pass through the flight tube of the mass analyzer — larger ions arrive later at the detector than the smaller ones. TOF can operate in a linear or reflectron mode. In the linear mode, the TOF resolution is low, but it has no specific upper *m/z* range limit for the measurement. On the contrary, TOF operating in the reflectron mode has high resolution but has limited *m/z* range.[85] The QIT captures the ions in the space between two endcap electrodes and a ring electrode, where they move on the sinusoidal path before sequential ejection to the detector according to their *m/z*. In the case of LIT, the release of the ions into the detector is preceded by their trapping by oscillation between two rods of the mass analyzer. The IT mass analyzers are also characterized by rather low resolution, although LIT has a higher capacity for ion trapping, contributing to better sensitivity and broader dynamic range.[85,86] In proteomics applications, ITs are often used in multistage MS/MS systems with, for example, Q, OT, TOF, or FT-ICR mass analyzers.[85,87] The FT-ICR and OT also belong to trapping mass analyzers. However, the generation of the mass spectrum, and so the *m/z* of the analytes, requires a Fourier transformation of the time domain transients produced by the image current resulting from the FT-ICR or OT analysis.[82] In conditions of high vacuum and high magnetic field, ions move in the FT-ICR in circular orbits with a characteristic frequency depending on *m/z* — ions of higher *m/z* oscillate with lower frequency. Although FT-ICR performance is characterized by the ultra-high resolution and the highest mass accuracy, its use in proteomics is limited due to the complicated maintenance and high costs (it requires supercooled superconductive magnet).[83] Instead, OT, performing also with a very high resolution, has become one of the most often used mass analyzers in proteomics. The ions are injected into OT by high-speed pulses and are set up in oscillatory motion around the inner spindle electrode and along its axis by an electromagnetic field. The oscillation frequencies of the ions in the OT depend on their *m/z*.[82,85,87] The type of ionization technique determines also the preferable mass analyzer(s) to be used for the most efficient ion separations. Taking that into account, ESI is generally compatible with various typically used mass analyzers in proteomics, such as OT, Q, IT, and FT-ICR. MALDI is most frequently coupled with the TOF mass analyzer.[85]

Mass spectrometers equipped with only one mass analyzer are generally used as balances providing the accurate weight of the investigated molecules. However, the mass of an intact peptide or protein is often not sufficient for their unbiased identification. The same amino acids can be arranged differently in a protein/peptide sequence, giving the same intact mass. Thus, the peptides and proteins are sequenced to obtain the order of amino acids in peptide chains for their confident detection. Multistage mass spectrometers, equipped with more than one mass analyzer, have high-throughput protein and peptide sequencing capabilities that rely on the fragment ion mass spectra (MS/MS spectra). Thus, the ability to generate information-rich MS/MS spectra is another important feature of mass analyzers,[2] and the vast majority of mass spectrometers used in proteomics nowadays are multistage instruments.[52] The fragment mass spectra are generated in MS/MS analysis. There, the *m/z* of specific precursor ion is selected in the first mass analyzer and fragmented in a collision cell (e.g., with the use of collisional-induced dissociation). The resulting fragment ions are separated according to their *m/z* ratio in the second mass analyzer and then counted by the detector.[86] Depending on the purpose of the proteomic experiment, different multistage mass spectrometers are employed. A *bottom-up* proteomic approach with extensive sample fractionation prior to MS/MS detection is usually applied for untargeted, discovery studies, where the focus is on the identification and relative quantification of the largest possible number of proteins in complex samples. Discovery proteomics requires high resolution, high sensitivity, and high mass accuracy of the measurements, which can be provided by a variety of hybrid OT mass spectrometers, such as ESI–Q–OT, ESI–LIT–OT, or ESI–Q–LIT–OT. The fast scanning ESI–Q–TOF instruments, which have lower than hybrid OT instruments but still high resolution and mass accuracy, can also be successfully applied to discovery proteomic studies.[87,88] MALDI–TOF/TOF can be applied to identify proteins in less complex samples, such as those in gel spots from SDS-PAGE separations.[84] Unlike in discovery experiments, in targeted proteomics, the focus is on the precise quantification of a selected, predefined number of proteins, often in large sample sets. Targeted proteomic assays must provide sensitive, accurate, and reproducible detection and quantification of analytes over a wide linear concentration range. These features can be provided by a triple quadrupole mass spectrometer, and it is most commonly used for targeted analyses.[89–91] In *top-down* proteomics, on the other hand, high-resolution instruments such as FT-ICR, hybrid OT, or Q-TOF systems with ESI or MALDI ion sources will provide the high mass accuracy needed, for example, to detect post-translational modifications of intact proteins or their complexes.[43,84]

References

1. Domon B, Aebersold R. Mass Spectrometry and Protein Analysis. *Science (1979)*. 2006;312(5771):212–217. doi: 10.1126/science.1124619.
2. Aebersold R, Mann M. Mass Spectrometry-Based Proteomics. *Nature*. 2003;422(6928):198–207. doi: 10.1038/nature01511.
3. Switzar L, Giera M, Niessen WMA. Protein Digestion: An Overview of the Available Techniques and Recent Developments. *J. Proteome Res.* 2013;12(3):1067–1077. doi: 10.1021/pr301201x.

4. Murray KK, Boyd RK, Eberlin MN, *et al.* Definitions of Terms Relating to Mass Spectrometry (IUPAC Recommendations 2013). *Pure Appl. Chem.* 2013;85(7):1515–1609. doi: 10.1351/PAC-REC-06-04-06.
5. Siuzdak G. *An Introduction to Mass Spectrometry Ionization: An Excerpt from the Expanding Role of Mass Spectrometry in Biotechnology*, 2nd ed. MCC Press: San Diego, 2005. *JALA*. 2004;9(2):50–63. doi: 10.1016/j.jala.2004.01.004.
6. Fenn JB, Mann M, Meng CK, *et al.* Electrospray Ionization for Mass Spectrometry of Large Biomolecules. *Science (1979)*. 1989;246(4926): 64–71. doi: 10.1126/science.2675315.
7. Michael K, Franz H. Laser Desorption Ionization of Proteins with Molecular Masses Exceeding 10,000 Daltons. *Anal. Chem.* 1988; 60(20):2299–2301. doi: 10.1021/ac00171a028.
8. Fenn JB. *John B.* Fenn — Nobel Lecture. *NobelPrize.org*. Nobel Prize Outreach AB 2023. Published May 17, 2023. Accessed May 17, 2023. https://www.nobelprize.org/prizes/chemistry/2002/fenn/lecture/.
9. Tanaka K. Koichi Tanaka — Nobel Lecture. *NobelPrize.org*. Nobel Prize Outreach AB 2023. Published May 17, 2023. Accessed May 17, 2023. https://www.nobelprize.org/prizes/chemistry/2002/tanaka/lecture/.
10. Blades AT, Ikonomou MG, Paul K. Mechanism of Electrospray Mass Spectrometry: Electrospray as an Electrolysis Cell. *Anal. Chem.* 1991;63(19):2109–2114. doi: 10.1021/ac00019a009.
11. Konermann L, Ahadi E, Rodriguez AD, *et al.* Unraveling the Mechanism of Electrospray Ionization. *Anal. Chem.* 2013;85(1):2–9. doi: 10.1021/ac302789c.
12. Cech NB, Enke CG. Practical Implications of Some Recent Studies in Electrospray Ionization Fundamentals. *Mass Spectrom. Rev.* 2001; 20(6):362–387. doi: 10.1002/mas.10008.
13. Rohner TC, Lion N, Girault HH. Electrochemical and Theoretical Aspects of Electrospray Ionisation. *Phys. Chem. Chem. Phys.* 2004;6(12):3056. doi: 10.1039/b316836k.
14. Wilm M. Principles of Electrospray Ionization. *Mol. Cell. Proteomics.* 2011;10(7):M111.009407. doi: 10.1074/mcp.M111.009407.

15. Aliyari E, Konermann L. Formation of Gaseous Proteins via the Ion Evaporation Model (IEM) in Electrospray Mass Spectrometry. *Anal. Chem.*2020;92(15):10807–10814.doi:10.1021/acs.analchem.0c02290.
16. McNaught AD, Wilkinson A. *The IUPAC Compendium of Chemical Terminology*, 2nd ed. (Gold V, ed.). International Union of Pure and Applied Chemistry (IUPAC), 2019. doi: 10.1351/goldbook.
17. Steckel A, Schlosser G. An Organic Chemist's Guide to Electrospray Mass Spectrometric Structure Elucidation. *Molecules.* 2019;24(3):611. doi: 10.3390/molecules24030611.
18. Xu G, Stupak J, Yang L, *et al.* Deconvolution in Mass Spectrometry Based Proteomics. *Rapid Commun. Mass Spectrom.* 2018;32(10): 763–774. doi: 10.1002/rcm.8103.
19. Carr SA, Annan RS. Overview of Peptide and Protein Analysis by Mass Spectrometry. *Curr. Protoc. Protein Sci.* 1996;4(1). doi: 10.1002/0471140864.ps1601s04.
20. Banerjee S, Mazumdar S. Electrospray Ionization Mass Spectrometry: A Technique to Access the Information Beyond the Molecular Weight of the Analyte. *Int. J. Anal. Chem.* 2012;2012:1–40. doi: 10.1155/2012/282574.
21. Wilm M, Mann M. Analytical Properties of the Nanoelectrospray Ion Source. *Anal. Chem.* 1996;68(1):1–8. doi: 10.1021/ac9509519.
22. Juraschek R, Dülcks T, Karas M. Nanoelectrospray — More than Just a Minimized-Flow Electrospray Ionization Source. *J. Am. Soc. Mass Spectrom.* 1999;10(4):300–308. doi: 10.1016/S1044-0305(98)00157-3.
23. Lee TD, Moore RE, Young MK. Introducing Samples Directly into Electrospray Ionization Mass Spectrometers Using a Nanospray Interface. *Curr. Protoc. Protein Sci.* 2000;22(1). doi: 10.1002/0471140864.ps1608s22.
24. Hillenkamp F, Karas M, Beavis RC, *et al.* Matrix-Assisted Laser Desorption/Ionization Mass Spectrometry of Biopolymers. *Anal. Chem.* 1991;63(24):1193A–1203A. doi: 10.1021/ac00024a002.
25. Gross JH. Matrix-Assisted Laser Desorption/Ionization. In: *Mass Spectrometry: A Textbook.* Springer Berlin Heidelberg, 2011:507–559. doi: 10.1007/978-3-642-10711-5_11.

26. Clark AE, Kaleta EJ, Arora A, *et al.* Matrix-Assisted Laser Desorption Ionization–Time of Flight Mass Spectrometry: A Fundamental Shift in the Routine Practice of Clinical Microbiology. *Clin. Microbiol. Rev.* 2013;26(3):547–603. doi: 10.1128/CMR.00072-12.
27. Knochenmuss R. Ion Formation Mechanisms in UV-MALDI. *Analyst.* 2006;131(9):966. doi: 10.1039/b605646f.
28. Zenobi R, Knochenmuss R. Ion Formation in MALDI Mass Spectrometry. *Mass Spectrom. Rev.* 1998;17:337–366.
29. Liu BH, Charkin OP, Klemenko N, *et al.* Initial Ionization Reaction in Matrix-Assisted Laser Desorption/Ionization. *J. Phys. Chem. B.* 2010;114(33):10853–10859. doi: 10.1021/jp104178m.
30. Knochenmuss R, Zenobi R. MALDI Ionization: The Role of In-Plume Processes. *Chem. Rev.* 2003;103(2):441–452. doi: 10.1021/cr0103773.
31. Jaskolla TW, Karas M. Compelling Evidence for Lucky Survivor and Gas Phase Protonation: The Unified MALDI Analyte Protonation Mechanism. *J. Am. Soc. Mass Spectrom.* 2011;22(6):976–988. doi: 10.1007/s13361-011-0093-0.
32. Israr MZ, Bernieh D, Salzano A, *et al.* Matrix-Assisted Laser Desorption Ionisation (MALDI) Mass Spectrometry (MS): Basics and Clinical Applications. *Clin. Chem. Lab. Med.* 2020;58(6):883–896. doi: 10.1515/cclm-2019-0868.
33. Karas M, Glückmann M, Schäfer J. Ionization in Matrix-Assisted Laser Desorption/Ionization: Singly Charged Molecular Ions Are the Lucky Survivors. *J. Mass Spectrom.* 2000;35(1):1–12. doi: 10.1002/(SICI)1096-9888(200001)35:1<1::AID-JMS904>3.0.CO;2-0.
34. Macklin A, Khan S, Kislinger T. Recent Advances in Mass Spectrometry Based Clinical Proteomics: Applications to Cancer Research. *Clin. Proteomics.* 2020;17(1):17. doi: 10.1186/s12014-020-09283-w.
35. Dupree EJ, Jayathirtha M, Yorkey H, *et al.* A Critical Review of Bottom-Up Proteomics: The Good, the Bad, and the Future of this Field. *Proteomes.* 2020;8(3):14. doi: 10.3390/proteomes8030014.
36. Brown KA, Melby JA, Roberts DS, *et al.* Top-Down Proteomics: Challenges, Innovations, and Applications in Basic and Clinical Research. *Expert Rev. Proteomics.* 2020;17(10):719–733. doi: 10.1080/14789450.2020.1855982.

37. Furey A, Moriarty M, Bane V, *et al.* Ion Suppression: A Critical Review on Causes, Evaluation, Prevention and Applications. *Talanta.* 2013;115:104–122. doi: 10.1016/j.talanta.2013.03.048.
38. Loo RRO, Dales N, Andrews PC. Surfactant Effects on Protein Structure Examined by Electrospray Ionization Mass Spectrometry. *Protein Sci.* 1994;3(11):1975–1983. doi: 10.1002/pro.5560031109.
39. Kachuk C, Doucette AA. The Benefits (and Misfortunes) of SDS in Top-Down Proteomics. *J. Proteomics.* 2018;175:75–86. doi: 10.1016/j.jprot.2017.03.002.
40. Wiśniewski JR. Filter Aided Sample Preparation — a Tutorial. *Anal. Chim. Acta.* 2019;1090:23–30. doi: 10.1016/j.aca.2019.08.032.
41. Yu YQ, Gilar M, Lee PJ, *et al.* Enzyme-Friendly, Mass Spectrometry-Compatible Surfactant for In-Solution Enzymatic Digestion of Proteins. *Anal. Chem.* 2003;75(21):6023–6028. doi: 10.1021/ac0346196.
42. Saveliev SV, Woodroofe CC, Sabat G, *et al.* Mass Spectrometry Compatible Surfactant for Optimized In-Gel Protein Digestion. *Anal. Chem.* 2013;85(2):907–914. doi: 10.1021/ac302423t.
43. Barth M, Schmidt C. Native Mass Spectrometry — a Valuable Tool in Structural Biology. *J. Mass Spectrom.* 2020;55(10). doi: 10.1002/jms.4578.
44. Breslow R, Guo T. Surface Tension Measurements Show that Chaotropic Salting-in Denaturants Are not Just Water-Structure Breakers. *Proc. Natl. Acad. Sci.* 1990;87(1):167–169. doi: 10.1073/pnas.87.1.167.
45. Donnelly DP, Rawlins CM, DeHart CJ, *et al.* Best Practices and Benchmarks for Intact Protein Analysis for Top-Down Mass Spectrometry. *Nat. Methods.* 2019;16(7):587–594. doi: 10.1038/s41592-019-0457-0.
46. Thomas SL, Thacker JB, Schug KA, *et al.* Sample Preparation and Fractionation Techniques for Intact Proteins for Mass Spectrometric Analysis. *J. Sep. Sci.* 2021;44(1):211–246. doi: 10.1002/jssc.202000936.
47. Bladergroen MR, van der Burgt YEM. Solid-Phase Extraction Strategies to Surmount Body Fluid Sample Complexity in High-Throughput Mass Spectrometry-Based Proteomics. *J. Anal. Methods Chem.* 2015;2015:1–8. doi: 10.1155/2015/250131.
48. Wang C, Qian L, Ji L, *et al.* Affinity Chromatography Assisted Comprehensive Phosphoproteomics Analysis of Human Saliva for Lung

Cancer. *Anal. Chim. Acta.* 2020;1111:103–113. doi: 10.1016/j.aca.2020.03.043.

49. Gundry RL, White MY, Murray CI, *et al.* Preparation of Proteins and Peptides for Mass Spectrometry Analysis in a Bottom-Up Proteomics Workflow. In: *Current Protocols in Molecular Biology.* John Wiley & Sons, Inc., 2009. doi: 10.1002/0471142727.mb1025s88.
50. Rappsilber J, Ishihama Y, Mann M. Stop and Go Extraction Tips for Matrix-Assisted Laser Desorption/Ionization, Nanoelectrospray, and LC/MS Sample Pretreatment in Proteomics. *Anal. Chem.* 2003;75(3):663–670. doi: 10.1021/ac026117i.
51. Goldman AR, Beer LA, Tang H, *et al.* Proteome Analysis Using Gel-LC-MS/MS. *Curr. Protoc. Protein Sci.* 2019;96(1). doi: 10.1002/cpps.93.
52. Darie-Ion L, Whitham D, Jayathirtha M, *et al.* Applications of MALDI-MS/MS-Based Proteomics in Biomedical Research. *Molecules.* 2022;27(19):6196. doi: 10.3390/molecules27196196.
53. Jensen C, Haebel S, Andersen SO, *et al.* Towards Monitoring of Protein Purification by Matrix-Assisted Laser Desorption Ionization Mass Spectrometry. *Int. J. Mass Spectrom. Ion Process.* 1997;160(1–3):339–356. doi: 10.1016/S0168-1176(96)04499-0.
54. Zhang N, Li L. Effects of Common Surfactants on Protein Digestion and Matrix-Assisted Laser Desorption/Ionization Mass Spectrometric Analysis of the Digested Peptides Using Two-Layer Sample Preparation. *Rapid Commun. Mass Spectrom.* 2004;18(8):889–896. doi: 10.1002/rcm.1423.
55. Padliya ND, Wood TD. A Strategy to Improve Peptide Mass Fingerprinting Matches Through the Optimization of Matrix-Assisted Laser Desorption/Ionization Matrix Selection and Formulation. *Proteomics.* 2004;4(2):466–473. doi: 10.1002/pmic.200300567.
56. Vorm O, Mann M. Improved Mass Accuracy in Matrix-Assisted Laser Desorption/Ionization Time-of-Flight Mass Spectrometry of Peptides. *J. Am. Soc. Mass Spectrom.* 1994;5(11):955–958. doi: 10.1016/1044-0305(94)80013-8.
57. Kriegsmann J, Kriegsmann M, Casadonte R. MALDI TOF Imaging Mass Spectrometry in Clinical Pathology: A Valuable Tool for Cancer

Diagnostics (Review). *Int. J. Oncol.* 2015;46(3):893–906. doi: 10.3892/ijo.2014.2788.

58. Ščupáková K, Balluff B, Tressler C, *et al.* Cellular Resolution in Clinical MALDI Mass Spectrometry Imaging: The Latest Advancements and Current Challenges. *Clin. Chem. Lab. Med.* 2020;58(6):914–929. doi: 10.1515/cclm-2019-0858.
59. Wang H, Wong CH, Chin A, *et al.* Integrated Mass Spectrometry–Based Analysis of Plasma Glycoproteins and Their Glycan Modifications. *Nat. Protoc.* 2011;6(3):253–269. doi: 10.1038/nprot.2010.176.
60. Pitt JJ, Goman JJ. Matrix-Assisted Laser Desorption Ionization Time-of-Flight Mass Spectrometry of Sialylated Glycopeptides and Proteins Using 2,6-Dihydroxyacetophenone as a Matrix. *Rapid Commun. Mass Spectrom.* 1996;10:1786–1788.
61. Goman JJ, Ferguson BL, Nguyen TB. Use of 2,6-Dihydroxyacetophenone for Analysis of Fragile Peptides, Disulphide Bonding and Small Proteins by Matrix-Assisted Laser Desorption Ionization. *Rapid Commun. Mass Spectrom.* 1996:529–536.
62. Hou J, Xie Z, Xue P, *et al.* Enhanced MALDI-TOF MS Analysis of Phosphopeptides Using an Optimized DHAP/DAHC Matrix. *J. Biomed. Biotechnol.* 2010;2010:1–12. doi: 10.1155/2010/759690.
63. Řehulka P, Šalplachta J, Chmelík J. Improvement of Quality of Peptide Mass Spectra in Matrix-Assisted Laser Desorption/Ionization Time-of-Flight Mass Spectrometry and Post-Source Decay Analysis of Salty Protein Digests by Using On-Target Washing. *J. Mass Spectrom.* 2003;38(12):1267–1269. doi: 10.1002/jms.548.
64. Laugesen S, Roepstorff P. Combination of Two Matrices Results in Improved Performance of MALDI MS for Peptide Mass Mapping and Protein Analysis. *J. Am. Soc. Mass Spectrom.* 2003;14(9):992–1002. doi: 10.1016/S1044-0305(03)00262-9.
65. Fukuyama Y, Funakoshi N, Takeyama K, *et al.* 3-Aminoquinoline/p-Coumaric Acid as a MALDI Matrix for Glycopeptides, Carbohydrates, and Phosphopeptides. *Anal. Chem.* 2014;86(4):1937–1942. doi: 10.1021/ac4037087.
66. Chen Y, Mehok AR, Mant CT, *et al.* Optimum Concentration of Trifluoroacetic Acid for Reversed-Phase Liquid Chromatography of

Peptides Revisited. *J. Chromatogr. A.* 2004;1043(1):9–18. doi: 10.1016/j.chroma.2004.03.070.

67. Lewandowska AE, Macur K, Czaplewska P, *et al.* Human Follicular Fluid Proteomic and Peptidomic Composition Quantitative Studies by SWATH-MS Methodology. Applicability of High pH RP-HPLC Fractionation. *J. Proteomics.* 2019;191:131–142. doi: 10.1016/j.jprot.2018.03.010.
68. Yeung D, Mizero B, Gussakovsky D, *et al.* Separation Orthogonality in Liquid Chromatography–Mass Spectrometry for Proteomic Applications: Comparison of 16 Different Two-Dimensional Combinations. *Anal. Chem.* 2020;92(5):3904–3912. doi: 10.1021/acs.analchem.9b05407.
69. Wolters DA, Washburn MP, Yates JR. An Automated Multidimensional Protein Identification Technology for Shotgun Proteomics. *Anal. Chem.* 2001;73(23):5683–5690. doi: 10.1021/ac010617e.
70. Wang Z, Ma H, Smith K, *et al.* Two-Dimensional Separation Using High-pH and Low-pH Reversed Phase Liquid Chromatography for Top-Down Proteomics. *Int. J. Mass Spectrom.* 2018;427:43–51. doi: 10.1016/j.ijms.2017.09.001.
71. Yang Z, Shen X, Chen D, *et al.* Improved Nanoflow RPLC-CZE-MS/MS System with High Peak Capacity and Sensitivity for Nanogram Bottom-up Proteomics. *J. Proteome Res.* 2019;18(11):4046–4054. doi: 10.1021/acs.jproteome.9b00545.
72. McCool EN, Lubeckyj RA, Shen X, *et al.* Deep Top-Down Proteomics Using Capillary Zone Electrophoresis-Tandem Mass Spectrometry: Identification of 5700 Proteoforms from the *Escherichia coli* Proteome. *Anal. Chem.* 2018;90(9):5529–5533. doi: 10.1021/acs.analchem.8b00693.
73. Hedges JB, Vahidi S, Yue X, *et al.* Effects of Ammonium Bicarbonate on the Electrospray Mass Spectra of Proteins: Evidence for Bubble-Induced Unfolding. *Anal. Chem.* 2013;85(13):6469–6476. doi: 10.1021/ac401020s.
74. Kašička V. Peptide Mapping of Proteins by Capillary Electromigration Methods. *J. Sep. Sci.* 2022;45(23):4245–4279. doi: 10.1002/jssc.20g2200664.
75. Li Y, Champion MM, Sun L, *et al.* Capillary Zone Electrophoresis-Electrospray Ionization-Tandem Mass Spectrometry as an Alternative Proteomics Platform to Ultraperformance Liquid Chromatography-

Electrospray Ionization-Tandem Mass Spectrometry for Samples of Intermediate Complexity. *Anal. Chem.* 2012;84(3):1617–1622. doi: 10.1021/ac202899p.

76. Ramautar R, Heemskerk AAM, Hensbergen PJ, *et al.* CE–MS for Proteomics: Advances in Interface Development and Application. *J. Proteomics.* 2012;75(13):3814–3828. doi: 10.1016/j.jprot.2012.04.050.
77. Chen D, McCool EN, Yang Z, *et al.* Recent Advances (2019–2021) of Capillary Electrophoresis-Mass Spectrometry for Multilevel Proteomics. *Mass Spectrom. Rev.* 2023;42(2):617–642. doi: 10.1002/mas.21714.
78. Harper S, Speicher DW. Comparing Complex Protein Samples Using Two-Dimensional Polyacrylamide Gels. *Curr. Protoc. Protein Sci.* 2019;96(1). doi: 10.1002/cpps.87.
79. Kondo T. Cancer Biomarker Development and Two-Dimensional Difference Gel Electrophoresis (2D-DIGE). *Biochimica. et Biophysica. Acta (BBA) — Proteins Proteomics.* 2019;1867(1):2–8. doi: 10.1016/j.bbapap.2018.07.002.
80. Weng CY, Wang CE, Xie WB, *et al.* Comparative Proteome Analysis of Actinoplanes Utahensis Grown on Various Saccharides Based on 2D-DIGE and MALDI-TOF/TOF-MS. *J. Proteomics.* 2021;239:104193. doi: 10.1016/j.jprot.2021.104193.
81. Jacqueline C, Dracz M, Boothman S, *et al.* Identification of Cell Surface Molecules That Determine the Macrophage Activation Threshold Associated with an Early Stage of Malignant Transformation. *Front. Immunol.* 2021;12. doi: 10.3389/fimmu.2021.749597.
82. Scigelova M, Hornshaw M, Giannakopulos A, *et al.* Fourier Transform Mass Spectrometry. *Mol. Cell. Proteomics.* 2011;10(7):M111.009431. doi: 10.1074/mcp.M111.009431.
83. Li C, Chu S, Tan S, *et al.* Towards Higher Sensitivity of Mass Spectrometry: A Perspective from the Mass Analyzers. *Front. Chem.* 2021;9. doi: 10.3389/fchem.2021.813359.
84. Han X, Aslanian A, Yates JR. Mass Spectrometry for Proteomics. *Curr. Opin. Chem. Biol.* 2008;12(5):483–490. doi: 10.1016/j.cbpa.2008.07.024.
85. Noor Z, Ahn SB, Baker MS, *et al.* Mass Spectrometry–Based Protein Identification in Proteomics — a Review. *Brief Bioinform.* 2021;22(2):1620–1638. doi: 10.1093/bib/bbz163.

86. Domon B, Aebersold R. Mass Spectrometry and Protein Analysis. *Science (1979)*. 2006;312(5771):212–217. doi: 10.1126/science.1124619.
87. Savaryn JP, Toby TK, Kelleher NL. A Researcher's Guide to Mass Spectrometry-Based Proteomics. *Proteomics.* 2016;16(18):2435–2443. doi: 10.1002/pmic.201600113.
88. Lin L, Lin H, Zhang M, *et al.* Types, Principle, and Characteristics of Tandem High-Resolution Mass Spectrometry and Its Applications. *RSC Adv*. 2015;5(130):107623–107636. doi: 10.1039/C5RA22856E.
89. Macur K, Hagen L, Ciesielski TM, *et al.* A Targeted Mass Spectrometry Immunoassay to Quantify Osteopontin in Fresh-Frozen Breast Tumors and Adjacent Normal Breast Tissues. *J. Proteomics.* 2019;208:103469. doi: 10.1016/j.jprot.2019.103469.
90. Macur K, Zieschang S, Lei S, *et al.* SWATH-MS and MRM: Quantification of Ras-Related Proteins in HIV-1 Infected and Methamphetamine-Exposed Human Monocyte-Derived Macrophages (hMDM). *Proteomics.* 2021;21(15):2100005. doi: 10.1002/pmic.202100005.
91. Vidova V, Spacil Z. A Review on Mass Spectrometry-Based Quantitative Proteomics: Targeted and Data Independent Acquisition. *Anal. Chim. Acta.* 2017;964:7–23. doi: 10.1016/j.aca.2017.01.059.

https://doi.org/10.1142/9789811283376_0003

Chapter 3

Basic Proteomics Protocols

Paulina Czaplewska* and Marc Muller

Intercollegiate Faculty of Biotechnology UG&MUG,
University of Gdańsk, Gdańsk, Poland

Contents

---

*Corresponding author.
Laboratory of Mass Spectroscopy, Intercollegiate Faculty of Biotechnology, University of Gdansk, Gdansk, Poland Paulina Czaplewska, Email: Paulina.Czaplewska@ug.edu.pl, Tel +48 585236437.

Abstract

The development of proteomics has made it possible to take a big step forward in terms of identification and qualitative and quantitative analyses of proteins. For a novice scientist, basic protocols such as digestion in a solution or digestion of proteins separated in a gel are enough to obtain the first information about the test sample. They are helpful not only to scientists dealing with proteomics but also to people who want to use them to check the correctness of their work in a biological, microbiological, or chemical laboratory. After reading the following chapter, any person starting their scientific work will be able to start their adventure with proteomics, choose the optimal combination of reagents, and enjoy working with proteomics.

1. Source of sample in proteomics research

The main focus in proteomics is proteins and their qualitative and quantitative analyses. Therefore, this technique allows us to identify proteins, define their sequences and post-translational modifications, and quantitatively determine the proteome response to introduced or pathological changes. It is currently one of the main approaches with high sensitivity and throughput that allow identifying and validating biomarkers or molecular targets for new drugs. It is not a simple task, and we often encounter problems mainly generated by the biological material in proteomics. In proteomics, work can be done on any provided material containing proteins. The most popular are animal and human tissues (fresh, frozen, embedded in paraffin or formalin) and body fluids (blood, saliva, cerebrospinal fluid, synovial fluid, follicular fluid, saline, and urine). It applies to both animal and bacterial or yeast cells. The cells are excellent material for protein analysis, where it is easy to make the intended changes directly in the genome and observe the result of these changes at the protein level. Among the samples mentioned above, sometimes there are fewer standards in the form of deposits such as amyloid deposits, stones (sialoliths, bile, urinary), bone tissue, teeth, and fat. Also, do not forget about the analysis of the protein preparations of plant preparations, from which

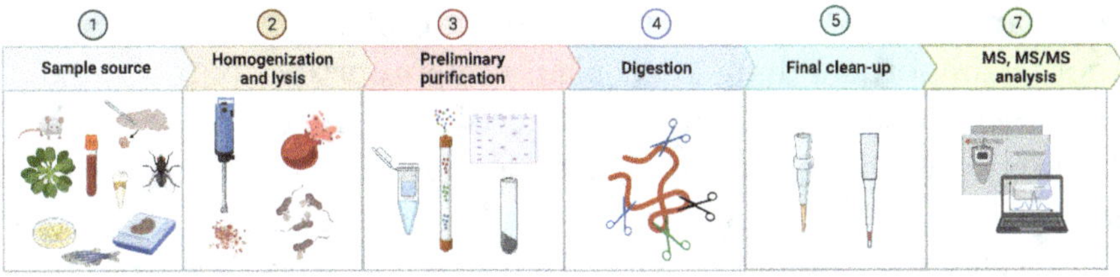

Figure 1. Basic steps in proteomics sample preparation. MS, mass spectrometry; MS/MS, tandem mass spectrometry.

you can analyze each part of the plant, such as leaves, stems, roots, as well as flowers, fruit, and seeds.

The list of potential samples is enormous; therefore, it is hard to talk about one standard sample preparation method. Essential steps can be distinguished as shown in Figure 1, and they are applied to each type of material. However, the detailed course of perception must be matched individually, based on literature or refined by the experimenter for a given material, and additional information such as the scientific question, the amount and homogeneity of material to conduct proteomics, location of the protein that interests us, or the protein fraction in the cell. Both the first stages, as the tissue homogenization or the choice of the suitable buffer for the cell lysis, preliminary purification and extraction of proteins, the efficient digestion process itself, the selection of the appropriate enzyme, and the final clean-up will significantly affect the final result. Therefore, the workflow developed during preliminary tests will be crucial for the success of the complete proteomics analysis.

2. Sample homogenization and lysis

When starting preparations for proteomics analysis with the use of a selected biological matrix, first of all, the possibility of collecting an appropriate number of samples should be taken into account. According to good practice, three biological and three technical repetitions should be carried out to eliminate fluctuations resulting from the repeatability of the experiment itself and the registration of mass spectrometry (MS) and tandem mass spectrometry (MS/MS) spectra. Another important thing is to plan an appropriate control for our experiment, that is, to obtain material that reflects the baseline (healthy) state of the tested sample as much as possible. The main goal of this step is to fragment the material and break the continuity of the cell wall/membrane, which in turn leads to the release of proteins, nucleic acids, and any biomolecules in the cells. In this way, obtaining the soluble form of the selected analytes allows for further analysis.

Focusing on different types of samples (tissue, body fluid, and cells), we will briefly present possible approaches to the first stage of protein extraction from material by homogenization and cellular lysis. Considering the diversity of the material undergoing proteomics analysis, it is always best to search for the most effective homogenization methods and cell lysis in the current literature.

When working with *cells*, the situation is relatively straightforward. After receiving the material for work, make sure that the cells have been washed and the remains of the breeding medium have been removed, which in some pads can be a problem in further processing (hesitating detergents) or cause the presence of additional signals on MS and MS/MS spectra (e.g., medium with albumin or other protein additions). In most cases, it is sufficient to use an adequately selected lysis buffer (Table 1) or a commercially available kit dedicated to a specific cell type. Still, in the case of more resistant cells, they can also use properly selected mechanical homogenization. The non-mechanical methods include introducing cells into osmotic shock, thermolysis, or adding chemicals such as detergents, non-polar solvents, chaotropic substances, bases, and chelating agents.

When working with *animal tissue*, we often do not influence the collection stage performed by a surgeon or person working, for example, with animals. If the work concerns pathological changes in the tissue like cancer or amyloid, the collected fragment must contain altered tissue, but at the same time, as minor of the surrounding healthy tissue or, for example, fat as possible. If possible, fresh tissue should be rinsed with cold saline to remove residual blood or other surrounding fluids. If our task is to study global changes, not local

Table 1. Examples of lysis buffers used in proteomics, depending on the organism as the sample source.

Sample type	Lysis buffer composition	Conditions	Ref.
Algae	**Urea buffer** • 9 M urea, 4% w/v CHAPS (3-[(3-Cholamidopropyl) dimethylammonio]-1-propane sulfonate), 60 mM dithiothreitol (DTT), 40 mM Tris-HCl, pH 7.4, 1% w/v carrier ampholytes, pH 3–10 • 8 M urea, 4% w/v 3-CHAPS, 40 mM Tris, 0.2% Bio-Lyte 4–7, 2 mM TBP **Sodium dodecyl sulfate (SDS) buffer** • 50 mM Tris-HCl (pH 7.8), 150 mM NaCl, 1% SDS **Sucrose buffer** • 30% sucrose, 0.1% KCl, 2% SDS, 50 mM EDTA, 0.1 M Tris, pH 8	Mechanical disruption Sonication	9–12
Bacteria	**Urea buffer** • 8 M urea, 2 M thiourea, 4% (w/v) CHAPS, 20 mM DTT, 2% Pharmalyte • 8 M urea, 4% w/v CHAPS, 50 mM Tris-HCl, 65 mM DTE **SDS buffer** • 0.3% SDS, 200 mM DTT, 50 mM Tris (pH = 8.0) • 5% SDS, 50 mM Tris-HCl, pH 8.5, 0.15 M NaCl, 0.1 mM EDTA, 1 mM $MgCl_2$; 50 mM DTT **Alkaline lysis** • 0.2 N NaOH, 1% (w/v) SDS, de-ionized water **Lysozyme buffer** • 10 mM Na-phosphate containing 10 mg/mL lysozyme • 50 mM Tris-HCl, pH 8, 20% sucrose, 4 mM EDTA, 0.5 mg/mL lysozyme, 1 mM PMSF, 10 mM $MgCl_2$	30 min in 4°C 30-min sonication 60 min 30°C water bath	13–17

(*Continued*)

Table 1. (*Continued*)

Sample type	Lysis buffer composition	Conditions	Ref.
Yeast	**Lysis buffer** • 50 mM Tris, pH 7.5, 100 mM NaCl, 0.5% NP40 • (pH 7.4) 3.8 mM $NaH_2PO_4 \times H_2O$, 49.4 mM $Na_2HPO_4 \times 2H_2O$, 48.4 mM NaCl, 5 mM KCl, 20% glycerol, 1% 2-mercaptoethanol, 0.3% IGEPAL CA-630 • 0.1 M NaOH, 0.05 M EDTA, 2% SDS, 2% 2-mercaptoethanol • 8 M Urea, 0.1 N NaOH, 0.05 M EDTA, 2% 2-mercaptoethanol	Mechanical disruption	18–22
	Lyticase lysis buffer • 50 mM Tris-HCl, pH 8.0, 10 mM EDTA, 1.2 M sorbitol, 10 mM β-mercaptoethanol containing 2 U of lyticase/μL	Incubation 1 h at 37°C	
	Proteinase K lysis buffer • 50 mM Tris-HCl, pH 8.0, 10 mM EDTA, 1.2 M sorbitol, 10 mM β-mercaptoethanol, 100 μL proteinase K		
Plants	**O'Farrell lysis buffer** • 9 M urea, 2% Nonidet P-40 (NP-40), 2% 2-ME, 2% carrier ampholytes (any desired pH interval) **Urea buffer** • 8 M urea, 2 M thiourea, 2 mM disodium EDTA salt, 4% CHAPS, 65 mM DTT, 2% ampholytes (pH 3–10), 1% TBP (tributylphosphine) **Other buffers** • 0.5 M EDTA, pH 8.0, 1 M Tris-HCl, pH 6.8, 10% SDS, 100% 2-ME, 100% glycerol, bromophenol blue powder • 50 mM Tris-HCl, 15 mM EDTA, 1 mM NaF, 0.5 mM Na_3VO_3, 15 mM β-glycerophosphate, 1 mM PMSF, 1 mM DTT, 2 μg/mL pepstatin, 2 μg/mL aprotinin, 2 μg/mL leupeptin, pH 7.5	Mechanical disruption	23,24

Human cells/tissue/FFPE tissue	**GnHCl lysis buffer** • 6 M GdnHCl and 0.1 M Tris-HCl, pH 8.5, 5 mM tris(2-carboxyethyl) phosphine (TCEP), 10 mM CAA	Boiled at 95°C for 10 min	3,25–28
	SDS lysis buffer: • 4% SDS and 0.1 M Tris-HCl, pH 8.5, 5 mM TCEP, 10 mM CAA • 4% SDS, 0.1 M Tris-HCl, pH 8.0, 0.1 M DTT, 0.5% (w/v) polyethylene glycol 20,000	Pressure-assisted lysis	
	Urea buffer • 4 M urea/50 mM NH_4HCO_3/10 mM Tris, TCEP/40 mM iodoacetamide		
	RIPA buffer • 25 mM Tris/HCl, 150 mM NaCl, 1% NP40, 1% SDS **GlyNP40 buffer** • 50 mM HEPES, 150 mM KCl, 1 mM $MgCl_2$, 10% glycerol, 0.5% NP-40	Mechanical disruption	

ones in the resected tissue, then such tissue should be homogenized to reduce intratissue heterogeneity, for example, in solid tumors. If we are interested in changes in the subsections of the delivered material, and the size of the tissue allows it, then before homogenization, the tissue should be divided into fragments and treated separately.

In the laboratory, mechanical methods are often used for homogenization are bead mills, manual rotor stator homogenizers with different tips, high-pressure homogenization [pressure cycling technology (PCT)], ultrasound or microwave, and mortar. The main problem is the homogenization method's adjustment to the material's size. The most common practice is to place the tissue or its fragments in liquid nitrogen and then mechanically crush it (in a mortar or mill). The resulting powder is then transferred to the Eppendorf′ tube and flooded with the selected lysis buffer. Now all that remains is to support the action of the buffer by heating it or placing it in ultrasound, which will release the entire content of the cells.

Tissues collected from patients during surgery or biopsy most often go to pathomorphological laboratories, which are fixed with formalin and embedded in paraffin. It allows for prolonged storage of samples and their banking. This way, it is possible to obtain samples from various centers worldwide. The scraps are cut from the block and stored in this way with the help of a cryoton, which can be further processed. The first step is to get rid of paraffin using hot water or xylene/ethanol, which has been extensively described in many studies.[1,2] Here, attention should be paid to the possible occurrence of protein modifications caused by formaldehyde.[3,4] The problem is the formation of methylene bridges, which limit the availability of trypsin during digestion, resulting in incomplete digestion and cross-linked peptides. In turn, is a problem in the identification of proteins? It can be prevented by applying the correct temperature and pH to paraffin removal. It is, therefore, important that the stage of preparing the sections for homogenization is carried out as precisely and scrupulously as possible. Due to the size of the material after deparaffinization for homogenization, the PCT method, which consists of cyclic pressure changes (high to low) to break down the tissue, works perfectly well.[5–7]

In plant tissues requiring direct mechanical forces using homogenization, the most common method is crushing the tissue in liquid nitrogen in a mortar, pestle, or adjusted bead-beating machine, followed by extraction of proteins in a dedicated lysis buffer.[8]

3. Additives in sample preparation

In proteomics analysis, enzymatic digestion is the critical step in identifying proteins. For this process to be efficient, it is necessary to provide full enzyme access to the protein, which must be deprived of any interactions with partners, that is, dissociate potential protein–protein, peptide–protein, lipid–protein complexes, and so on. It is also necessary to deprive the protein of its structure, that is, carry out the denaturation process while keeping it soluble. It is a tremendous challenge for a scientist, especially when working with a complex and protein-rich matrix.[29] Different chemical agents are an obligatory addition to buffers already at the stage of cell lysis, protein extraction, or during enzymatic digestion itself. The most commonly used additives include chaotropes, detergents, reducing agents, surfactants, or salts. At the same time, the other less commonly helpful additives are organic solvents, ampholytes, or strong bipolar molecules.[30] However, not all of the commonly known and used additives are fully compatible with the liquid chromatography–mass spectrometry (LC–MS) technique. Therefore, it is crucial at the beginning of planning experiments to make an excellent choice to ensure the maximum increase in protein identification to the effort and costs needed to remove additives before the final LC–MS analysis. When choosing the right set of additives for planned work, it is good to read the literature to find the collected information concerning the type of sample or analysis you want to perform (qualitative analysis, quantitative analysis, analysis of interactions in complexes).[31,32]

The most commonly used chaotropic additives are urea and sodium dodecyl sulfate (SDS), and popular components of the lysis buffers are used in the preparation of samples for analysis using electrophoresis (Table 2). The first mentioned compound is a potent protein denaturant, disrupting non-covalent and ionic bonds between amino acid residues, and the result is unfolded proteins.[33] In addition to denaturation, it also helps to dissolve proteins.[34] Urea can be used in concentrations ranging from 5 to 8 M, freshly prepared before

Table 2. Examples of chemical additives used during sample preparation for proteomics studies.

Subtype	Chemical name and formula	Additive name	Purpose	Compatibility with LC–MS	Ref.
Chaotropic	Carbonyldiamide CH_4N_2O or NH_2CONH_2	**Urea**	Non-ionic chaotropic agent–denaturing proteins, applied concentrations up to 8 M, compatible with other additives, temperature sensitive, stable up to about 30°C, and removed by dialysis or membrane filtration.	No	[29,32,39,40]
	Sodium dodecyl sulfate $C_{12}H_{25}NaO_4S$	**SDS**	Ionic, micellar [critical micelle concentration (CMC) 6–8 mM]; the most potent used detergent for protein solubilization breaks protein–lipid and lipid–lipid interactions; popular component of lysis buffers (1%–4% concentration); and can be removed by precipitation (with KCl), strong cation exchange chromatography, or membrane filtration (like in FASP approach).	No	[41,44,87,88]

Detergent	4-Nonylphenyl-polyethene glycol $HO(CH_2CH_2O)_{nC}6H_4C_9H_{19}$	**NP-40**	Non-ionic detergent (CMC 0.05–0.3 mM) used for the isolation and purification of functional membrane proteins and removed by commercial kits and dedicated detergent removal resins.	No	[89,90]
	Octylphenolpoly(ethylene glycol ether)$_x$ $C_{16}H_{26}O_2$	**Triton X-100**	Non-ionic detergent, used as 10% or less solution, solubilize membrane proteins, CMC 0.22–0.24 mM, retains protein activity, interferes with protein quantification by absorption, and removed by polymeric resins like Amberlite.	No	[91–93]
	Polyethylene glycol sorbitan monooleate $C_{32}H_{60}O_{10}$	**Tween 80**	Non-ionic detergent, essential for protein solubilization, a polymer composed of PEGylated sorbitan, where the total number of poly (ethylene glycol) units is 20 (w + x + y + z = 20) and an oleoyl group caps a single terminal. Aggregation protector prevents surface absorption, solubilizes membrane proteins and removed by commercial kits and detergent removal resins.	No	[89,94]

(*Continued*)

Table 2. (*Continued*)

Subtype	Chemical name and formula	Additive name	Purpose	Compatibility with LC–MS	Ref.
	3-[(3-Cholamidopropyl) dimethylammonio]-1-propane sulfonate $C_{32}H_{58}N_2O_7S$	**CHAPS**	Non-denaturing zwitterionic detergent with mild denaturing properties, well soluble (CMC 8–10 mM), solubilizes membrane proteins, breaks protein–protein interactions and removed by dialysis.	No	95–98
	$C_{22}H_{42}O_{11}$	**n-Decyl-beta-D-maltoside**	Non-ionic, mild, and non-denaturing detergent used to disrupt protein–lipid, lipid–lipid interactions and used in the extraction of membrane proteins in their native state, not compatible with ultraviolet measurements at 280 nm.	No	99,100
Reducing agents	Dithiothreitol $C_4H_{10}O_2S_2$	**DTT**	Redox molecule, Cleland's reagent, used to reduce disulfide bonds of proteins/peptides, not stable for a long time in reduced form — used with urea or SDS.	No	101,102
	2-Mercaptoethanol $HOCH_2CH_2SH$	**BME**	Effectively cleaves and reduces disulfide bonds, disrupts tertiary and quaternary protein structures, and unfolds native proteins.	No	103,104

	Tris (2-carboxyethyl) phosphine $C_9H_{15}O_6P$	**TCEP**	Extremely potent, selective reductant in acid media, not reactive with other functional groups, and impermeable to the hydrophobic core of the protein.	No	101,102
Surfactants	3-[(2-methyl-2-undecyl-1,3-dioxolan-4-yl) methoxy]-1-propane sulfonate	**RapiGest**	An acid-cleavable anionic detergent facilitates digestion; hydrolyzed 0.5% (v/v) trifluoroacetic acid (TFA) for 45 min at 37°C; and used for plasma, cells, and tissues, and plant proteomics.	Yes	31,56,105–108
	Sulfonate sodium 3-((1-(furan-2-yl) undecyloxy) carbonylamino) propane-1-sulfonate)	**ProteaseMAX**	Hydrophobic, anionic, acid-labile surfactant; in concentration 0.1%–0.2% (w/v) hydrolyses simultaneously with enzymatic proteolysis; degraded completely in 0.5 TFA for 15 min at 37°C or neutral pH incubation at 95°C.	Yes	31,54,59
	Surfactant 3192	**SoluMAX**	Acid-labile surfactant.	Yes	55

(*Continued*)

Table 2. (*Continued*)

Subtype	Chemical name and formula	Additive name	Purpose	Compatibility with LC–MS	Ref.
	3-[3-(1,1-bisalkyloxyethyl) pyridin-1-yl] propane-1-sulfonate	**PPS Silent**	Acid-labile, a zwitterionic surfactant, is applied to cells, tissues, and plasma; degraded at low pH; and the product does not interfere with mass spectrometry.	YES	31,57
	3-[3-(1,1-bisalkyloxyethyl)-1-yl]propane-1-sulfonate	**AALS II**	Anionic; acid-labile surfactant II improves protein solubilization, cell lysis, and enzymatic digestion; reduces non-specific interactions of the protein with the surface; hydrolyze on the proteolysis, at pH 2.5–3 with 1% TFA incubation for 10 min at 37°C.	YES	31,58,60
	Sodium deoxycholate	**SDC or DOC**	Ionic bile salt surfactants' polar and non-polar parts, mild detergent; under acidic conditions forms insoluble deoxycholic acid and precipitate; and removed by centrifugation.	YES	77,78,109–111

use, but can also be stored at −80°C for up to several months. It perfectly tolerates the presence of other additives. It should be remembered that heating the urea solution leads to the release of isocyanate, leading to protein carbamylation (α- and ε-groups of lysine, arginine, *N*-terminus of peptides) and charge heterogeneities of the samples.[35–37] Consequently, this causes identification problems and isotopic labeling attempts in quantification.[38] The solution to the problem of carbamylation can be, for example, using ammonium-containing buffers[39] or sample processing and digestion with a temperature not exceeding 37°C.[40] Urea is incompatible with MS and must be removed from the sample before analysis. Commonly known clean-up methods such as acetone precipitation, membrane filtration, or desalination can be used for this purpose.

Another chaotropic agent combined with gel electrophoresis is SDS (Table 2). It plays a vital role in the denaturation and solubilization of proteins and affects the lysis and binding of proteins. The usage of SDS in proteomics protocols comes with specific advantages and disadvantages. The most significant benefit of SDS in the case of in-gel digestion is the possibility to separate proteins by molecular weight during gel electrophoresis, allowing for an easy, fast, and cheap way to separate proteins.[41] Detergents or chaotropes can disrupt hydrogen bonds in the solution by exerting chaotropic activity, thereby increasing the solubility of hydrophobic particles.[42] They can also desaturate macromolecules (e.g., protein) by weakening hydrophobic effects by creating perturbation in the hydrophobic network. Direct binding to proteins can weaken hydrophobic interactions between non-polar amino acids by high pressure, which destabilizes the native state of proteins and desaturates proteins.[43]

On the other hand, a significant downside of SDS usage is its interference with chromatography. When the concentration is above 0.01% and reaches above 0.1%, removal is necessary before any further step can be taken.[41] In the case of SDS, precipitation with warm acetone can be used to remove it,[44] and precipitation with potassium chloride (KCl) is more accurate and effective.[45] Currently, the use of filter-aided sample preparation (FASP) etching is prevalent, which allows for the efficient use of other aids, thanks to which washing centrifugal spin filters with a membrane with a mass cut-off of 10 or 30 kDa combined with centrifugation is obtained.[46]

The use of detergents in molecular and cell biology goes beyond their use as a niche agent. It is used for various biological experiments, where its main activity is that of a surfactant (surface-acting agent). These agents are used in MS to lower surface tension and increase access to unfolded proteins. Unfortunately, different detergents are better suitable for different proteins, and all of them need to be removed because they interfere with chromatography and ionization in the MS ion source. Detergents can inhibit the evaporation of solvents in the ESI ion source and can also be ionized, leading to strong signals in detector systems (suppressing peptide signals).[47–49] Removing these detergents is therefore vital for successful sample preparation and can be achieved using different techniques, reaching from protein precipitation to protein-absorbing beads in commercial ready-to-use kits. Various detergents are available to suit the needs of different research fields and sample types (Table 1). The detergents most often used and best described are the following: NP-40, Triton X-100, CHAPS (3-[(3-Cholamidopropyl) dimethylammonio]-1-propane sulfonate), Tween 80, and n-dodecyl-β-D-maltoside, but this does not mean that these are the only ones used, and several others have been used and described in various publications. Most detergents work similarly by denaturing proteins when forming micelles.[47] Using detergents could be considered a double-edged sword, as on the one hand, they need to be removed and adding detergents should be done only after solid considerations. On the other hand, they enable higher performance in proteomics and precision medicine.[50]

Another group of regularly used substances is reducing agents. They induce the reduction of disulfide bonds in proteins, leading to denaturation, solubilization, and protection of thiols against forming disulfide bonds again.[51] They are often used with thiol-binding reagents, like iodoacetamide (IAA), to prevent their reformation effectively.[30] This allows the enzymes to digest the protein more efficiently and reduces the sample loss.[52] The most commonly used reducing agents in proteomics are dithiothreitol (DTT), tris (2-carboxyethyl)phosphine, and β-mercaptoethanol. All are highly efficient at making amino acids accessible for enzymes like trypsin, the most widely applied enzyme in enzymatic digestion, which could be considered a golden standard in sample preparation.[53]

Due to the high sensitivity of both chromatography and MS to the presence of additives, a new generation of compounds, surfactants, are being developed that are fully compatible and do not require additional steps in the preparation of samples for analysis. Most of these are compounds that disintegrate in an acidic environment — acid-labile surfactants (ALS) like ProteaseMAX,[54] SoluMAX (Promega),[55] RapiGest (Waters),[56] PPS Silent (Agilent),[57] and AALS II (Progenta).[58] Using any compound from this commercially available group is an excellent alternative, especially for novice scientists in proteomics. For the decomposition of ALS, it is enough to use 0.5%–1% of trifluoroacetic acid (v/v, TFA) and briefly warm the sample (15–45 min 37°C).[59–61] They are perfect for working with biological matrices, such as bacterial[62] or human cells,[63] tissues,[64–66] serum[67] and plasma,[68–70] or plant samples.[49,71–73] The formulae and a brief description of each of them are presented in Table 1. An example of an equally helpful last-group surfactant that yields insoluble precipitates at low pH is sodium deoxycholate. A steroid similar in structure to SDS is also an ionic bile salt isolated from the gastrointestinal tract.[74] It belongs to the mild detergents facilitating the dissolution and increasing enzyme activity during the digestion of proteins.[75] To remove it prior to analysis, the pH of the solution must be lowered to form insoluble deoxycholic acid, which can be easily removed by centrifuging sample.[76] Alternatively, the precipitated decomposition product can be removed by phase transfer (PTS) to the organic phase.[75] It is a very popular additive that works well with all kinds of samples in proteomics.[77,78]

Organic solvents such as acetonitrile, methanol or ethanol, and trifluoroethanol (or its combination) can also be added to the buffers intended for digestion.[79–81] They increase protein solubility by the stabilization of hydrophobic fragments of molecules, and their use does not require a particular method of sample purification prior to LC–MS analysis, as is the case with standard detergents. There is also no risk of contamination of the instrument. Many publications report a significant improvement in digestion efficiency and an increased number of identifications, especially in the case of proteolytic-resistant proteins, when using trypsin together with adding an organic solvent or a mixture of solvents.[82–86] The percentage of the organic solvent or mixture concerning the water phase varies depending on the experiment, from 10% to 40%–60%, and even 80% in the case of acetonitrile. Mixed organic-aqueous systems are used in the case of in-solution and in-gel digestion and with enzymes immobilized on a solid support.

4. Preliminary purification

All cell components, including proteins and genetic material, are released into the solution during the first stage of sample preparation — homogenization and cell disruption. Enzymatic digestion can be carried out straightaway on such material, but it often introduces many contaminants and interferences into the results due to the complicated matrix composition and incompatible MS additives or other impurities. Therefore, different methods can be used to prepurify the material and to extract only or mostly proteins that are the main interest in proteomics.

The most commonly used methods of preliminary purification are different types of protein precipitation with an organic solvent (e.g., cold acetone, acetonitrile, or methanol),[87] salt-like ammonium sulfate,[112,113] polyethylenimine,[114] or a more targeted approach like immunoprecipitation (Figure 2).[115] In the first case, turbidity is observed after adding the organic solvent to the sample, and centrifugation at 15,000–20,000×*g* for 30 min allows the sediment to be effectively separated from the solution. Another approach to the disruption material and simultaneous protein extraction is using TCA/acetone precipitation[116] together with phenol or SDS extraction. In all these cases, pellet contains mostly proteins present in the used material. If a more targeted approach is needed, specific antibodies can bind the protein of interest for further targeted proteomics analysis.

Filtration: In the case of body fluids, the most frequently used diagnostic material, the main problem is its composition, namely

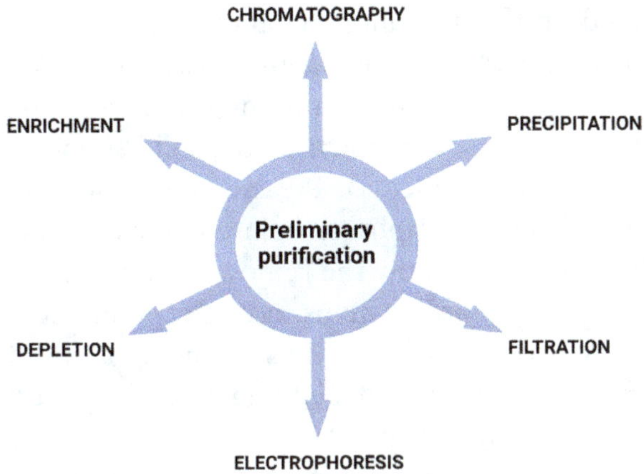

Figure 2. Techniques used in the preliminary purification of protein mixtures in proteomics analysis.

proteins in high concentration. Of course, we are talking about, among others, albumin, immunoglobulins, transthyretin, apolipoproteins, and the C3 component. Unfortunately, most of the proteins of interest to us with biomarker potential are low-concentration proteins. The simplest way is using mass cut-off filters like Amicon with available membranes in the range of 5–150 kDa can be fractionated according to their mass. The selection of the size of the membranes used and the number of fractions depend on the operator and the complexity of the mixture. This approach can be used, for example, to remove proteins of a certain mass from a mixture, such as removing antibodies from plasma samples, which is sufficient to centrifuge on a previously prepared Amicon with a 100-kDa membrane. Another variant of filtration is dialysis, which allows you to get rid of low-molecular-weight compounds, such as reducing agents, and residues of compounds are used to label proteins or preservatives (sodium azide, thimerosal). It effectively allows you to get rid of impurities and replace the buffer before the next steps of the procedure. The most commonly used are membranes with a mass cut-off of 2–10 kDa.[117]

A special type of preliminary purification is protein depletion, which is used when working with body fluids such as serum. This method is based on the immunoprecipitation of proteins using commercially available kits. In this case, antibodies against proteins in high concentration, such as albumin, immunoglobulins, complement C, transthyretin, and others, are attached to a solid support (sepharose, magnetic beads, or other media). Preassembled columns for attachment to liquid chromatographs or spin-columns are available with various antibody sets that allow the removal of only albumin or albumin and Ig as well as more proteins in one experiment. An example is the MARS column (Agilent), which removes 7–14 of the most common proteins in the serum depending on the number of antibodies applied.[118–121] The system is perfect for automation in chromatographic columns and works on a smaller number of samples in the form of a spin cartridge. Another one working on the same rules is the Aurum Affi-Gel Blue kit (Bio-Rad), the ProteoPrep immunoaffinity albumin, Seppro-IgY-14, and IgG depletion kit (Sigma).[122,123] The reverse idea is used in the ProteMiner kit from Bio-Rad. This set can enrich the sample in low abundant proteins.[124–126] In this case, peptides that specifically bind proteins present at low serum concentrations are attached to the solid matrix. In this way, proteins that do not interact with the peptides are washed away first, and only at the elution stage, those bound to the peptide linker are selectively released. The enrichment strategy can also be used when analyzing proteins with post-translational modifications, such as phosphorylation and glycosylation. The use of specific solid-phase extraction (SPE) dedicated to particular modifications, such as TiO_2, allows for capturing only modified proteins for further digestion. All kinds of specific interactions can also be used for this approach, such as binding proteins containing tags (His-tag) and metal ion affinity.

If the laboratory ***chromatographic*** systems are in use, then size exclusion chromatography, flash chromatography, or even high-performance liquid chromatography can be used for the first purification. Size exclusion chromatography and ion exchange chromatography are most commonly used, but other types, such as, affinity (dye–ligand, metal–ion, and antigen–antibody) and hydrophobic interaction chromatography, can also be used.

Typically, SDS-PAGE is the first choice with preferred in MS Coomassie blue staining. After separation, the band of choice can be cut, or the whole gel can be divided into pieces and digested. For more precise separation, two-dimensional gels are often used. Also, routinely used electrophoresis can be used as a tool for preliminary fractionation.

If you reached to get the final protein solution fractionated or you get your proteins as gel bands, you can start preparation for the crucial step — digestion of the proteins.

5. Standard digestion methods

5.1. *In-solution digestion*

In-solution digestion is the most straightforward technique for protein digestion into peptides in bottom-up proteomics (Figure 3). It allows for more control over the outcome, compared to in-gel digestion (see Chapter x.x.x), as the conditions (i.e., pH, digestion buffer, additives, and enzyme/substrate ratio) can be altered and controlled more easily.[127] Furthermore, it does not need other specific equipment, like filters or gels, to be performed. In-solution digestion consists of three significant steps. First, the sample set-up with steps before the digestion, like denaturation of the proteins and alkylation. The second step is the digestion itself, which can be enzymatic digestion or chemical cleavage of the proteins. The third and last step is the sample clean-up, followed by identification and quantification measurements.

Sample set-up before digestion heavily depends on the starting material and the purification protocol and needs to be adjusted accordingly to suit the researcher's needs. The resulting protein solution should not contain anything interfering with the following digestion to ensure high specificity and efficiency. Traditionally, complex buffers with high salt concentrations and additives were used to disperse proteins and disrupt their tertiary structure to warrant quality digestion.[127] The most often used reagents are urea, guanidine

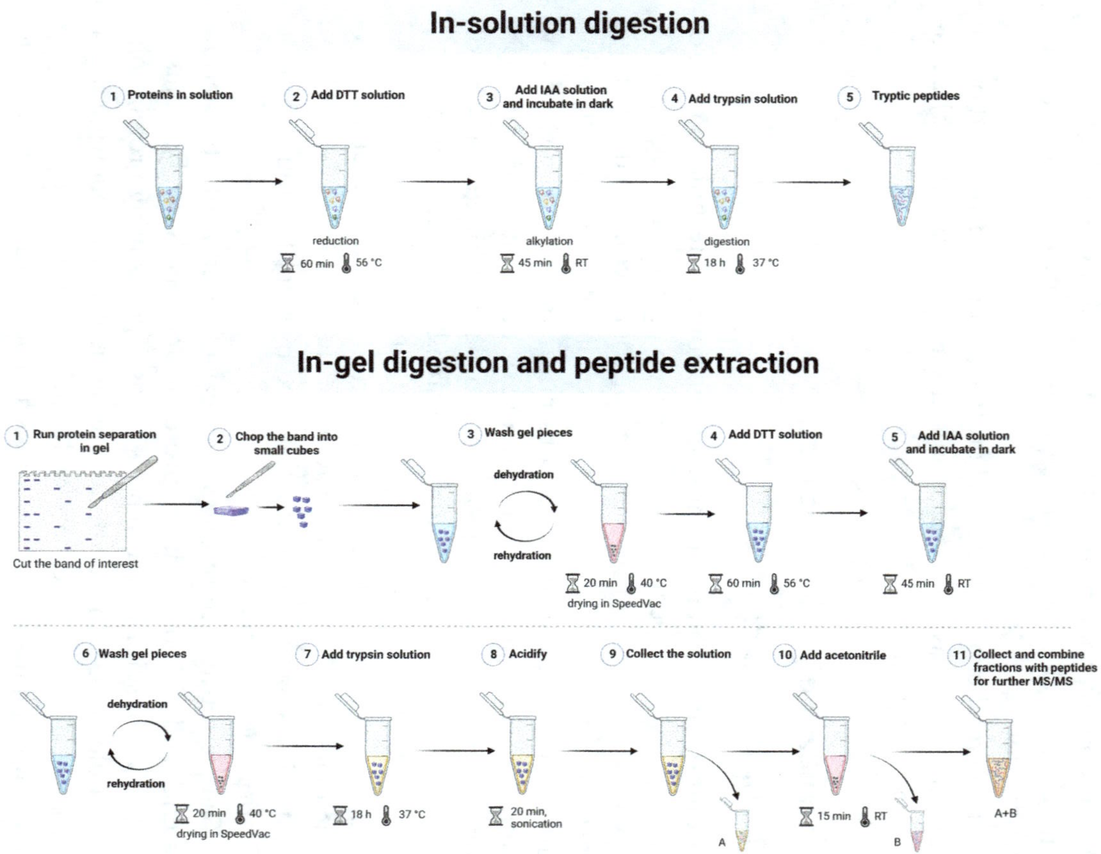

Figure 3. Graphical comparison of the procedure of digestion in solution and digestion of proteins separated in gel, in-gel digestion.

hydrochloride, or organic solvents to denature the proteins in the solution.[52]

After the sample set-up is performed, the proteins need to be digested, either by cleaving specific bonds, commonly found in proteins, or by the usage of enzymes that can "cut" specific bonds or by chemical cleavage, dissolving specific bonds, to create the peptides, desired for analysis. For enzymatic digestion, various endoproteases are used to cleave the proteins. Each protease has its optimal digestion conditions, and the parameters (e.g., temperature, pH) must be adjusted accordingly. The performed digestion protocol also needs to consider salts and detergents in the solution after the sample set-up, as they might interfere with the enzymes.

The most commonly used enzyme in enzymatic digestion is trypsin, as it is available for low cost and cleaves from the *C*-terminal of the basic residues of Lys and Arg, which frequently appear in most proteins.[128] Other often used alternatives with high efficiency and specificity are Arg-C, Asp-N, Glu-C, Lys-C, and Lys-N, with all having different optimal conditions, as well as their advantages and disadvantages in usage.[53]

Chemical cleavage is the second alternative to enzymatic digestion, using diluted solutions of different chemicals with high cleavage specificity. Only a handful of chemicals are helpful to create predictable cleavage products, with each needing a different experimental set-up and each having its ups and downsides when used.[127]

Chemicals commonly used for chemical cleavage of proteins are formic acid (FA) cleaving Asp-Xxx bonds; 2-M hydroxylamine cleaving Asn-Gly bonds; hydrochloric acid (HCl_{aqu}) cleaving Asp-Xxx bonds; 2-nitro-5-thiocyanobenzoate cleaving Cys-Xxx bonds, and cyanogen bromide cleaving Met Xxx bonds.[129–133]

5.2. *In-gel digestion*

In-gel digestion was one of the first methods developed for the proteomics identification of proteins.[134] Although it is no longer considered as the most sophisticated separation tool, gel electrophoresis has been used as a fast, inexpensive, high-resolution protein separation method.[135,136] Over the recent years, several changes have been introduced to improve the quantity and the quality of the peptide yield, while the main components remained the same. Most basic protocols (Figure 3) start with the solubilization of the sample and the separation of the proteins via gel electrophoresis. Essential steps in this procedure to obtain high-quality spectra consist of destaining, reduction, and alkylation of cysteines, enzymatic digestion of proteins in gel piece, and peptide extraction from the gel.[137,138] The quality of resulting mass spectra depends on the amount of sample introduced. It is critical for in-gel digestion, as it is sensitive to sample loss due to insufficient separation, destaining, or adsorption on surfaces, where losses can get up to 50% of the sample introduced.[139,140]

The first crucial step after the proteins have been separated by gel electrophoresis and before the in-gel digestion is the visualization of the proteins for isolation. The staining technique depends on the type of electrophoresis used and the amount of protein in the sample. If the targeted gel spot contains approximately 30 ng of protein, the organic dye Coomassie blue can be used as the most convenient staining method.[141] For the detection and visualization of lower protein amounts until the low femtomole range, silver staining methods or fluorescent staining can be utilized.[142,143] Furthermore, radioactive labeling, staining, and visualization are possible, but it requires equipment suited for that labeling.

To generate samples, which can be analyzed by MS, the target protein separated by gel electrophoresis needs to be digested. It can also be done after the gel has been dried out by rehydration of the gel.[144] Proteins that have been digested in-gel leave defined peptides with specific molecular masses, depending on the type of gel separation and digestion. These peptides can be used in MS to identify the protein. In-gel digestion usually uses either chemical cleavage or enzymatic digestion to generate the peptides.

Chemical cleavage is the most undemanding digestion type, performed under heat and acidic conditions.[145,146] Although acid is a universal digester, its disadvantage is low and unspecific cleavage under mild and acidic conditions. It leads to complex mixtures of peptides and the generation of a more random sample, as chemical cleavage is also more error-prone, leading to the creation of peptide artifacts by modification of amino acids and random cleavage.[147]

Trypsin is the most commonly used enzyme for enzymatic cleavage, cleaving peptide bonds specifically after the carboxyl group of arginine and lysine. Therefore, cleavage with trypsin leads to predictable cleavage and results in peptides of masses fit for analysis by MS.[148] Besides, trypsin, Lys-C, Glu-C, and Asp-N are enzymes frequently used for in-gel digestion. Since the proteins are embedded in the gel matrix and the protease must enter the gel, rehydration of the gel improves the directional process, even more so rehydration with acetonitrile.[144,149] Additionally, reducing the size of the gel piece when performing in-gel digestion can significantly increase the process's efficiency. Standard gel-digestion procedures with enzymes need overnight incubation of 12–18 h for complete digestion, which can be reduced significantly by various steps.[150] The optimization of biochemical and physical parameters (e.g., pH, temperature, enzyme concentration, and surface area) can lead to sufficient digestion yield after 30 min.[151] Coupling optimization steps using an ultrasonic field can reduce incubation time to 60 s.[152] Also, microwave irradiation can reduce the time needed for complete digestion, independent of other optimization steps.[153]

After the in-gel digestion of the protein, the peptides must be released from the gel matrix. Therefore, two different extraction steps are often performed to ensure the extraction of peptides with different physiochemical properties.[140] The first step usually is extraction with ammonium bicarbonate followed by extraction with FA for subsequent electrospray ionization or with TFA for subsequent matrix-assisted laser desorption/ionization analysis. Peptides are extracted by incubation and the diffusion of the peptides into the buffer. Isolated peptides from different extraction steps can be combined. A vacuum centrifugation step is performed before the MS analysis is performed. It ensures the removal of organic solvents and concentrates the diluted peptides.[154,155]

Before sample analysis by MS, the peptides must be concentrated and desalted to reduce the impurities and increase the general performance of the analysis. The sample can be concentrated using a vacuum centrifuge to reduce the general sample volume while simultaneously removing a part of the salt from the buffer. Most often, the go-to method for desalting and concentration is using microcolumns containing alkyl chains (C_4–C_{18}), which adsorb the peptides.[156] After washing the matrix and eluting the peptides from the columns, the sample is ready to be analyzed via mass spectrometer.

6. Final clean-up

The last stage before recording fragmentation spectra is the final clean-up, also called desalting. It is related to removing substances from the digested solution that may interfere with the recorded spectrum, suppress peptide signals, or damage the mass spectrometer itself. If the used LC–MS system can conduct such a process online, this step is already carried out directly on the chromatographic column. In this case, non-volatile compounds are washed out during the first stage of chromatographic separation by the aqueous phase and transferred to the waste, and the rest of the peptides are gradually released from the column directed to the mass spectrometer. If the LC system is not possible, the purification process should be carried out offline before placing the sample in the autosampler. Most often, all additives that facilitate or increase digestion efficiency, that is, SDS, urea, detergents such as Triton X-100 or NP-40, and all non-volatile buffers in which the digestion process takes place, such as phosphate-buffered saline, are subjected to attention.

The first choice is run on a mini-scale SPE performed on commercially available pipette tips, microcolumns, or cartridges filled with C4–C18. The most popular ones are C18 ZipTip (Millipore), where the process is carried out manually with an automatic pipette, and the size of the phase-filled tip can be selected according to the sample size purified. StageTips are another form where the process is carried out with a centrifuge. Regardless of the form or the tool used (pipe, centrifuge), the steps are the same:

1. preparation/wetting of the bed (organic solvent line acetonitrile or methanol),
2. phase conditioning (acidic water — TFA/FA or acetic acid),

3. sample introduction,
4. washing off the impurities (acidic water), and
5. elution of the peptides (water with an organic solvent, 50%–80% acetonitrile or methanol).

The combination of solutions used for the preparation of tips is usually recommended by manufacturers so that the process is carried out as efficiently as possible. It is, therefore, always worth getting acquainted with the manual attached to the set. However, it is crucial to use solvents with LC–MS purity class at this stage.

Tips with stationary phase can also be prepared manually in the laboratory according to the procedure described by Rappsilber, where, using a specially prepared needle, discs are cut from the C18 phase discs and then placed in an empty pipette tip.[157] The entire procedure is then carried out using a tip centrifuge. The preparation of such desalination tips in a laboratory has the advantage that the various phases can be arbitrarily combined to increase the efficiency of the extraction of peptides with different chemical properties. The most common phase combinations are C18–SCX, C18–SCX–C18, SDB–SAX, or metal oxide–C8. Finally, in all these cases, we obtain peptides suspended in a solution of 50%–70% acetonitrile or methanol with acidic water (up to 1% FA, acetic acid, or TFA). So it only remains to concentrate the sample, evaporate the organic solvent, and start the analysis.

Final clean-up can also be used for the specific enrichment of the sample with low-concentration peptides, like in the case of post-translational modification analysis, such as phosphorylation or glycosylation. The use of beds specifically binding selected modifications significantly improves the quality of the sample and allows for complete analysis. This step is also suitable for high throughput by using multiwell plates with extraction phases inserted therein. Commercial kits based on the SPE method are also used to select and efficiently remove detergents, such as Triton X-100 or NP-40.

References

1. Mansour A, Chatila R, Bejjani N, *et al.* A Novel Xylene-Free Deparaffinization Method for the Extraction of Proteins from Human

Derived Formalin-Fixed Paraffin Embedded (FFPE) Archival Tissue Blocks. *MethodsX*. 2014;1:90–95. doi: 10.1016/j.mex.2014.07.006.

2. Mitsa G, Guo Q, Goncalves C, *et al.* A Non-Hazardous Deparaffinization Protocol Enables Quantitative Proteomics of Core Needle Biopsy-Sized Formalin-Fixed and Paraffin-Embedded (FFPE) Tissue Specimens. *Int. J. Mol. Sci.* 2022;23(8):4443. doi: 10.3390/ijms23084443.
3. Magdeldin S, Yamamoto T. Toward Deciphering Proteomes of Formalin-Fixed Paraffin-Embedded (FFPE) Tissues. *Proteomics.* 2012;12(7):1045–1058. doi: 10.1002/pmic.201100550.
4. Nirmalan N, Harnden P, Selby P, *et al.* Mining the Archival Formalin-Fixed Paraffin-Embedded Tissue Proteome: Opportunities and Challenges. *Mol. Biosyst.* 2008;4:712–720. doi: 10.1039/b800098k.
5. Zhu Y, Weiss T, Zhang Q, *et al.* High-Throughput Proteomic Analysis of FFPE Tissue Samples Facilitates Tumor Stratification. *Mol. Oncol.* 2019;13(11):2305–2328. doi: 10.1002/1878-0261.12570.
6. Gao H, Zhang F, Liang S, *et al.* Accelerated Lysis and Proteolytic Digestion of Biopsy-Level Fresh-Frozen and FFPE Tissue Samples Using Pressure Cycling Technology. *J. Proteome Res.* 2020;19(5):1982–1990. doi: 10.1021/acs.jproteome.9b00790.
7. Handa T, Sasaki H, Takao M, *et al.* Proteomics-Based Investigation of Cerebrovascular Molecular Mechanisms in Cerebral Amyloid Angiopathy by the FFPE-LMD-PCT-SWATH Method. *Fluids Barriers CNS.* 2022;19(1):1–23. doi: 10.1186/s12987-022-00351-x.
8. He C-F, Wang Y-M. Protein Extraction from Leaves of Aloe Vera L., a Succulent and Recalcitrant Plant, for Proteomic Analysis. *Plant Mol. Biol. Report.* 2008;26(4):292–300. doi: 10.1007/s11105-008-0040-9.
9. Contreras L, Ritter A, Dennett G, *et al.* Two-Dimensional Gel Electrophoresis Analysis of Brown Algal Protein Extracts[1]. *J. Phycol.* 2008;44(5):1315–1321. doi: 10.1111/j.1529-8817.2008.00575.x.
10. Stuhr M, Blank-Landeshammer B, Reymond CE, *et al.* Disentangling Thermal Stress Responses in a Reef-Calcifier and Its Photosymbionts by Shotgun Proteomics. *Sci. Rep.* 2018;8(1):1–13. doi: 10.1038/s41598-018-21875-z.
11. Karthikaichamy A, Deore P, Rai V, *et al.* Time for Multiple Extraction Methods in Proteomics? A Comparison of Three Protein Extraction

Methods in the Eustigmatophyte Alga *Microchloropsis Gaditana* CCMP526. *OMICS J. Integr. Biol.* 2017;21(11):678–683. doi: 10.1089/omi.2017.0128.

12. Wong P-F, Tan L-J, Nawi H, *et al.* Proteomics of the Red Alga, *Gracilaria changii* (Gracilariales, Rhodophyta)1. *J. Phycol.* 2006; 42(1):113–120. doi: 10.1111/j.1529-8817.2006.00182.x.
13. Tanca A, Abbondio M, Pisanu S, *et al.* Critical Comparison of Sample Preparation Strategies for Shotgun Proteomic Analysis of Formalin-Fixed, Paraffin-Embedded Samples: Insights from Liver Tissue. *Clin. Proteomics.* 2014;11(1):28. doi: 10.1186/1559-0275-11-28.
14. Wild P, Gabrieli A, Schraner EM, *et al.* Reevaluation of the Effect of Lysoyzme on Escherichia Coli Employing Ultrarapid Freezing Followed by Cryoelectronmicroscopy or Freeze Substitution. *Microsc. Res. Tech.* 1997;39(3):297–304. doi: 10.1002/(SICI)1097-0029 (19971101)39:3<297::AID-JEMT8>3.0.CO;2-H.
15. Encheva V, Gharbia SE, Wait R, *et al.* Comparison of Extraction Procedures for Proteome Analysis of Streptococcus Pneumoniae and a Basic Reference Map. *Proteomics.* 2006;6(11):3306–3317. doi: 10.1002/pmic.200500744.
16. Imperi F, Ciccosanti F, Perdomo AB, *et al.* Analysis of the Periplasmic Proteome of Pseudomonas Aeruginosa, a Metabolically Versatile Opportunistic Pathogen. *Proteomics.* 2009;9(7):1901–1915. doi: 10.1002/pmic.200800618.
17. Chourey K, Jansson J, VerBerkmoes N, *et al.* Direct Cellular Lysis/ Protein Extraction Protocol for Soil Metaproteomics. *J. Proteome Res.* 2010;9(12):6615–6622. doi: 10.1021/pr100787q.
18. Gao Q, Madian AG, Liu X, *et al.* Coupling Protein Complex Analysis to Peptide Based Proteomics. *J. Chromatogr. A.* 2010;1217(49): 7661–7668. doi: 10.1016/j.chroma.2010.09.071.
19. Yoo B-S, Regnier FE. Proteomic Analysis of Carbonylated Proteins in Two-Dimensional Gel Electrophoresis Using Avidin-Fluorescein Affinity Staining. *Electrophoresis.* 2004;25(9):1334–1341. doi: 10.1002/elps.200405890.
20. Rodríguez A, Vaneechoutte M. Comparison of the Efficiency of Different Cell Lysis Methods and Different Commercial Methods for

RNA Extraction from Candida Albicans Stored in RNAlater. *BMC Microbiol.* 2019;19(1):94. doi: 10.1186/s12866-019-1473-z.

21. Haar, T. von der. Optimized Protein Extraction for Quantitative Proteomics of Yeasts. *PLoS One.* 2007;2(10):e1078. doi: 10.1371/journal.pone.0001078.
22. Goldschmidt P, Degorge S, Merabet L, *et al.* Enzymatic Treatment of Specimens Before DNA Extraction Directly Influences Molecular Detection of Infectious Agents. *PLoS One.* 2014;9(6):e94886. doi: 10.1371/journal.pone.0094886.
23. O'Farrell PH. High Resolution Two-Dimensional Electrophoresis of Proteins. *J. Biol. Chem.* 1975;250(10):4007–4021.
24. Alvarez S, Naldrett MJ. Plant Structure and Specificity — Challenges and Sample Preparation Considerations for Proteomics. In *Modern Proteomics — Sample Preparation, Analysis and Practical Applications*; Mirzaei H, Carrasco M, Eds.; *Adv. Exp. Med. Biol.* Cham: Springer International Publishing, 2016;919:63–81. doi: 10.1007/978-3-319-41448-5_4.
25. Neset L, Takayidza G, Berven FS, *et al.* Comparing Efficiency of Lysis Buffer Solutions and Sample Preparation Methods for Liquid Chromatography–Mass Spectrometry Analysis of Human Cells and Plasma. *Molecules.* 2022;27(11):3390. doi: 10.3390/molecules27113390.
26. Winter D, Steen H. Optimization of Cell Lysis and Protein Digestion Protocols for the Analysis of HeLa S3 Cells by LC-MS/MS. *Proteomics.* 2011;11(24):4726–4730. doi: 10.1002/pmic.201100162.
27. Pirog A, Faktor J, Urban-Wojciuk Z, *et al.* Comparison of Different Digestion Methods for Proteomic Analysis of Isolated Cells and FFPE Tissue Samples. *Talanta.* 2021;233:122568. doi: 10.1016/j.talanta.2021.122568.
28. Wiśniewski JR. Proteomic Sample Preparation from Formalin Fixed and Paraffin Embedded Tissue. *J. Vis. Exp.* 2013;(79). doi: 10.3791/50589.
29. Ngoka LCM. Sample Prep for Proteomics of Breast Cancer: Proteomics and Gene Ontology Reveal Dramatic Differences in Protein Solubilization Preferences of Radioimmunoprecipitation Assay and Urea Lysis Buffers. *Proteome Sci.* 2008;6:1–24. doi: 10.1186/1477-5956-6-30.
30. Hughes CS, Moggridge S, Müller T, *et al.* Single-Pot, Solid-Phase-Enhanced Sample Preparation for Proteomics Experiments. *Nat. Protoc.* 2019;14(1):68–85. doi: 10.1038/s41596-018-0082-x.

31. Danko K, Lukasheva E, Zhukov VA, *et al.* Detergent-Assisted Protein Digestion — On the Way to Avoid the Key Bottleneck of Shotgun Bottom-Up Proteomics. *Int. J. Mol. Sci.* 2022;23(22):13903. doi: 10.3390/ijms232213903.
32. Zhang X. Less Is More: Membrane Protein Digestion Beyond Urea-Trypsin Solution for Next-Level Proteomics. *Mol. Cell. Proteomics.* 2015;14(9):2441–2453. doi: 10.1074/mcp.R114.042572.
33. Shaw MM, Riederer BM. Sample Preparation for Two-Dimensional Gel Electrophoresis. *Proteomics.* 2003;3(8):1408–1417. doi: 10.1002/pmic.200300471.
34. Herbert B. Advances in Protein Solubilisation for Two-Dimensional Electrophoresis. *Electrophoresis.* 1999;20(4–5):660–663. doi: 10.1002/(SICI)1522-2683(19990101)20:4/5<660::AID-ELPS660>3.0.CO;2-Q.
35. Peach M, Marsh N, Miskiewicz EI, *et al.* Solubilization of Proteins: The Importance of Lysis Buffer Choice. In *Western Blotting*; Kurien BT, Scofield RH, Eds.; *Methods Mol. Biol.* New York: Springer, 2015;1312:49–60. doi: 10.1007/978-1-4939-2694-7_8.
36. Kollipara L, Zahedi RP. Protein Carbamylation: In Vivo Modification or in Vitro Artefact? *Proteomics.* 2013;13(6):941–944. doi: 10.1002/pmic.201200452.
37. McCarthy J, Hopwood F, Oxley D, *et al.* Carbamylation of Proteins in 2-D Electrophoresis — Myth or Reality? *J. Proteome Res.* 2003;2(3):239–242. doi: 10.1021/pr025564b.
38. Lo A, Tang Y, Chen L, *et al.* Automation of Dimethylation After Guanidination Labeling Chemistry and Its Compatibility with Common Buffers and Surfactants for Mass Spectrometry-Based Shotgun Quantitative Proteome Analysis. *Anal. Chim. Acta.* 2013;788:81–88. doi: 10.1016/j.aca.2013.05.060.
39. Sun S, Zhou J-Y, Yang W, *et al.* Inhibition of Protein Carbamylation in Urea Solution Using Ammonium-Containing Buffers. *Anal. Biochem.* 2014;446:76–81. doi: 10.1016/j.ab.2013.10.024.
40. Betancourt LH, Sanchez A, Pla I, *et al.* Quantitative Assessment of Urea In-Solution Lys-C/Trypsin Digestions Reveals Superior Performance at Room Temperature over Traditional Proteolysis at 37°C. *J. Proteome Res.* 2018;17(7):2556–2561. doi: 10.1021/acs.jproteome.8b00228.

41. Kachuk C, Doucette AA. The Benefits (and Misfortunes) of SDS in Top-Down Proteomics. *J. Proteomics.* 2018;175:75–86. doi: 10.1016/j.jprot.2017.03.002.
42. Russo D. The Impact of Kosmotropes and Chaotropes on Bulk and Hydration Shell Water Dynamics in a Model Peptide Solution. *Chem. Phys.* 2008;345(2–3):200–211. doi: 10.1016/j.chemphys.2007.08.001.
43. Salvi G, De Los Rios P, Vendruscolo M. Effective Interactions Between Chaotropic Agents and Proteins. *Proteins Struct. Funct. Genet.* 2005;61(3):492–499. doi: 10.1002/prot.20626.
44. Botelho D, Wall MJ, Vieira DB, *et al.* Top-Down and Bottom-Up Proteomics of SDS-Containing Solutions Following Mass-Based Separation. *J. Proteome Res.* 2010;9(6):2863–2870. doi: 10.1021/pr900949p.
45. Zhou J-Y, Dann GP, Shi T, *et al.* Simple Sodium Dodecyl Sulfate-Assisted Sample Preparation Method for LC-MS-Based Proteomics Applications. *Anal. Chem.* 2012;84(6):2862–2867. doi: 10.1021/ac203394r.
46. Wiśniewski JR. Filter Aided Sample Preparation — a Tutorial. *Anal. Chim. Acta.* 2019;1090:23–30. doi: 10.1016/j.aca.2019.08.032.
47. Linke D. *Chapter 34 Detergents: An Overview*, 1st ed., Vol. 463. Elsevier Inc., 2009. doi: 10.1016/S0076-6879(09)63034-2.
48. Dupree EJ, Jayathirtha M, Yorkey H, *et al.* A Critical Review of Bottom-Up Proteomics: The Good, the Bad, and the Future of this Field. *Proteomes.* 2020;8(3):14. doi: 10.3390/proteomes8030014.
49. Smolikova G, Gorbach D, Lukasheva E, *et al.* Bringing New Methods to the Seed Proteomics Platform: Challenges and Perspectives. *Int. J. Mol. Sci.* 2020;21(23):9162. doi: 10.3390/ijms21239162.
50. Zhang X. Detergents: Friends Not Foes for High-Performance Membrane Proteomics Toward Precision Medicine. *Proteomics.* 2017. doi: 10.1002/pmic.201600209.
51. Mthembu SN, Sharma A, Albericio F, *et al.* Breaking a Couple: Disulfide Reducing Agents. *ChemBioChem.* 2020;21(14):1947–1954. doi: 10.1002/cbic.202000092.
52. Sung CK, Chen Y, Mirza S, *et al.* A Clean, More Efficient Method for in-Solution Digestion of Protein Mixtures Without Detergent or Urea. *J. Proteome Res.* 2006;5(12):3446–3452. doi: 10.1021/pr0603396.

53. Switzar L, Giera M, Niessen WMA. Protein Digestion: An Overview of the Available Techniques and Recent Developments. *J. Proteome Res.* 2013;12(3):1067–1077. doi: 10.1021/pr301201x.
54. *ProteaseMAX™ Surfactant, Trypsin Enhancer.* https://pl.promega.com/en/products/mass-spectrometry/proteases-and-surfactants/proteasemax-surfactant_-trypsin-enhancer/ (accessed 2023-01-20).
55. *Maximizing-Protein-Recovery-and-Digestion-Efficiency.pdf.* https://promega.widen.net/s/dmfpktzksn/maximizing-protein-recovery-and-digestion-efficiency (accessed 2023-01-20).
56. RapiGest SF Surfactant|Reagent for Enzymatic Protein Digestion. *Waters.* https://www.waters.com/nextgen/be/en/products/standards-and-reagents/rapigest-sf-surfactant.html?xcid=ppc-ppc_25933&gclid=Cj0KCQiAlKmeBhCkARIsAHy7WVtHX_Hie6S1bYwyMEWTO5L3jgVN4CZeUWtBIPyLAe1kMhZfiFThI_UaAukSEALw_wcB (accessed 2023-01-20).
57. *Agilent_PPS_SilentSurfactant.Pdf.* https://proteomicsresource.washington.edu/docs/protocols03/Agilent_PPS_SilentSurfactant.pdf (accessed 2023-01-20).
58. Progenta™ Anionic Acid Labile Surfactant II (AALS II), Protea. *VWR.* https://us.vwr.com/store/product/14450751/progentatm-anionic-acid-labile-surfactant-ii-aals-ii-protea (accessed 2023-01-20).
59. Ji Y, Liu M, Bachschmid MM, *et al.* Surfactant-Induced Artifacts During Proteomic Sample Preparation. *Anal. Chem.* 2015;87(11):5500–5504. doi: 10.1021/acs.analchem.5b00249.
60. Li M, Powell MJ, Razunguzwa TT, *et al.* A General Approach to Anionic Acid-Labile Surfactants with Tunable Properties. *J. Org. Chem.* 2010;75(18):6149–6153. doi: 10.1021/jo100954q.
61. Wu F, Sun D, Wang N, *et al.* Comparison of Surfactant-Assisted Shotgun Methods Using Acid-Labile Surfactants and Sodium Dodecyl Sulfate for Membrane Proteome Analysis. *Anal. Chim. Acta.* 2011;698(1):36–43. doi: 10.1016/j.aca.2011.04.039.
62. Chooneea D, Karlsson R, Encheva V, *et al.* Elucidation of the Outer Membrane Proteome of Salmonella Enterica Serovar Typhimurium Utilising a Lipid-Based Protein Immobilization Technique. *BMC Microbiol.* 2010;10(1):44. doi: 10.1186/1471-2180-10-44.

63. Murphy S, Zweyer M, Henry M, *et al.* Proteomic Profiling of Liver Tissue from the Mdx-4cv Mouse Model of Duchenne Muscular Dystrophy. *Clin. Proteomics.* 2018;15(1):34. doi: 10.1186/s12014-018-9212-2.
64. Waanders LF, Chwalek K, Monetti M, *et al.* Quantitative Proteomic Analysis of Single Pancreatic Islets. *Proc. Natl. Acad. Sci.* 2009;106(45):18902–18907. doi: 10.1073/pnas.0908351106.
65. Miyauchi E, Furuta T, Ohtsuki S, *et al.* Identification of Blood Biomarkers in Glioblastoma by SWATH Mass Spectrometry and Quantitative Targeted Absolute Proteomics. *PLoS One.* 2018;13(3): e0193799. doi: 10.1371/journal.pone.0193799.
66. Dapic I, Uwugiaren N, Kers J, *et al.* Evaluation of Fast and Sensitive Proteome Profiling of FF and FFPE Kidney Patient Tissues. *Molecules.* 2022;27(3):1137. doi: 10.3390/molecules27031137.
67. Li Y, Huang Y, Ferrant J, *et al.* Assessing in Vivo Dynamics of Multiple Quality Attributes from a Therapeutic IgG4 Monoclonal Antibody Circulating in Cynomolgus Monkey. *mAbs.* 2016;8(5):961–968. doi: 10.1080/19420862.2016.1167298.
68. Ting YS, Egertson JD, Bollinger JG, *et al.* PECAN: Library-Free Peptide Detection for Data-Independent Acquisition Tandem Mass Spectrometry Data. *Nat. Methods.* 2017;14(9):903–908. doi: 10.1038/nmeth.4390.
69. Miroshnychenko O, Chalkley RJ, Leib RD, *et al.* Proteomic Analysis of Platelet-Rich and Platelet-Poor Plasma. *Regen. Ther.* 2020;15:226–235. doi: 10.1016/j.reth.2020.09.004.
70. Kramer G, Woolerton Y, Straalen JP van, *et al.* Accuracy and Reproducibility in Quantification of Plasma Protein Concentrations by Mass Spectrometry Without the Use of Isotopic Standards. *PLoS One.* 2015;10(10):e0140097. doi: 10.1371/journal.pone.0140097.
71. Robison FM, Heuberger AL, Brick MA, *et al.* Proteome Characterization of Leaves in Common Bean. *Proteomes.* 2015;3(3):236–248. doi: 10.3390/proteomes3030236.
72. Aryal UK, Krochko JE, Ross ARS. Identification of Phosphoproteins in Arabidopsis Thaliana Leaves Using Polyethylene Glycol Fractionation, Immobilized Metal-Ion Affinity Chromatography, Two-Dimensional Gel Electrophoresis and Mass Spectrometry. *J. Proteome Res.* 2012;11(1):425–437. doi: 10.1021/pr200917t.

73. Kaspar-Schoenefeld S, Merx K, Jozefowicz AM, *et al.* Label-Free Proteome Profiling Reveals Developmental-Dependent Patterns in Young Barley Grains. *J. Proteomics.* 2016;143:106–121. doi: 10.1016/j.jprot.2016.04.007.
74. Mukhopadhyay S, Maitra U. Chemistry and Biology of Bile Acids. *Curr. Sci.* 2004;87(12):1666–1683.
75. Masuda T, Tomita M, Ishihama Y. Phase Transfer Surfactant-Aided Trypsin Digestion for Membrane Proteome Analysis. *J. Proteome Res.* 2008;7(2):731–740. https://pubs.acs.org/doi/10.1021/pr700658q (accessed 2023-01-20).
76. Lin Y, Liu Y, Li J, *et al.* Evaluation and Optimization of Removal of an Acid-Insoluble Surfactant for Shotgun Analysis of Membrane Proteome. *Electrophoresis.* 2010;31(16):2705–2713. doi: 10.1002/elps.201000161.
77. Lin Y, Lin H, Liu Z, *et al.* Improvement of a Sample Preparation Method Assisted by Sodium Deoxycholate for Mass-Spectrometry-Based Shotgun Membrane Proteomics†. *J. Sep. Sci.* 2014;37(22):3321–3329. doi: 10.1002/jssc.201400569.
78. Lau BYC, Othman A. Evaluation of Sodium Deoxycholate as Solubilization Buffer for Oil Palm Proteomics Analysis. *PLoS One.* 2019;14(8):e0221052. doi: 10.1371/journal.pone.0221052.
79. Waas M, Bhattacharya S, Chuppa S, *et al.* Combine and Conquer: Surfactants, Solvents, and Chaotropes for Robust Mass Spectrometry Based Analyses of Membrane Proteins. *Anal. Chem.* 2014;86(3):1551–1559. doi: 10.1021/ac403185a.
80. Proc JL, Kuzyk MA, Hardie DB, *et al.* A Quantitative Study of the Effects of Chaotropic Agents, Surfactants, and Solvents on the Digestion Efficiency of Human Plasma Proteins by Trypsin. *J. Proteome Res.* 2010;9(10):5422–5437. doi: 10.1021/pr100656u.
81. Russell WK, Park ZY, Russell DH. Proteolysis in Mixed Organic-Aqueous Solvent Systems: Applications for Peptide Mass Mapping Using Mass Spectrometry. *Anal. Chem.* 2001;73(11):2682–2685. doi: 10.1021/ac001332p.
82. Laštovičková M, Bobál P, Strouhalová D, *et al.* Acetonitrile-Assisted Enzymatic Digestion Can Facilitate the Bottom-Up Identification of

Proteins of Cancer Origin. *Anal. Biochem.* 2019;570:1–4. doi: 10.1016/j.ab.2019.01.004.

83. Direct Acetonitrile-Assisted Trypsin Digestion Method Combined with LC–MS/MS-Targeted Peptide Analysis for Unambiguous Identification of Intact Ricin. *J. Proteome Res.* https://pubs-1acs-1org-19fb477cy01ee.hansolo.bg.ug.edu.pl/doi/10.1021/acs.jproteome.0c00458 (accessed 2023-01-20).
84. Wu S, Zhang H, Yang K, *et al.* A Rapid Protein Sample Preparation Method Based on Organic-Aqueous Microwave Irradiation Technique. *Sci. China Chem.* 2015;58(3):526–531. doi: 10.1007/s11426-014-5163-2.
85. Strader MB, Tabb DL, Hervey WJ, *et al.* Efficient and Specific Trypsin Digestion of Microgram to Nanogram Quantities of Proteins in Organic–Aqueous Solvent Systems. *Anal. Chem.* 2006;78(1):125–134. doi: 10.1021/ac051348l.
86. Ranasinghe A, Mehl J, D'Arienzo C, *et al.* Fucosyl Monosialoganglioside: Quantitative Analysis of Specific Potential Biomarkers of Lung Cancer in Biological Matrices Using Immunocapture Extraction/Tandem Mass Spectrometry. *Rapid Commun. Mass Spectrom.* 2018;32(17):1481–1490. doi: 10.1002/rcm.8194.
87. Santa C, Anjo SI, Manadas B. Protein Precipitation of Diluted Samples in SDS-Containing Buffer with Acetone Leads to Higher Protein Recovery and Reproducibility in Comparison with TCA/Acetone Approach. *Proteomics.* 2016;16(13):1847–1851. doi: 10.1002/pmic.201600024.
88. Crowell AMJ, MacLellan DL, Doucette AA. A Two-Stage Spin Cartridge for Integrated Protein Precipitation, Digestion and SDS Removal in a Comparative Bottom-Up Proteomics Workflow. *J. Proteomics.* 2015;118:140–150. doi: 10.1016/j.jprot.2014.09.030.
89. Shimazaki Y, Manabe T. Detection of Activity and Mass Spectrometric Identification of Mouse Liver Carboxylesterase and Aldehyde Dehydrogenase Separated by Non-Denaturing Two-Dimensional Electrophoresis After Extraction with Detergents. *Biochim. Biophys. Acta BBA — Proteins Proteomics.* 2005;1749(1):95–101. doi: 10.1016/j.bbapap.2005.02.011.

90. Ahsan N, Komatsu S. Comparative Analyses of the Proteomes of Leaves and Flowers at Various Stages of Development Reveal Organ-Specific Functional Differentiation of Proteins in Soybean — Ahsan — 2009. *Proteomics.* 2009. Wiley Online Library. https://analyticalsciencejournals.onlinelibrary.wiley.com/doi/10.1002/pmic.200900308 (accessed 2023-01-23).
91. Casey TM, Meade JL, Hewitt EW. Organelle Proteomics: Identification of the Exocytic Machinery Associated with the Natural Killer Cell Secretory Lysosome*. *Mol. Cell. Proteomics.* 2007;6(5):767–780. doi: 10.1074/mcp.M600365-MCP200.
92. Sun N-N, Wang Z-J, Gao X, *et al.* Comparative Study of Three Methods of Sample Preparation for Proteomics Research. *Chin. J. Anal. Chem.* 2015;43(6):808–813. doi: 10.1016/S1872-2040(15)60828-3.
93. Pham MD, Wen T-C, Li H-C, *et al.* Streamlined Membrane Proteome Preparation for Shotgun Proteomics Analysis with Triton X-100 Cloud Point Extraction and Nanodiamond Solid Phase Extraction. *Materials.* 2016;9(5):385. doi: 10.3390/ma9050385.
94. Jäpelt KB, Johnsen LG, Christensen JH. Removal of Polysorbate 80 by Complexation Prior to LC–MS Analysis. *Anal. Bioanal. Chem.* 2016;408(9):2303–2307. doi: 10.1007/s00216-016-9326-1.
95. Bhadauria V, Peng Y-L. Optimization of a Protein Extraction Technique for Fungal Proteomics. *Indian J. Microbiol.* 2010;50(1):127–131. doi: 10.1007/s12088-010-0072-3.
96. Partearroyo MA, Goñi FM, Katime I, *et al.* Micellar Properties of the Zwitterionic Bile Derivative CHAPS. *Biochem. Int.* 1988;16(2):259–265.
97. Chaiyarit S, Thongboonkerd V. Comparative Analyses of Cell Disruption Methods for Mitochondrial Isolation in High-Throughput Proteomics Study. *Anal. Biochem.* 2009;394(2):249–258. doi: 10.1016/j.ab.2009.07.026.
98. Henningsen R, Gale BL, Straub KM, *et al.* Application of Zwitterionic Detergents to the Solubilization of Integral Membrane Proteins for Two-Dimensional Gel Electrophoresis and Mass Spectrometry. *Proteomics.* 2002;2(11):1479–1488. doi: 10.1002/1615-9861(200211)2:11<1479::AID-PROT1479>3.0.CO;2-A.

99. Liu J, Wang F, Mao J, *et al.* High-Sensitivity N-Glycoproteomic Analysis of Mouse Brain Tissue by Protein Extraction with a Mild Detergent of N-Dodecyl β-D-Maltoside. *Anal. Chem.* 2015;87(4):2054–2057. https://pubs.acs.org/doi/10.1021/ac504700t (accessed 2023-01-23).
100. Katayama H, Tabata T, Ishihama Y, *et al.* Efficient In-Gel Digestion Procedure Using 5-Cyclohexyl-1-Pentyl-β-D-Maltoside as an Additive for Gel-Based Membrane Proteomics. *Rapid Commun. Mass Spectrom.* 2004;18(20):2388–2394. doi: 10.1002/rcm.1637.
101. Santarino IB, Oliveira SCB, Oliveira-Brett AM. Protein Reducing Agents Dithiothreitol and Tris(2-Carboxyethyl)Phosphine Anodic Oxidation. *Electrochem. Commun.* 2012;23:114–117. doi: 10.1016/j.elecom.2012.06.027.
102. Getz EB, Xiao M, Chakrabarty T, *et al.* A Comparison Between the Sulfhydryl Reductants Tris(2-Carboxyethyl)Phosphine and Dithiothreitol for Use in Protein Biochemistry. *Anal. Biochem.* 1999;273(1):73–80. doi: 10.1006/abio.1999.4203.
103. Shin S, Lee A, Lee S, *et al.* Microwave-Assisted Extraction of Human Hair Proteins. *Anal. Biochem.* 2010;407(2):281–283. doi: 10.1016/j.ab.2010.08.021.
104. Lametsch R, Lonergan S, Huff-Lonergan E. Disulfide Bond Within M-Calpain Active Site Inhibits Activity and Autolysis. *Biochim. Biophys. Acta BBA — Proteins Proteomics.* 2008;1784(9):1215–1221. doi: 10.1016/j.bbapap.2008.04.018.
105. Chen EI, Cociorva D, Norris JL, *et al.* Optimization of Mass Spectrometry-Compatible Surfactants for Shotgun Proteomics. *J. Proteome Res.* 2007;6(7):2529–2538. doi: 10.1021/pr060682a.
106. Mosen PR, Hardt R, Winter D. RapiGest Precipitation Depends on Peptide Concentration. *Proteomics.* 2021;21(20):2100129. doi: 10.1002/pmic.202100129.
107. Chen W, Kirui D, Millenbaugh NJ. Optimized Peptide Extraction Method for Analysis of Antimicrobial Peptide Kn2-7/DKn2-7 Stability in Human Serum by LC–MS. *Future Sci. OA.* 2022;8(6):FSO807. doi: 10.2144/fsoa-2022-0013.
108. Föll MC, Fahrner M, Oria VO, *et al.* Reproducible Proteomics Sample Preparation for Single FFPE Tissue Slices Using Acid-Labile Surfactant

and Direct Trypsinization. *Clin. Proteomics.* 2018;15(1):1–15. doi: 10.1186/s12014-018-9188-y.

109. Zhu R, Zhou S, Peng W, *et al.* Enhanced Quantitative LC-MS/MS Analysis of N-Linked Glycans Derived from Glycoproteins Using Sodium Deoxycholate Detergent. *J. Proteome Res.* 2018;17(8):2668–2678. doi: 10.1021/acs.jproteome.8b00127.
110. Pasing Y, Colnoe S, Hansen T. Proteomics of Hydrophobic Samples: Fast, Robust and Low-Cost Workflows for Clinical Approaches. *Proteomics.* 2017;17(6):1500462. doi: 10.1002/pmic.201500462.
111. Serra A, Zhu H, Gallart-Palau X, *et al.* Plasma Proteome Coverage Is Increased by Unique Peptide Recovery from Sodium Deoxycholate Precipitate. *Anal. Bioanal. Chem.* 2016;408(7):1963–1973. doi: 10.1007/s00216-016-9312-7.
112. Duong-Ly KC, Gabelli SB. Salting Out of Proteins Using Ammonium Sulfate Precipitation. *Methods Enzymol.* 2014;541:85–94. doi: 10.1016/B978-0-12-420119-4.00007-0.
113. Wingfield P. Protein Precipitation Using Ammonium Sulfate. *Curr. Protoc. Protein Sci.* 2001, *Appendix 3*, Appendix 3F. doi: 10.1002/0471140864.psa03fs13.
114. Burgess RR. Protein Precipitation Techniques. *Methods Enzymol.* 2009;463:331–342. doi: 10.1016/S0076-6879(09)63020-2.
115. Bonifacino JS, Gershlick DC, Dell'Angelica EC. Immunoprecipitation. *Curr. Protoc. Cell Biol.* 2016;71. doi: 10.1002/cpcb.3.
116. Méchin V, Damerval C, Zivy M. Total Protein Extraction with TCA-Acetone. *Methods Mol. Biol. Clifton NJ.* 2007;355:1–8. doi: 10.1385/1-59745-227-0:1.
117. Andrew SM, Titus JA, Zumstein L. Dialysis and Concentration of Protein Solutions. *Curr. Protoc. Toxicol.* 2001;10(1):A.3H.1–A.3H.5. doi: 10.1002/0471140856.txa03hs10.
118. Tu C, Rudnick PA, Martinez MY, *et al.* Depletion of Abundant Plasma Proteins and Limitations of Plasma Proteomics. *J. Proteome Res.* 2010;9(10):4982–4991. https://pubs.acs.org/doi/10.1021/pr100646w (accessed 2022-11-03).
119.. Maccarrone G, Milfay D, Birg I, *et al.* Mining the Human Cerebrospinal Fluid Proteome by Immunodepletion and Shotgun

Mass Spectrometry. *Electrophoresis.* 2004;25(14):2402–2412. doi: 10.1002/elps.200305909.

120. Jankovska E, Svitek M, Holada K, *et al.* Affinity Depletion Versus Relative Protein Enrichment: A Side-by-Side Comparison of Two Major Strategies for Increasing Human Cerebrospinal Fluid Proteome Coverage. *Clin. Proteomics.* 2019;16:9. doi: 10.1186/s12014-019-9229-1.
121. Ahn SB, Kamath KS, Mohamedali A, *et al.* Use of a Recombinant Biomarker Protein DDA Library Increases DIA Coverage of Low Abundance Plasma Proteins. *J. Proteome Res.* 2021;20(5):2374–2389. doi: 10.1021/acs.jproteome.0c00898.
122. Fratantoni SA, Piersma SR, Jimenez CR. Comparison of the Performance of Two Affinity Depletion Spin Filters for Quantitative Proteomics of CSF: Evaluation of Sensitivity and Reproducibility of CSF Analysis Using GeLC-MS/MS and Spectral Counting. *Proteomics — Clin. Appl.* 2010;4(6–7):613–617. doi: 10.1002/prca.200900179.
123. Wetterhall M, Zuberovic A, Hanrieder J, *et al.* Assessment of the Partitioning Capacity of High Abundant Proteins in Human Cerebrospinal Fluid Using Affinity and Immunoaffinity Subtraction Spin Columns. *J. Chromatogr. B.* 2010;878(19):1519–1530. doi: 10.1016/j.jchromb.2010.04.003.
124. Boschetti E, Righetti PG. The ProteoMiner in the Proteomic Arena: A Non-Depleting Tool for Discovering Low-Abundance Species. *J. Proteomics.* 2008;71(3):255–264. doi: 10.1016/j.jprot.2008.05.002.
125. Murphy S, Dowling P. DIGE Analysis of ProteoMinerTM Fractionated Serum/Plasma Samples. *Methods Mol. Biol. Clifton NJ.* 2018;1664:109–114. doi: 10.1007/978-1-4939-7268-5_10.
126. Jahangir S, John P, Bhatti A, *et al.* LC-MS/MS-Based Serum Protein Profiling for Identification of Candidate Biomarkers in Pakistani Rheumatoid Arthritis Patients. *Life Basel Switz.* 2022;12(3):464. doi: 10.3390/life12030464.
127. Medzihradszky KF. In-Solution Digestion of Proteins for Mass Spectrometry. *Methods in Enzymology.* 2005:50–65. doi: 10.1016/S0076-6879(05)05003-2.
128. Burkhart JM, Schumbrutzki C, Wortelkamp S, *et al.* Systematic and Quantitative Comparison of Digest Efficiency and Specificity Reveals

the Impact of Trypsin Quality on MS-Based Proteomics. *J. Proteomics.* 2012;75(4):1454–1462. doi: 10.1016/j.jprot.2011.11.016.

129. Li A, Sowder RC, Henderson LE, *et al.* Chemical Cleavage at Aspartyl Residues for Protein Identification. *Anal. Chem.* 2001;73(22):5395–5402. doi: 10.1021/ac010619z.
130. Park HB, Pyo SH, Hong SS, *et al.* Optimization of the Hydroxylamine Cleavage of an Expressed Fusion Protein to Produce a Recombinant Antimicrobial Peptide. *Biotechnol. Lett.* 2001;23(8):637–641. doi: 10.1023/A:1010358131124.
131. Lo L-H, Huang T-L, Shiea J. Acid Hydrolysis Followed by Matrix-Assisted Laser Desorption/Ionization Mass Spectrometry for the Rapid Diagnosis of Serum Protein Biomarkers in Patients with Major Depression. *Rapid Commun. Mass Spectrom.* 2009;23(5):589–598. doi: 10.1002/rcm.3908.
132. Tang HY, Speicher DW. Identification of Alternative Products and Optimization of 2-Nitro-5-Thiocyanatobenzoic Acid Cyanylation and Cleavage at Cysteine Residues. *Anal. Biochem.* 2004;334(1):48–61. doi: 10.1016/j.ab.2004.08.008.
133. Patwa TH, Wang Y, Simeone DM, *et al.* Enhanced Detection of Autoantibodies on Protein Microarrays Using a Modified Protein Digestion Technique. *J. Proteome Res.* 2008;7(6):2553–2561. doi: 10.1021/pr800023g.
134. Rosenfeld J, Capdeville J, Guillemot JC, *et al.* In-Gel Digestion of Proteins for Internal Sequence Analysis After One- or Two-Dimensional Gel Electrophoresis. *Anal. Biochem.* 1992;203:173–179.
135. Rogers JC, Bomgarden RD. Sample Preparation for Mass Spectrometry-Based Proteomics from Proteomes to Peptides. *Adv. Exp. Med. Biol.* 2016;919:3–62. doi: 10.1007/978-3-319-41448-5_3.
136. Rabilloud T, Chevallet M, Luche S, *et al.* Two-Dimensional Gel Electrophoresis in Proteomics: Past, Present and Future. *J. Proteomics.* 2010;73(11):2064–2077. doi: 10.1016/j.jprot.2010.05.016.
137. Granvogl B, Reisinger V, Eichacker LA. Mapping the Proteome of Thylakoid Membranes by de Novo Sequencing of Intermembrane Peptide Domains. *Proteomics.* 2006;6(12):3681–3695. doi: 10.1002/pmic.200500924.

138. Washburn MP, Wolters D, Yates JR. Large-Scale Analysis of the Yeast Proteome by Multidimensional Protein Identification Technology. *Nat. Biotechnol.* 2001;19(3):242–247. doi: 10.1038/85686.
139. Speicher KD, Kolbas O, Harper S, *et al.* Systematic Analysis of Peptide Recoveries from In-Gel Digestions for Protein Identifications in Proteome Studies. *J. Biomol. Tech.* 2000;11(2):74–86.
140. Granvogl B, Plöscher M, Eichacker LA. Sample Preparation by In-Gel Digestion for Mass Spectrometry-Based Proteomics. *Anal. Bioanal. Chem.* 2007;389(4):991–1002. doi: 10.1007/s00216-007-1451-4.
141. Tal M, Silberstein A, Nusser E. Why Does Coomassie Brilliant Blue R Interact Differently with Different Proteins? A Partial Answer. *J. Biol. Chem.* 1985;260(18):9976–9980. doi: 10.1016/s0021-9258(17)39198-6.
142. Mortz E, Krogh TN, Vorum H, *et al.* Improved Silver Staining Protocols for High Sensitivity Protein Identification Using Matrix-Assisted Laser Desorption/Ionization-Time of Flight Analysis. *Proteomics.* 2001;1(11):1359–1363. doi: 10.1002/1615-9861(200111)1:11<1359::aid-prot1359>3.3.co;2-h.
143. Marouga R, David S, Hawkins E. The Development of the DIGE System: 2D Fluorescence Difference Gel Analysis Technology. *Anal. Bioanal. Chem.* 2005;382(3):669–678. doi: 10.1007/s00216-005-3126-3.
144. Hellman U, Wernstedt C, Góñez J, *et al.* Improvement of an “in-Gel” Digestion Procedure for the Micropreparation of Internal Protein Fragments for Amino Acid Sequencing. *Analytical Biochemistry.* 1995:451–455. doi: 10.1006/abio.1995.1070.
145. Kowit JD, Maloney J. Protein Cleavage by Boiling in Sodium Dodecyl Sulfate Prior to Electrophoresis. *Anal. Biochem.* 1982;123(1):86–93. doi: 10.1016/0003-2697(82)90627-3.
146. Hill C. Phytochrome Structure: Peptide Fragments from the Amino-Terminal Domain. *Planta.* 1989;178(2):147–156.
147. Gobom J. Use of Vapor-Phase Acid Hydrolysis for Mass Spectrometric Peptide Mapping and Protein Identification. *Anal. Chem.* 1999;71(5):919–927. doi: 10.1021/ac981239p.
148. Olsen JV, Ong SE, Mann M. Trypsin Cleaves Exclusively C-Terminal to Arginine and Lysine Residues. *Mol. Cell. Proteomics.* 2004;3(6):608–614. doi: 10.1074/mcp.T400003-MCP200.

149. Harpel PC, Hayes MB, Hugli TE. Heat-Induced Fragmentation of Human A2-Macroglobulin. *J. Biol. Chem.* 1979;254(17):8669–8678. doi: 10.1016/S0021-9258(19)86944-2.
150. Jenö P, Mini T, Moes S, *et al.* Internal Sequences from Proteins Digested in Polyacrylamide Gels. *Anal. Biochem.* 1995:75–82. doi: 10.1006/abio.1995.1010.
151. Havliš J, Thomas H, Šebela M, *et al.* Fast-Response Proteomics by Accelerated in-Gel Digestion of Proteins. *Anal. Chem.* 2003;75(6):1300–1306. doi: 10.1021/ac026136s.
152. López-Ferrer D, Capelo JL, Vázquez J. Ultra Fast Trypsin Digestion of Proteins by High Intensity Focused Ultrasound. *J. Proteome Res.* 2005;4(5):1569–1574. doi: 10.1021/pr050112v.
153. Sun W, Gao S, Wang L, *et al.* Microwave-Assisted Protein Preparation and Enzymatic Digestion in Proteomics. *Mol. Cell. Proteomics.* 2006;5(4):769–776. doi: 10.1074/mcp.T500022-MCP200.
154. Castellanos-Serra L, Ramos Y, Huerta V. An In-Gel Digestion Procedure that Facilitates the Identification of Highly Hydrophobic Proteins by Electrospray Ionization-Mass Spectrometry Analysis. *Proteomics.* 2005;5(11):2729–2738. doi: 10.1002/pmic.200401164.
155. Erdjument-Bromage H, Lui M, Lacomis L, *et al.* Examination of Micro-Tip Reversed-Phase Liquid Chromatographic Extraction of Peptide Pools for Mass Spectrometric Analysis. *J. Chromatogr. A.* 1998;826(2):167–181. doi: 10.1016/S0021-9673(98)00705-5.
156. Donnelly DP, Rawlins CM, DeHart CJ, *et al.* Best Practices and Benchmarks for Intact Protein Analysis for Top-Down Mass Spectrometry. *Nat. Methods.* 2019;16(7):587–594. doi: 10.1038/s41592-019-0457-0.
157. Rappsilber J, Mann M, Ishihama Y. Protocol for Micro-Purification, Enrichment, Pre Fractionation and Storage of Peptides for Proteomics Using StageTips. *Nat. Protoc.* 2007;2(8):1896–1906. doi: 10.1038/nprot.2007.261.

https://doi.org/10.1142/9789811283376_0004

Chapter 4

Improving the Basic Workflow

Paulina Czaplewska*, Natalia Musiał, and Michał Puchalski

Intercollegiate Faculty of Biotechnology UG&MUG, University of Gdańsk, Gdańsk, Poland

Contents

*Corresponding author.
Laboratory of Mass Spectroscopy, Intercollegiate Faculty of Biotechnology, University of Gdansk, Gdansk, Poland Paulina Czaplewska, Email: Paulina.Czaplewska@ug.edu.pl, Tel +48 585236437.

Abstract

With the development of proteomics and the growing demand for faster, more efficient and high-throughput methods, new methods of digesting biological materials have been developed and proposed to scientists. Some proposed solutions did not impress the users, but some were adopted and are currently used routinely in many laboratories worldwide. One such method is methods with the use of membranes, among them the filter-aided sample preparation approach, commonly known as FASP, which enjoys a large group of supporters. Using a simple vessel with a membrane with a selected mass cutoff point made it possible to use reagents incompatible with mass spectrometers, getting rid of detergents and other low-molecular-weight impurities. It was also perfect for further modification, as described in the following chapter. The reader can get acquainted with the basics of the FASP method, possible variants of membrane methods as well as some handy methods to improve sample preparation for the final tandem mass spectrometry analysis. Reading this chapter will allow beginners to become familiar with the currently most popular and effective approaches to protein digestion, allowing you to quickly decide and choose the optimal procedure for your research or encourage the reader to research further or optimize the approaches presented here.

1. Introduction

The most significant step in the proteomic analysis is the sample preparation, including extracting proteins from the biological material and the enzymatic digestion process. The classical in-solution digestion method has undergone numerous modifications to obtain the most effective and complete digestion process, giving maximum coverage of the analyzed protein sequences in the mass spectrometry (MS) image in a complex biological matrix. Emphasis is now also placed on developing methods for achieving high throughput for clinical trial analyses. It is possible due to the rapid development of technology and new instruments facilitating the work with many samples in a short time. On the chemical side, it is facilitated by additives, such as detergents or chaotropic salts. Another aspect is introducing an easy way to remove mass-incompatible substances, that is, all non-volatile compounds or ion suppressors in clinical analyses, where the speed of the analyses is essential concerning a large number of samples, the implementation of high-throughput protocols.

One approach that is intensively used and constantly modified is the method based on introducing the mass cutoff membrane into the reaction vessel. It can include here all the so-called membrane-based methods. The most popular is the filter-aided sample preparation approach, shortly called FASP. It meets all the requirements of modern sample preparation and can be easily used for various biological materials, starting from cell extracts, homogenized tissues, and tissues fixed in formalin or paraffin. It also gives great opportunities to analyze many samples by introducing multiwell plates with a membrane and shortening the time of individual stages by using microwave or ultrasonic reactors. In this chapter, we will present an overview of the most popular membrane method — FASP and its modifications, the equivalent methods using beads, and some attractive methods of modern proteomics.

2. Membrane-based approaches

2.1. *Filter-aided sample preparation (FASP) and multiple enzyme digestion FASP (MED-FASP)*

Filter-aided sample preparation (FASP) is currently one of the most popular sample preparation methods in bottom-up proteomics. It allows performing three basic steps in one pot: protein extraction, enzymatic digestion, and peptide isolation after digestion, leading to sample preparation for MS analysis. It was proposed in 2009 by Wiśniewski and colleagues, and the standard workflow is presented in Figure 1.[1,2] They combined the best elements from the existing methods, that is, the popular sodium dodecyl sulfate (SDS) and 8 M urea with ultrafiltration. SDS is a popular component in lysis buffers facilitating the solubilization of proteins. Unfortunately, even in small concentrations, it can preclude enzymatic digestion or effectively disrupt the registration of mass spectra. Therefore, the removal of SDS is critical for efficient analysis in proteomics. In FASP, detergents are removed from the sample using 8 M urea. This chaotropic agent reduces the size of SDS micelles by decreasing concentration below its critical micelle concentration and dissociates SDS from proteins that allow for its flow through the pores of the filter membrane. Urea also keeps proteins denatured and soluble in solution after removing SDS. According to that, the protein samples are applied to ultrafiltration concentrators (the most popular and efficient are flat bottom, e.g., Microcons with a 10- or 30-kDa membrane) as reactors and are washed with urea, so low-molecular-weight compounds flow through the filter and can be discarded.[3] Centrifugation is an element that occurs after each step of the procedure, and the use of ultrafiltration concentrators with an adequately selected membrane ensures effective removal, and proteins remain in solution. It is possible to carry out the process in filters with a conic bottom membrane of the Amicon type, but it has been proven that they are not sufficiently effective in removing SDS, which ultimately lowers the sensitivity of the analysis.[3] The protein amount that can be efficiently processed with this procedure is 10–100 μg range.[4] Since the reduction of cystine to cysteine

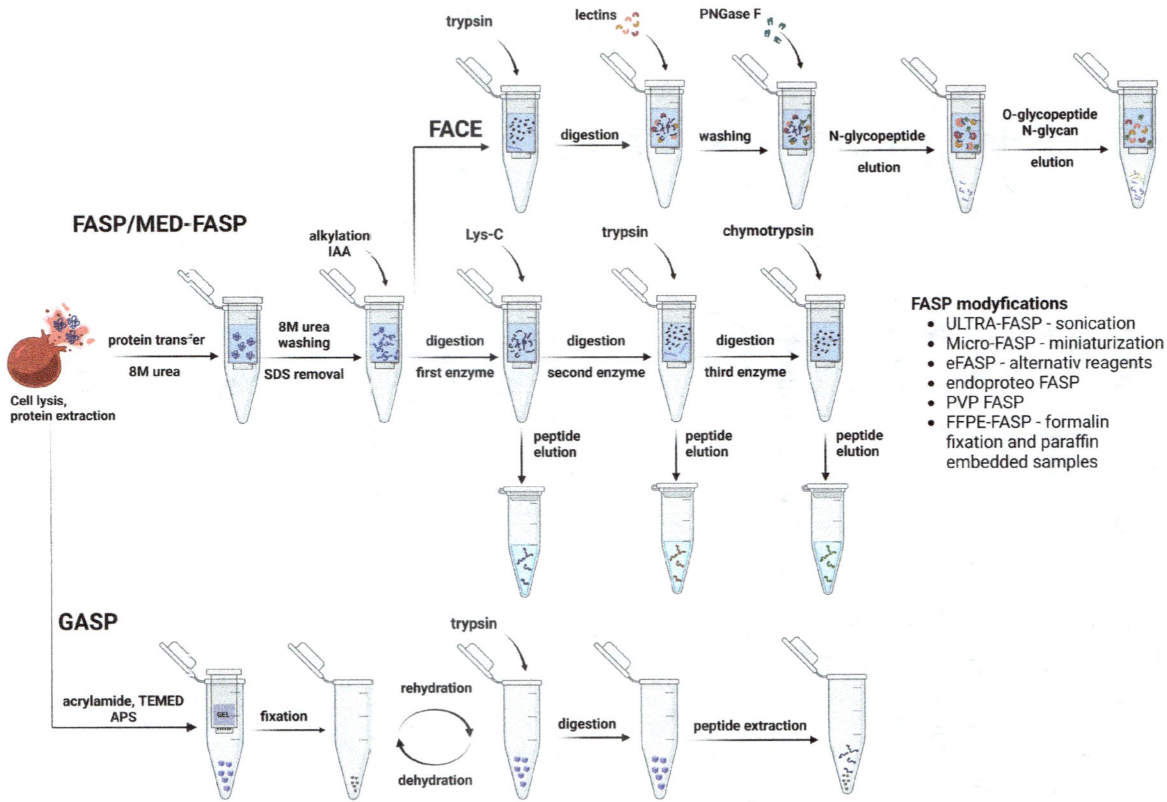

Figure 1. Standard workflow of filter-aided sample preparation (FASP)/multiple enzyme digestion FASP (MED-FASP), filter-aided capture and elution (FACE), and gel-aided sample preparation (GASP).

is carried out at the stage of cell lysis [reducing agents such as dithiothreitol [DTT], tris(2-carboxyethyl) phosphine (TCEP), or β-mercaptoethanol are components of buffers], the next step is quenching its reactivity by alkylation with the iodoacetamide (IAA). The mixture is now ready for enzymatic digestion, and the enzyme of the first choice is usually trypsin, but the FASP approach does not exclude the use of every other enzyme available. LysC and ArgC endoproteinases perform best, with high-sequence coverage in liquid chromatography–tandem mass spectrometry (LC–MS/MS) analysis. After the enzymatic digestion of proteins, peptides can be eluted from the vessel by centrifugation.

The detailed procedure, all reagents and equipment used, and the key steps have been described and explained by its creator in the publication — tutorial.[2] In 2020, Wisniewski and colleagues proposed to modify the protocol by omitting the thiol-alkylation step. They showed that while maintaining a low concentration of DTT (0.1–1 mM DTT), thiols remain in a reduced form, enzymes such as trypsin and LysC are still effective, and the identification of proteins and peptides increases by 10%–20%. Although it is widely used in protein digestion procedures, it is not cysteine-thiol specific and causes changes in, for example, methionine, histidine, lysine, and tyrosine with prolonged storage.[5,6] Eliminating an integral part of the bottom-up routine facilitates and shortens the workflow and reduces the likelihood of artificially modified peptides.[7]

FASP can be successfully used for many applications thanks to its combination of in-gel and in-solution digestion advantages, compatibility with different detergents, and high purity of eluted peptides. However, the time-consuming nature of the FASP protocol and the aim to get improved sensitivity, recovery, and proteomic coverage caused the development of new filter-based approaches.[4]

The most popular modification of the FASP protocol is *multiple enzyme digestion FASP (MED-FASP)*.[8,9] In this method, digestion is carried out with the use of at least two different enzymes during separate steps. The most commonly applied enzyme combination is endoproteinase Lys-C, trypsin, and chymotrypsin, but GluC or ArgC can also be used. MED-FASP is more efficient in proteomic analysis than

FASP because, during multiple enzyme digestion, more different peptides are generated, increasing by 20%–40% in the identified proteins. What is more, the number of peptides with missed cleavages decreases.[3] The efficiency of the entire process depends on the enzyme used, the type of the processed sample (bacterial and yeast cultures, tissues, body fluids), the initial amount of the material, and the correctly determined protein concentration in the material (the presence of detergents, reducing agents is not compatible with colorimetric measurements).

3. FASP modifications

Since its popularization, the FASP method has gained new features in the form of modifications aimed at achieving high throughput, shortening the procedure time, or adapting to the analysis of specific proteins and peptide modifications. Several have already gained a large group of supporters, and the results confirmed the entire approach's effectiveness. There is also the possibility to accelerate chemical reactions, like reduction or alkylation, and enzymatic digestion of proteins during FASP using various tools, including heating, immobilized enzymes, and different kinds of energy sources: microwave, ultrasonic, or infrared. Thanks to these modifications, the time of the whole procedure, as the reagents, sample and energy consumption, generation of by-products, and the cost of the entire process is reduced. FASP accelerated by ultrasonic energy is the most advanced method to speed up sample preparation. The time needed to prepare the sample can be reduced from overnight to 2.5 h. Moreover, this tool introduces high throughput crucial in studies on large sample batches.[10] Below we present some of the most popular modifications in contemporary literature.

3.1. *Filter-aided capture and elution (FACE)*

Zielińska and colleagues used the specific binding of lectins to carbohydrate moiety to create an effective method of N-glycopeptide analysis.[11] The basis of the procedure is the FASP protocol, the modification of which was introduced after enzymatic digestion with trypsin (Figure 1). One of the first modifications added to the original FASP procedure was the enrichment of the sample with glycoproteins. To enrich the sample with modified peptides, lectins that specifically bind glycopeptides are added. Thanks to this, the first elution allows the separation of the two types of peptides — the modified ones (in a complex with a lectin) will remain on the membrane, and the non-glycosylated ones will go into the filtrate. The enzyme PNGase can then be added to release the glycopeptides in the last elution. Thus, a sample enriched with modified peptides is obtained after the final centrifugation, ready for LC–MS/MS analysis. The downside to this procedure is that it is limited to N-glycopeptides. In 2016, Arul modified the procedure by treating the peptides after tryptic digestion with a mixture of concanavalin A (ConA) and wheat germ agglutinin (WGA) lectins, following PNGase F. Released peptides were discharged, and after washing the rest of the mixture on the filter was treated with 80% acetonitrile to release N-glycans and O-glycopeptides.[11]

A variation of this modification of the FASP approach is the use of specific antibodies for material enrichment — filter-aided antibody capturing and elution. Wisniewski and his associates used it to analyze the phosphoproteome of murine brain tissue. In this study, the combination of size exclusion chromatography and FASP fraction analysis with direct digestion of the brain tissue lysate with FASP and enrichment of the FACE peptide fraction with antibody resulted in over 12,000 mapped phosphorylation sites.[12]

3.2. *Gel-aided sample preparation (GASP)*

The acronym gel-aided sample preparation in short GASP refers to the FASP because they share the same working principle and similar workflow. The difference is that placing white on the membrane has been replaced by binding them to a gel. It is an excellent alternative to digestion in solution, reduces sample contamination and sample loss, and is also robust, scalable, sensitive, and easy to use by inexperienced users. It is derived initially from the tube-gel digestion (TG) proposed by Lu and Zhu, developed for membrane proteins[13] was modified, simplified, and adapted to the analysis of each material — all kinds of extracts or intact cells, by Fischer and Kessler in 2015.[14] In 2016, Muller and colleagues applied TG to high-throughput quantitative proteomics,[15] and they are still developing this approach under the name Tube-gel,[16] checking the possibilities of using various detergents or various types of gel polymerization.[17] Regardless of the name used in the literature, the main steps are the same, as shown in Figure 1 and include:

i) release of proteins from biological material (lysis) and extraction of proteins in the presence of a reducing agent (e.g., DTT),
ii) binding the proteins in the gel by copolymerization with acrylamide (18%–20% acrylamide concentration is optimal),
iii) cutting the gel into small pieces to increase the surface area (in GASP, it is recommended to use pulse spinning in the SpinX-Corning filter without a membrane and in TG, manual cutting),
iv) removal of low-molecular-weight compounds and impurities (dehydration and rehydration of gel pieces),
v) enzymatic digestion in a gel, and
vi) peptide extraction.

Comparisons of the amounts of proteins identified by the FASP and GASP approaches indicate similar effectiveness. At the peptide level, they identify different peptides, indicating the methods' complementarity. Depending on the biological material, it is worth considering carrying out both protocols to obtain more detailed information at the peptide level. Jailani and co-workers compared the GASP approach to solution digestion (two methods, CMW and RapiGest) for the *Saccharomyces cerevisiae* extract showing more proteins identified and better peptide recovery using the gel method.[18] Davis *et al.* conducted for MCF-7 breast cancer cells, multienzymatic GASP (trypsin and elastase) combined with offline fractionation at high pH ended with CHarge Ordered Parallel Ion Analysis (CHOPIN) on Orbitrap Fusion (Lumos) expanding proteome coverage in total over 13,728 protein groups).[19]

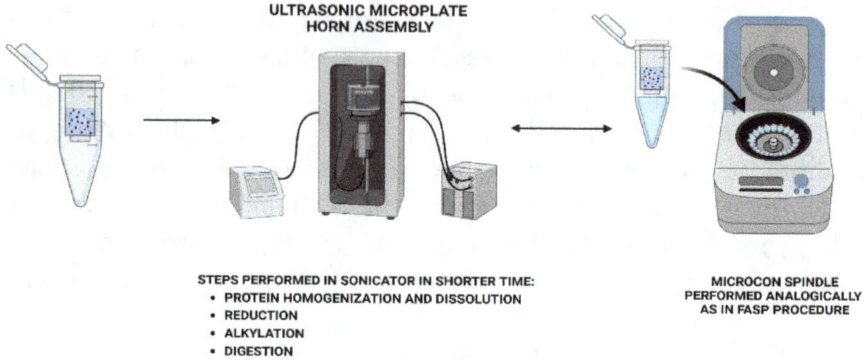

Figure 2. Ultrasonic FASP approach in sample preparation for proteomics.

3.3. *Ultrasonic-based filter-aided sample preparation (US-FASP)*

As the name suggests, this modification uses ultrasound in the standard FASP pipeline. It significantly increases the efficiency of each step and, above all, shortens the time of individual reactions. It was proposed by Carvalho and colleagues in 2020 and is based on a dedicated sonicator attachment in the Microplate Horn assembly (Figure 2).[20] Scientists tested the effect of ultrasound on the filters used, enzyme activity, time of individual reactions, applicability in quantitative and qualitative analysis, and the effectiveness of MH with other techniques. The optimal process parameters are 5.25 min with seven cycles: 30 s on and 15 s off ultrasonication time, 25% ultrasound amplitude, and 20 kHz ultrasound frequency. It reduces the time needed to complete the procedure from 18 to 2.5 h. The method applies to 96-well plates, which allows for the quick preparation of large sample sets. We get high throughput if we add a mass spectrometer with a suitably fast chromatographic system (15-min gradient for one sample).

3.4. *Microwave-assisted FASP*

The most time-consuming step in the FASP procedure is enzymatic digestion with trypsin incubation overnight at 37°C. This stage also makes it possible to introduce modifications to accelerate and increase this process's efficiency.

Yu *et al.*, in 2012, used this approach to analyze membrane proteins from the K562 leukemia cell line.[21] Two thousand ninety proteins were identified, and the procedure is described under GOFAST (GELFrEE optimized FASP technology). The membrane fraction was additionally separated in the GELFrEE system, which separates the proteins in the gel cartridge into 12 fractions, and after the separation is completed, the proteins are eluted from the bands. The material collected in this way was subjected to the FASP procedure with digestion in a microwave reactor.

One of the proposals is to use microwave-assisted trypsin digestion with an additional increase in its amount (trypsin:substrate 1:1) to accelerate and improve the digestion of low-abundant proteins. Such an alternative was proposed by Zhao *et al.* called imFASP. *E. coli* lysate was chosen as the material, and the proposed procedure was compared with the classical digestion in solution.[22] First of all, it was possible to speed up the procedure by using microwave etching and by reducing the amount of starting material to the level of nanograms, the identification of proteins and peptides increased by 30%–44% compared to the solution.

3.5. *Enhanced FASP (eFASP)*

eFASP allows the introduction of alternative reagents like mild detergent deoxycholic acid (DCA) or optionally even Tween 20 to improve sensitivity, recovery, and proteomic coverage when working with a small amount of sample.[23] Compared to the original FASP procedure, it was proposed to use 0.2% DCA instead of urea, which increased the trypsin digestion efficiency of cytosolic and membrane proteins. Another change is the introduction of passivation of Microcon and plastics with 5% T20, which reduced the losses associated with the non-specific binding of peptides to the surface of plastics. The authors claim that this ensures a detergent, reductant, and alkylator-free peptide, as well as salts, lipids, and nucleic acids. An additional proposition accelerating the process is a one-step reduction and alkylation using a combination of TCEP and 4-vinylpyridine

(4-VP). Alkylation with 4-VP is very cysteine specific and proceeds rapidly at neutral pH.[24]

3.6. *Miniaturized FASP (MICRO-FASP)*

One of the common challenges scientists face is the small amount of biological material dedicated to proteomic research. It requires a scale-down concerning the solutions' volumes and the reaction vessel itself. Zhang *et al.* proposed miniaturization in 2020 by moving the membrane to a 20-µL pipette tip and attaching it to the bottom and top support (Figure 3).[25] Thereby, the filtration area was reduced to 0.1 mm^2, the buffer volumes used were reduced to microliter values, and the sample amounts to 1 µg of proteins. The authors of the method performed tests on 1000 MCF-7 cells, obtaining 3069 protein groups for identification at 100 ng of the starting protein and 1895 groups of proteins at 10 ng of the material. The significant advantage of this modification is the lack of the need to purchase additional devices. The only thing is to master the creation of MICRO-FASP reactors, which the authors of the method support with detailed instructions. This approach is also suitable for high-throughput

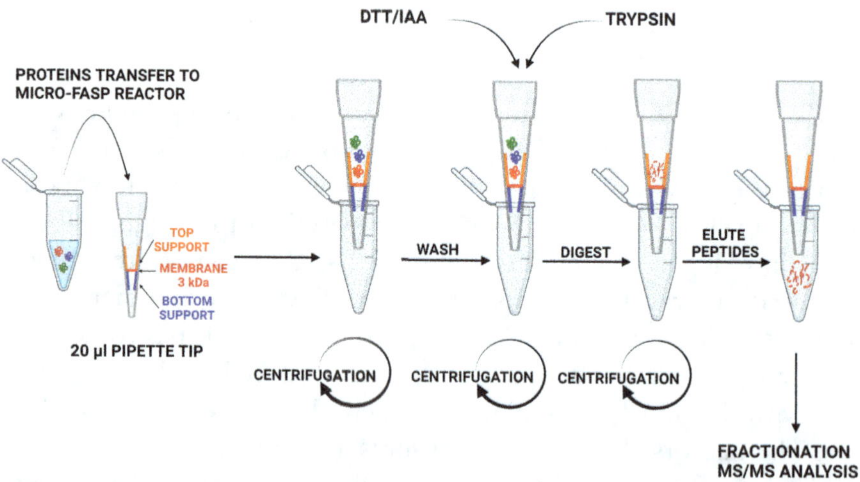

Figure 3. Schematic workflow for sample preparation by the MICRO-FASP method.

adaptation using polypropylene 96-well plates as microreactor support. The first tests from 2020 on 12 samples of *Xenopus laevis* lysate, each with 1 mg of lysate, processed in parallel in a polypropylene plate (2 mL, 96 wells) as support, resulted in the identification of 2701 groups of proteins.[25] The second described approach from 2022, where the system was refined for high throughput, allowed for the parallel processing of 113 single blastomeres from 50-cell stage *X. laevis* embryos. On average, 3468 groups of proteins were obtained per sample for each blastomer.[26] Inspired by Zhang's microplate idea, Sandbaumhüter and colleagues developed a new type of 96-well plate microtiter to analyze the single pancreatic islet proteome. It allows the insertion of filters with a diameter of 1 mm using a biopsy punch. On such a prepared microplate, they conducted proteomic and peptidomic studies for HeLA digest and the single islet of Langerhans.[27]

3.7. *EndoProteo FASP*

It is one of the specific adaptations of the FASP protocol to the biological material of saliva. Due to the easy accessibility and non-invasive method of sampling, it is gaining more and more supporters as an excellent source of information about pathological changes occurring in the human body. Trindade was the first to present a proof-of-concept on the possibility of using FASP for the simultaneous analysis of the proteome and salivary degradome — the bridge between salivary peptidome and salivary proteases.[28] The FASP approach guarantees simple sample clean-up but also works as a reactor for the natural degradation of saliva proteins by endogenous salivary protease pool and allows sequential elution of new peptides. The concept is simple and presented schematically in Figure 4. The first separation was carried out on a 30-kDa filter, allowing to obtain a natural saliva peptide fraction. Then the remaining high-mass fraction was incubated on the membrane in a bicarbonate solution for 18–115 h at 37°C. High-mass proteins remaining on the folate were analyzed using gel electrophoresis and zymography. After the wash sequence, the eluates were collected, pooled, and analyzed as newly generated peptides — degradome. This approach made it possible to track the salivary peptide's dynamic changes and determine the role of proteases in this process.[28]

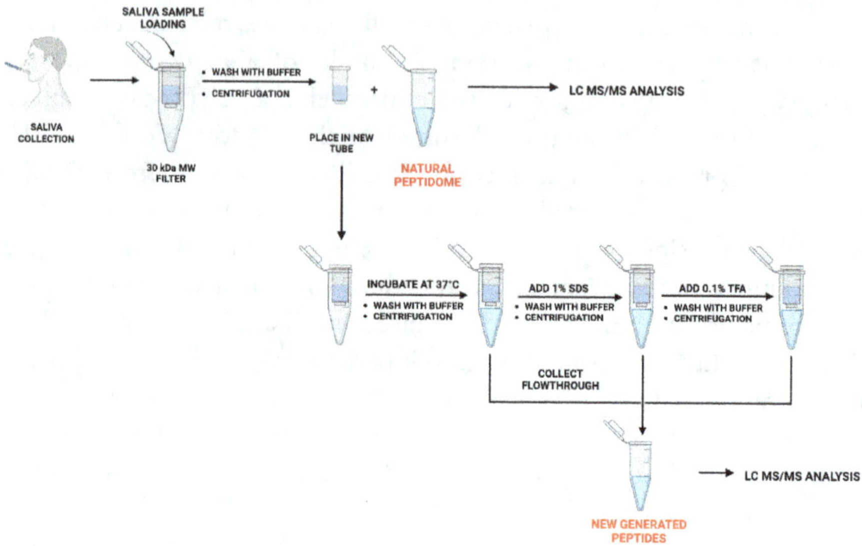

Figure 4. EndoProteo FASP workflow. TFA, trifluoroacetic acid.

3.8. *Formalin-fixed paraffin-embedded FASP (FFPE-FASP)*

The availability and quantity of biological material intended for proteomic research is one of the main limitations of the research work. In the case of clinical material, the procedures related to collecting and preserving samples are related to embedding the specimens in paraffin blocks. Of course, procedures have been developed to extract proteins from FFPE tissues. Cryosection of such material from which, on average, about 60,000 cells can be obtained utilizing laser-capture microdissection (LCM) allows for in-depth proteomic analysis. In 2010, Ostasiewicz *et al.* described the possibility of using FASP to analyze fragments of fresh-frozen and fixed tissues.[29] In the case of FFPE tissues, the first step is always to deparaffinize the material, which can be done with xylene or hot water. The material is then lyzed in a buffer containing DTT, ethylene glycol, and SDS. The presence of formalin in the original preparation causes the formation of chemical crosslinks that can be a problem in identifying proteins in proteomic analysis. High concentrations of detergents such as SDS and high temperatures help eliminate these protein modifications.

Acronym	Full name	Filter used	Modification	Reference
FASP	Filter-aided sample preparation	Cellulose membrane mass cutoff 10 or 30 kDa	Digestion on membrane	2
MED-FASP	Multiple enzyme digestion filter-aided sample preparation	Cellulose membrane mass cutoff 10 or 30 kDa	Sequential digestion of proteins with three enzymes, the most popular combination: LysC–trypsin–chymotrypsin	8
FACE	Filter-aided capture and elution		For preparation of N-glycans and O-glycopeptides	
US-FASP	Ultrasonic-based filter-aided sample preparation	Cellulose membrane mass cutoff 10 or 30 kDa	Placing the Microcon in the ultrasonic bath in the microplate horn device	20
MICRO-FASP	Miniaturized filter-aided sample preparation	Cellulose membrane mass cutoff 10 or 30 kDa	The membrane is cut and placed in the pipette tip, miniaturization	25
abFASP	Affinity-based filter-aided sample preparation	Cellulose membrane mass cutoff 10 or 30 kDa	Two-step affinity purifications before digestion with FASP protocol	36
pFASP	Plate-based filter-aided sample preparation	Cellulose membrane mass cutoff 10 or 30 kDa	96-Well plates variant for high throughput	37
eFASP	Enhanced filter-aided sample preparation	Cellulose membrane mass cutoff 10 or 30 kDa	Incorporation of alternative reagents to those of traditional FASP	23
EndoProteo FASP	Endogenous salivary proteases armamentarium FASP	Cellulose membrane mass cutoff 10 or 30 kDa	Effective enrichment of salivary peptidome by incubation on membrane and retention of high-molecular-weight material	28

(***Continued***)

(*Continued*)

Acronym	Full name	Filter used	Modification	Reference
GASP	Gel-aided sample preparation	20% acrylamide/ bis-acrylamide gel	Protein extracts or intact cells are copolymerized with acrylamide	14
GOFAST	GELFrEE optimized FASP technology	Cellulose membrane mass cutoff 10 or 30 kDa	Protein fractionation based on their molecular weights within gel-eluted liquid fraction entrapment electrophoresis (GELFrEE), digestion with the assistance of microwave reactor	21
MStern-blot	High-throughput polyvinylidene fluoride (PVDF) membrane-based proteomic sample preparation	96-Well plate with PVDF membrane	It uses a large-pore hydrophobic PVDF membrane that efficiently adsorbs proteins, resulting in fast liquid transfer through the membrane and significantly reduced sample processing times	38
iFASP	Isobaric mass tagging with filter-aided sample preparation	Cellulose membrane mass cutoff 10 or 30 kDa	Tryptic peptides are labeled (iTRAQ, TMT) on the membrane before elution	39
FFPE-FASP	Formalin-fixed and paraffin-embedded filter-aided sample preparation	Cellulose membrane mass cutoff 10 or 30 kDa	Combined a procedure of boiling FFPE sample in SDS buffer with subsequent loading onto 30-kDa filter units and digestion	29

After the tissue sample is treated with the lysis buffer, the material is ready for the classical FASP or Med-FASP procedure. According to the information provided by Wiśniewski *et al.*, from 100 to 200 nL of fixed tissue after LCM, the sample volume corresponds to 10–20 mm^2 of a 10-μm-thick slice, yields 10–30 μg of proteins.[30] Strong anion exchange (SAX) was used as an additional fractionation, as an optional step, which increased the number and dynamic range identification and extent of the protein coverage. Strong cation exchange step and StageTips final clean-up are performed using pipette-tip microcolumns.[31] Many studies in the literature compare the effectiveness of FFPE-FASP with other classical methods of fixed-tissue analysis. Depending on the type of tissue, its age (length of storage time in the biobank), and the methods compared (digestion in a solution, digestion in high pressure, digestion in the presence of RapiGest, or with the use of commercial kits), the comparisons are different.[32–35] The most important conclusion is that if you have enough samples, choose two compatible methods to increase the proteome identification coverage, width, and depth. If not, FFPE-FASP is a suitable starting method for performing qualitative and quantitative proteomic analyses on fixed tissues.

4. On-bead digestion-based methods

4.1. *Suspension trapping approach (S-Trap)*

The S-Trap simultaneous trapping method turned out to be an excellent alternative to FASP, which makes it possible to digest proteins dissolved in the detergent solution — initially proposed by Zougman in 2014, who used a pipette tip filled with sequentially placed quartz or borosilicate glass depth filters with a C18 membrane attached as a reaction vessel.[40] Next, a straightforward sequence was used where a suspension of previously denatured, reduced, and alkylated proteins in a concentrated SDS solution (5%), treated with the combination of phosphoric acid (12%) and methanol (90%), and introduced into a tip. The particles are trapped in a porous filter. Washing the entire system allows for the effective removal of detergent and impurities, and the trapped proteins are available for the next step, that is, enzymatic digestion. After digestion, peptides are trapped in the C18 bed, washed, and eluted. The method was very well received by scientists and was quickly commercialized. Currently, columns and 96-well plates with built-in filters are available, which can carry out the entire process on a small or large scale. A definite advantage over FASP is the short time of this procedure, less than 40 min for the total procedure, and the spinning time from 200 min to 15 min.[41]

Another advantage of using S-Trap is its compatibility with most detergents and lysis buffers: SDS, urea, NP-40, and RIPA.[43,44] Thanks to the effective trapping of proteins in the pores, intensive washing and removal of non-protein impurities can be used. The studies carried out so far have shown that the method is highly reproducible, the digestion provides high sequence coverage, and can be adapted to high throughput. Kuras and colleagues have shown that it also gives excellent results with FFPE tissue samples. In tissue studies, tonsil, lung adenocarcinoma and malignant melanoma used xylene-free deparaffinization and two protein extraction methods (SDS and urea) followed by S-Trap digestion. It gave outstanding results, the identification of the 9585 ID proteins. The protocol also proved compatible with quantitative proteomic TMT labeling,[45] iTRAQ,[46] label-free quantification,[47–49] and SWATH.[50,51]

4.2. *SiTrap*

In 2020, Zougman described a method based on S-Trap, which differs in the possibility of processing the sample without detergents.[48] It was created to respond to metabolomic needs because metabolites were lost along with the flow-through fraction during the classic S-Trap procedure (Figure 5b). Reduction and alkylation are performed in situ with triethylammonium bicarbonate, TCEP, and chloroacetamide (CAA) instead of the standard DTT/IAA, resulting in detergent-free flow-through. Only ammonium acetate or ammonium hydroxide and sonication are used at the cell lysis stage. Also, it was quickly commercialized, and commercial columns are available that allow for processing without detergents and thus enable obtaining samples from one material for metabolomic and proteomic analysis.

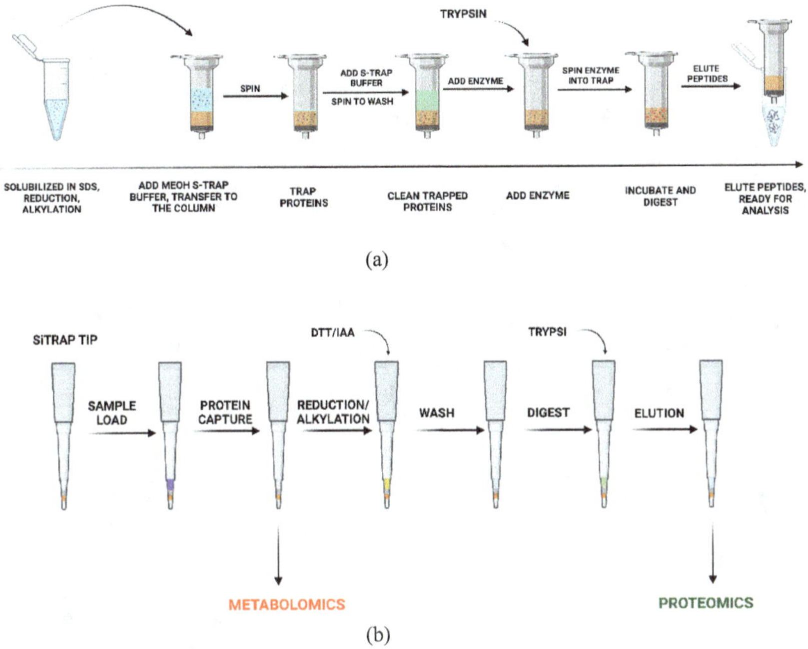

Figure 5. Schematic representation of the steps in the (a) S-Trap and (b) SiTrap workflow for metabolomics and proteomics based on Zougman.[42]

4.3. *Single-pot, solid-phase-enhanced sample preparation (SP-3)*

Single-pot, solid-phase-enhanced sample preparation, known as SP-3, relies on paramagnetic beads to bind and clean up proteins and peptides. The procedure requires the following steps: cell or tissue lysis and nucleic acids shearing (e.g., benzonase digestion, sonication, or bead beating) (Figure 6a). The beads are added to the protein-containing solution in an optimal SP3 beads/protein ratio of 1:10. The surface of the beads is covered with carboxylate functional groups, which enables the binding of proteins and peptides through hydrophilic interactions. After binding, the beads are attracted to the magnet, and all contamination can be removed by rinsing with ethanol. When clean-up is over, the whole proteins can be eluted from the

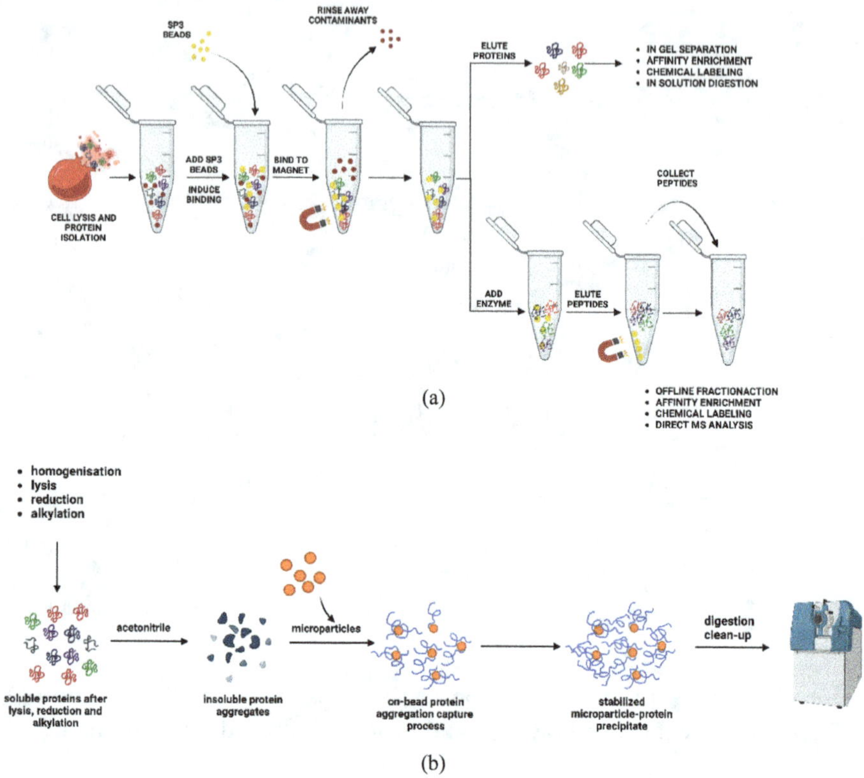

Figure 6. (a) General SP3 workflow.[52] (b) Hypothetical model of the protein aggregation on microparticles in protein aggregation capture, based on Ref. [54].

surface of the beads. If proteolysis is planned, it could proceed in the same tube by adding enzyme and digestion buffer to the beads with bounded proteins, followed by elution of the peptides ready for MS analysis.[52,53]

Johnston and colleagues noticed that in the SP3 method, the binding of proteins to the magnetic bed is related to solvent-induced denaturation, aggregation, and subsequent precipitation independent of bead surface chemistry (Figure 6b), So they proposed using glass spheres (GB) or entirely omitting solid beads (BF).[55] Named SP4 or solvent precipitation SP3 by the developers, this approach is a rapid single-pot, minimalistic, low-cost, and easy to apply even for an inexperienced scientist. In addition, it does not require special equipment or reagents, both with and without beads. Compared to the original method, which is excellent at low protein concentrations, SP4 is superior to it at high concentrations, as reported by Johnston.

The procedure is similar to those presented above, with the difference that proteins dissolved in the solution (after reduction and alkylation) are precipitated by adding an organic solvent, the most common 80% acetonitrile. Then, glass beads are added to the precipitate, on which protein aggregates are captured (glass beads SP4, SP4-GB). Further steps are the same, and there is no need to use a magnet as in the SP3 method. The authors also propose a version with complete omission of the beads and, after precipitation of the aggregates, further processing of the pellet (bead-free, SP4-BF).

4.4. *Protein aggregation capture (PAC)*

Protein aggregation is, in most cases, a disorderly and disease-related process or a process that makes working with proteins difficult. However, it was cleverly used in a method called by the developers as PAC. Batth *et al.* used the process of artificially generating protein aggregates in a cell lysis solution by adding an organic solvent.[54] They noticed that the thus obtained insoluble forms of proteins bind perfectly to the microparticles, which allows them to be captured and processed further. The basis of this approach is the phenomenon of the non-specific interaction of precipitated and aggregated proteins on any type of microparticle. This interaction is independent of the substrate chemistry of the particles used. The first step, as in any of the above-described procedures, is the homogenization and lysis of the material for proteomic analysis. Reduction and alkylation of the released proteins are carried out in the adhesive step. The solution obtained in this way is subjected to the first mandatory step in the PAC procedure, which is precipitation and formation of protein aggregates. Acetonitrile seems to be the most effective in this case. At this stage, microparticles with any chosen binding chemistry are introduced into the solution. It can be a standard bed used as a packing for C18 chromatographic columns and any magnetic bed.

The introduced and extruded microparticles act as nucleation seeds and induce the immobilization of protein aggregates. The last step is to add the selected enzyme and digest proteins associated with the resin. An important observation of the authors of the method is that high concentrations of proteins in solutions after lysis require a much smaller amount of microparticles than solutions with a low concentration of proteins to bind aggregates effectively. The procedure was described in 2019 and has already had a large group of users. It proved perfect in identifying cancer markers,[56–58] the characteristics of extracellular vesicles,[59] microbiomes,[60–62] and work on the proteome of archaeological samples[63,64] and quantitative research using isotope labeling, for example, TMT.[65]

5. On-pellet digestion (OPD)

The standard proteomic procedure assumes that the proteins in the mixture should be dissolved in the working buffer used and preferably unfolded before proteolytic digestion. It ensures the best access of the enzyme to the cleavage sites to make the process as effective as possible. Often, however, when working with complex biological matrices, there is a problem with the complete solubility of proteins. Similarly, after the precipitation of proteins with an organic solvent such as acetone, acetonitrile, or methanol, reintroducing proteins into the solution often causes many problems and the need to use an ultrasonic bath or surfactant additives. Therefore, to simplify the procedure, it was checked whether direct digestion of the pellet after precipitation with an organic solvent was possible and whether the obtained identification rates were similar to the results of traditional procedures. In 2009, such an approach was described by Duan *et al.* as an effective and repetitive precipitation/on-pellet digestion strategy[66] used for profiling the tissue proteome (Figure 7). In the case type, the first step was to extract proteins from the tissue using a detergent-containing buffer. The proteins were then precipitated with cold acetone, and after its removal, buffer and trypsin were added to the pellets. Alternatively, adding a surfactant called ProteasMAX was also used to ensure complete digestion. The reduction and alkylation

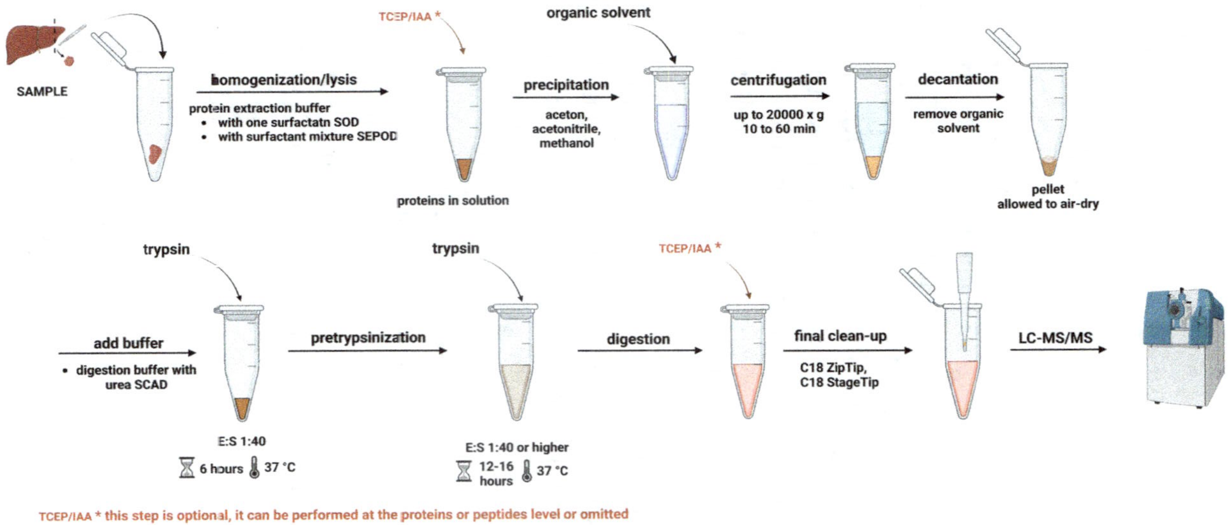

Figure 7. Schematic workflow of the on-pellet digestion procedure.

of cysteine residues were performed at the peptide stage, and then the samples were subjected to LC–MS/MS analysis. This approach has been successfully used for proteins from the tissue material described above and worked with plasma. The potential of organic solvents used for precipitation was checked, and apart from acetone, acetonitrile and methanol proved excellent.[67] The approach is also suitable for conducting quantitative high-throughput studies.[68]

The analysis of the literature data shows that the reduction and alkylation step can be performed immediately after protein extraction, just before precipitation with an organic solvent,[69] in the tryptic peptide step, or omitted altogether. Interestingly, this approach often neglects the reduction and alkylation steps, such as when analyzing monoclonal antibodies from serum as seen by Yuan and his colleagues. They showed that compared to the pretreatment approach (reduction and alkylation), the results of the on-pellet digestion analyses were the same or better than the classical approach.[70] Some papers describe procedures with two-phase trypsin use. The first phase, the so-called pellet-dissolving phase or pretrypsinization, is associated with the added trypsin in the enzyme:substrate ratio of 1:40 to predigest and dissolve the proteins. The solution is incubated at 37°C for 6 h. In the second step of the complete-cleavage phase, another aliquot of the enzyme at the same ratio or higher is added, and incubation is extended to 12–16 hours. It is to ensure the completeness and highest efficiency of the process.[71,72] Another modification of the original method is the introduction of surfactants or chaotropic substances during protein extraction or just before adding the enzyme. In the literature, therefore, one can find a procedure called surfactant-aided precipitation/on-pellet digestion (SOD), in which the extraction buffer contains one additive such as SDS, sodium deoxycholate (SDC), NP-40, or an MS-compatible surfactant like RapiGest.[71,73] Approach surfactant cocktail-aided extraction/precipitation/on-pellet digestion (SE-POD) introduces a mixture of surfactants (SDS, SDC, and IGEPAL CA-630) into the extraction buffer.[74,75] Another variant is a surfactant and chaotropic agent–assisted sequential extraction/on-pellet digestion (SCAD), where SDS was used in the extraction buffer, and urea was added to the digestion buffer.[76–78]

6. Other potent contemporary digestion approaches

6.1. *Sample preparation by easy extraction and digestion SPEED*

The first step in the proteomic analysis is obtaining proteins from the provided sample (tissues, cells, body fluid), and the presence of detergents or chaotropic agents increases the efficiency of this process. Some of them are also helpful in unfolding proteins, which facilitates the access of the enzyme to the entire structure. However, most of these additions are a problem for recording spectra. Therefore, new approaches are wanted, allowing work in an environment without substances incompatible with LC–MS. One solution might be sample preparation by easy extraction and digestion (SPEED), a fast and detergent-free sample preparation method (Figure 8). SPEED can apply to many sample types, even lysis-resistant or containing membrane and cytosolic proteins. This method is based on three key steps:

1) acidification with pure trifluoroacetic acid (TFA, 1–10 min): samples dissolve in minutes, and DNA is degraded,

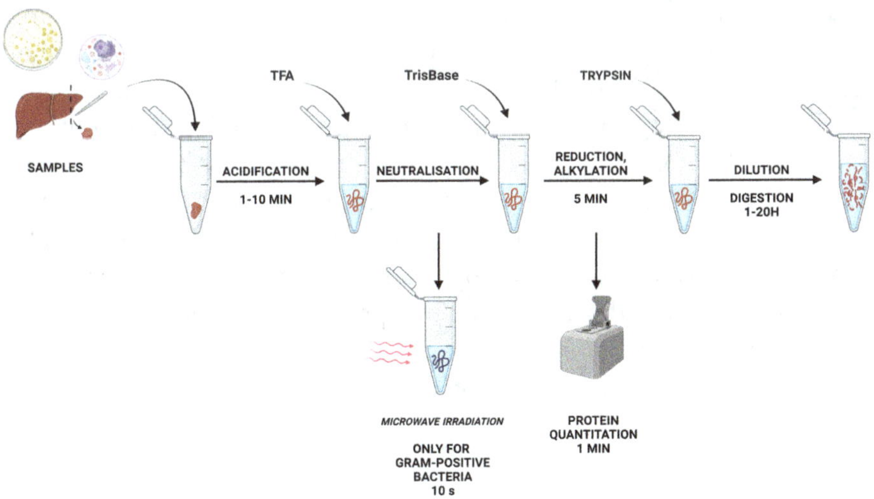

Figure 8. Workflow of sample preparation by easy extraction and digestion (SPEED).

2) neutralization (including reduction and alkylation, 5 min): neutralization is carried out with a weak base with a pKa corresponding to the optimal activity of the selected protease (e.g., 2 M Tris Base); the sample is heated (70°C–80°C) as a result of the exothermic process, which facilitates and shortens the following steps, reduction (TCEP) and alkylation (CAA);
3) digestion with the selected enzyme at optimal temperature (1–20 h): the sample must be diluted with water before adding the enzyme.

The developed procedure minimizes and significantly speeds up the steps needed to prepare the sample and perform the digestion. SPEED improves proteomic coverage and quantitative reproducibility. Besides, this approach is inexpensive, does not require special equipment, and has high automatization potential.[79]

Although TFA dissolves *Escherichia coli* cells and mouse liver tissue at room temperature entirely in a few minutes, lysis of Gram-positive bacteria *Staphylococcus aureus* requires an additional step of microwave irradiation for 10 s. After neutralization, the samples become slightly turbid as proteins precipitate. Proteins are quantified in the resulting dispersions by turbidity measurement at 360 nm and digested by adding trypsin. Afterward, peptides are desalted online or offline before LC–MS/MS measurements.[79]

The SPEED approach has been successfully applied, for example, for the study of transthyretin amyloidosis (ATTR) by Treitz and colleagues. In this case, samples from paraffin blocks containing carpal tunnel tissue after laser microdissection were dissolved in TFA.[80] The maximum dissolution of the amyloid sample is allowed in this case, not only to identify the amyloidogenic protein in the samples but also to perform an in-depth analysis of the entire proteome and relate its composition to the mechanism responsible for the disease.

6.2. *Pressure cycling technology (PCT)*

Pressure cycling technology (PCT), also known as barocycling, was introduced in the early 1990s; however, it did not find many supporters due to bulky equipment and limitations of the original highly specialized consumable sample containers called PULSE tubes. New instrumentation recently developed by Pressure BioSciences popularized high-pressure applications for sample preparation in Systems Biology fields such as proteomics or genomics. The main advantage of this procedure is the ability to homogenize, extract, solubilize, and enzymatically digest proteins from complex biological samples, solid and liquid, by using pressure in a much shorter time than traditional procedures. Up to 16 samples per batch can be placed in unique fluoropolymer PCT MicroTube containers and subjected to high-pressure process with or without adding a urea-containing lysis buffer (Figure 9).

The Barocycler 2320EXT generates short, alternating cycles of ultrahigh (up to 50,000 psi) and ambient hydrostatic pressure (Figure 10). The rapid change of pressure breaks down and

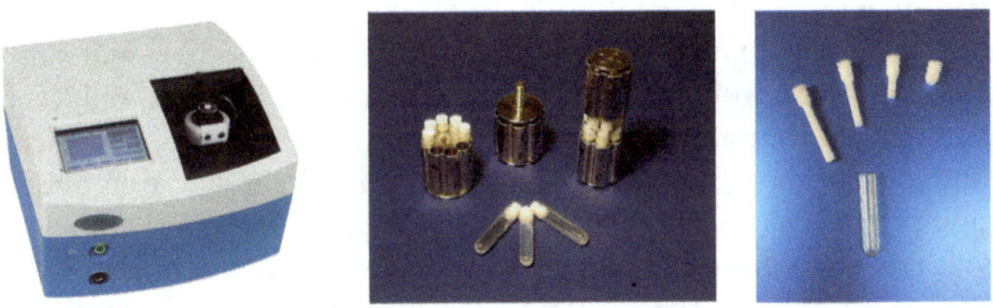

Figure 9. PCT Barocycler instrument with a unique screw holder for 16 samples and MicroPestle with various plugs.

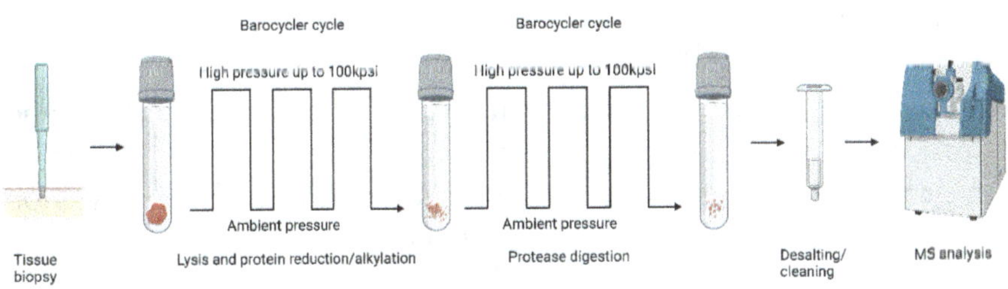

Figure 10. Schematic workflow of the sample preparation procedure for proteomic analysis using pressure cycling technology.

solubilizes samples to release their content (e.g., nucleic acids, proteins, and lipids). Using a water bath or a built-in electric heater allows the temperature to be controlled during the PCT procedure, preventing overheating and damaging the samples. After complete solubilization, it is possible to add enzymes and perform digestion during the next cycle. The hydrostatic pressure used in PCT improves the efficiency of enzymatic cleavage.[81,82] The peptides obtained in this manner can be subjected to final clean-up and analyzed directly, or in the case of a highly complex matrix, they can be fractionated, for example, at high pH, to increase protein identification coverage.

The further modification of the PCT procedure is the PCT-MicroPestle-based protocol, which combines the advantages of standard PCT and homogenization. It is based on using MicroPestle, a mechanical homogenizer placed inside the MicroTube. During high-pressure cycles, the tissue sample is ground between the MicroPestle and the tube walls, increasing solubilization efficiency. It results in 20%–40% more peptides after digestion than the classic PCT procedure.[83] The procedure with PCT is reproducible, effective, and suitable for high-throughput analyses, allowing its use not only for identification but also for quantitative analysis of proteins in clinical samples. As a first group, Lopez-Ferrer used protein digestion under high pressure in cyclin technology for proteomic analysis.[84] Guo and colleagues successfully combined the pressure technique with quantification by creating the PCT-SWATH protocol for biopsy-level tissues. The processing (lysis, protein extraction, and enzymatic digestion) was carried out in a Barocycler. It shortened the time from the initial material — biopsy tissue — to data collection to 12 h.[81]

Accelerated Barocycler lysis and extraction (ABLE) is a simple modification of the PCT procedure. The main difference is the replacement of urea in the lysis buffer with a combination of SDC and N-propanol. Thanks to that, solubilization and enzymatic digestion can proceed at higher temperatures without the risk of carbamylation and decreasing the final peptide outcome. It considerably speeds up the whole procedure, from 6 to 8 h in classic PCT to even 2 h in ABLE.[85] Its creators further adapted this method to simultaneously obtain proteomic, metabolomics, and lipidomic information as a multiomic Multi-ABLE method. Immediately after homogenization and lysis of the material in a Barocycler, the sample is divided into two parts, the first of which is further processed according to the ABLE protocol, and the second is subjected to extraction with organic and water phases allowing the separation of metabolites from lipids.[86] This technology has also been successfully applied to analyze frozen and FFPE tissues.[87–89]

Sample preparation in a Barocycler also allows for the combination of proteomic and microbiome analysis, for example, in the work of Bao *et al.*, where gingival tissue was processed according to the standard PCT protocol. Simultaneously with the isolation of proteins, DNA isolation of the microbiome was carried out. After digestion, proteins were analyzed based on fragmentation spectra, and isolated genetic material allowed for 16S rRNA sequencing and identification of the microbiome in the samples.[90]

An additional benefit of high pressure as a denaturant synergistic to chaotropic reagents is that proteases can efficiently digest proteins that have not been previously reduced, preserving disulfide bonds and thus allowing efficient multiattribute analytical methods for the quality control (QC) of biopharmaceuticals.[91] This approach has been adopted by several major biopharmaceutical companies in a regulated QC environment, thanks to the Good Manufacturing Practice–compliant features of the Barocycler software.

The ultimate adaptations of PCT sample preparation methods to study tumor microheterogeneity have been introduced by streamlining sample transfer from laser microdissection equipment directly into the PCT MicroTubes.[92–95] The resulting protocol incorporated Leica LMD tissue processing followed by Barocycler-assisted FFPE de-crosslinking and subsequent Barocycler-assisted enzymatic digestion without using chaotropic reagents, thus simplifying or eliminating the need for sample clean-up and preserving hydrophilic peptides otherwise lost during clean-up steps.

Pressure cycling technology presents several unique opportunities to streamline and improve the efficiency and throughput of biological sample preparation for proteomic workflows, protein characterization, and QC of biologics.

Acronym	Full name	Filter used	Modification	Reference(s)
SPEED	Sample preparation by easy extraction and digestion	In-solution digestion in Eppendorf tubes	Consists of three mandatory steps, acidification, neutralization, and digestion. SPEED is a universal method for peptide generation from various sources and is easily applicable even for lysis-resistant sample types, as pure TFA is used for highly efficient protein extraction by complete sample dissolution.	79
fa-SPEED	**Filter-aided sample preparation by easy extraction** and digestion	**Cellulose membrane mass cutoff 10 or 30 kDa**	**Combination of SPEED with FASP.**	79
SP3	Single-pot solid-phase-enhanced sample preparation	Requires the introduction of the solid matrix as beads (any bead with a hydrophilic coating and functionalized surface should be compatible with SP3)	"Single vessel" approach that aims at minimizing sample losses by simplifying the proteomic sample preparation workflow.	52
Urea-ISD	Urea-based in-solution digestion	Urea in-solution digestion at room temperature	The number and mass spectral signal intensity of peptides with carbamylation and pyroglutamic acid modifications are reduced, consequently reducing the negative effects of these modifications.	96

S-Trap	Suspension trapping	Required dedicated S-Trap microcolumns	Proteins are lyzed in 5% SDS, and a fine protein particulate suspension is created by adding phosphoric acid and a methanolic buffer solution. The suspension is trapped in a stack of filtration material, and residual SDS is washed away in one short wash step. Proteins are digested in the filter using a protease of choice before analysis via LC–MS/MS. The S-Trap possesses the advantages of FASP and other filter-based methods while decreasing the sample handling steps and time prior to MS analysis.	41,43
PCT	Pressure-assisted digestion	Use of dedicated vials and Barocycler instrument	Pressure cycling technology for rapid tissue homogenization, solubilization, and reliable, controlled proteolytic digestion.	97
iST	In-StageTip digestion	A pipette tip filled with an appropriately selected phase	The "single vessel" approach allows proteome-sample preparation in five pipetting steps and can readily be performed in a 96-well format.	98
SOD	Surfactant-aided-precipitation/on-pellet digestion procedure	In traditional Eppendorf tubes or other plastic tubes	Prior to the precipitation/on-pellet digestion process, a relatively high concentration of surfactant is employed to achieve extensive denaturation/reduction/alkylation of	73

(*Continued*)

(*Continued*)

Acronym	Full name	Filter used	Modification	Reference(s)
			proteins, to inactivate endogenous protease inhibitors, and to facilitate the removal of matrix components.	
SCAD	Surfactant and chaotropic agent–assisted sequential extraction/on-pellet digestion	In traditional Eppendorf tubes or other plastic tubes	Proteins were first solubilized by SDS and then subjected to precipitation to remove the detergent. The pellet was reconstituted in urea, and on-pellet digestion was performed.	76
SEPOD	Surfactant cocktail-aided extraction/precipitation/on-pellet digestion	In traditional Eppendorf tubes or other plastic tubes	Critical parameters of SEPOD: composition and concentrations of surfactants during extraction and denaturation, conditions of precipitation, and duration of proteolytic digestion.	73

References

1. Wiśniewski JR, Zougman A, Nagaraj N, *et al.* Universal Sample Preparation Method for Proteome Analysis. *Nat. Methods.* 2009;6(5): 359–362. doi: 10.1038/nmeth.1322.
2. Wiśniewski JR. Filter Aided Sample Preparation — a Tutorial. *Anal. Chim. Acta.* 2019;1090:23–30. doi: 10.1016/j.aca.2019.08.032.
3. Lipecka J, Chhuon C, Bourderioux M, *et al.* Sensitivity of Mass Spectrometry Analysis Depends on the Shape of the Filtration Unit Used for Filter Aided Sample Preparation (FASP). *Proteomics.* 2016;16(13):1852–1857. doi: 10.1002/pmic.201600103.
4. Wiśniewski JR. Filter-Aided Sample Preparation: The Versatile and Efficient Method for Proteomic Analysis. *Methods Enzymol.* 2017;585: 15–27. doi: 10.1016/bs.mie.2016.09.013.
5. Boja ES, Fales HM. Overalkylation of a Protein Digest with Iodoacetamide. *Anal. Chem.* 2001;73(15):3576–3582. doi: 10.1021/ac0103423.
6. Lapko VN, Smith DL, Smith JB. Identification of an Artifact in the Mass Spectrometry of Proteins Derivatized with Iodoacetamide. *J. Mass Spectrom.* 2000;35(4):572–575. doi: 10.1002/(SICI)1096-9888(200004)35:4<572::AID-JMS971>3.0.CO;2-2.
7. Wiśniewski JR, Zettl K, Pilch M, *et al.* 'Shotgun' Proteomic Analyses Without Alkylation of Cysteine. *Anal. Chim. Acta.* 2020;1100:131–137. Elsevier Enhanced Reader. doi: 10.1016/j.aca.2019.12.007.
8. Wiśniewski JR, Rakus D. Multi-Enzyme Digestion FASP and the 'Total Protein Approach'-Based Absolute Quantification of the Escherichia Coli Proteome. *J. Proteomics.* 2014;109:322–331. doi: 10.1016/j.jprot.2014.07.012.
9. Wiśniewski JR. Quantitative Evaluation of Filter Aided Sample Preparation (FASP) and Multienzyme Digestion FASP Protocols. *Anal. Chem.* 2016;88(10):5438–5443. doi: 10.1021/acs.analchem.6b00859.
10. Potriquet J, Laohaviroj M, Bethony JM, *et al.* A Modified FASP Protocol for High-Throughput Preparation of Protein Samples for Mass Spectrometry. *PLoS One.* 2017;12(7):e0175967. doi: 10.1371/journal.pone.0175967.

11. Arul A-B, Park J-M, Lee H, *et al.* Use of Filter-Aided Capture and Elution Protocol for Concurrent Preparation of N-Glycan and O-Glycopeptides for LC-MS/MS Analysis. *Curr. Proteomics.* 2016; 13(1):48–54. doi: 10.2174/1570164613666160413125041.
12. Wiśniewski JR, Nagaraj N, Zougman A, *et al.* Brain Phosphoproteome Obtained by a FASP-Based Method Reveals Plasma Membrane Protein Topology. *J. Proteome Res.* 2010;9(6):3280–3289. doi: 10.1021/pr1002214.
13. Lu XN, Zhu HN. Tube-Gel Digestion — a Novel Proteomic Approach for High Throughput Analysis of Membrane Proteins. *Mol. Cell. Proteomics.* 2005;4(12):1948–1958. doi: 10.1074/mcp.M500138-MCP200.
14. Fischer R, Kessler BM. Gel-Aided Sample Preparation (GASP) — a Simplified Method for Gel-Assisted Proteomic Sample Generation from Protein Extracts and Intact Cells. *Proteomics.* 2015;15(7):1224–1229. doi: 10.1002/pmic.201400436.
15. Muller L, Fornecker L, Van Dorsselaer A, *et al.* Benchmarking Sample Preparation/Digestion Protocols Reveals Tube-Gel Being a Fast and Repeatable Method for Quantitative Proteomics. *Proteomics.* 2016;16(23):2953–2961. doi: 10.1002/pmic.201600288.
16. Lu X, Zhu H. Tube-Gel Digestion. *Mol. Cell. Proteomics.* 2005; 4(12):1948–1958. Elsevier Enhanced Reader. doi: 10.1074/mcp. M500138-MCP200.
17. Muller L. Extended Investigation of Tube-Gel Sample Preparation: A Versatile and Simple Choice for High Throughput Quantitative Proteomics. *Sci. Rep.* 2018;9.
18. Jailani NM, Sani AAA, Fischer R, *et al.* Comparison of Gel-Aided Sample Preparation (GASP) and Two in-Solution Digestion Workflows for Proteomic Analysis Using Saccharomyces Cerevisiae Lysate. *Malays. J. Microbiol.* 2018;14(6):579–584. doi: 10.21161/mjm.1461817.
19. Davis S, Charles PD, He L, *et al.* Expanding Proteome Coverage with CHarge Ordered Parallel Ion ANalysis (CHOPIN) Combined with Broad Specificity Proteolysis. *J Proteome Res.* 2017;12.
20. Carvalho LB, Capelo-Martínez J-L, Lodeiro C, *et al.* Ultrasonic-Based Filter Aided Sample Preparation as the General Method to Sample Preparation in Proteomics. *Anal. Chem.* 2020;92(13):9164–9171. doi: 10.1021/acs.analchem.0c01470.

21. Yu Y, Xie L, Gunawardena HP, *et al.* GOFAST: An Integrated Approach for Efficient and Comprehensive Membrane Proteome Analysis. *Anal. Chem.* 2012;84(21):9008–9014. doi: 10.1021/ac300134e.
22. Zhao Q, Fang F, Wu C, *et al.* ImFASP: An Integrated Approach Combining in-Situ Filter-Aided Sample Pretreatment with Microwave-Assisted Protein Digestion for Fast and Efficient Proteome Sample Preparation. *Anal. Chim. Acta.* 2016;912:58–64. doi: 10.1016/j.aca.2016.01.049.
23. Erde J, Loo RRO, Loo JA. Enhanced FASP (EFASP) to Increase Proteome Coverage and Sample Recovery for Quantitative Proteomic Experiments. *J. Proteome Res.* 2014;13(4):1885–1895. doi: 10.1021/pr4010019.
24. Erde J, Loo RRO, Loo JA. Improving Proteome Coverage and Sample Recovery with Enhanced FASP (EFASP) for Quantitative Proteomic Experiments. In *Proteomics: Methods and Protocols*; Comai L, Katz JE, Mallick P, Eds.; Methods Mol. Biol. New York: Springer, 2017:11–18. doi: 10.1007/978-1-4939-6747-6_2.
25. Zhang Z, Dubiak KM, Huber PW, *et al.* Miniaturized Filter-Aided Sample Preparation (MICRO-FASP) Method for High Throughput, Ultrasensitive Proteomics Sample Preparation Reveals Proteome Asymmetry in Xenopus Laevis Embryos. *Anal. Chem.* 2020;92(7):5554–5560. doi: 10.1021/acs.analchem.0c00470.
26. Zhang Z, Dubiak KM, Shishkova E, *et al.* High-Throughput, Comprehensive Single-Cell Proteomic Analysis of *Xenopus Laevis* Embryos at the 50-Cell Stage Using a Microplate-Based MICROFASP System. *Anal. Chem.* 2022;94(7):3254–3259. doi: 10.1021/acs.analchem.1c04987.
27. Sandbaumhüter FA, Nezhyva M, Eriksson O, *et al.* Well-Plate MFASP for Proteomic Analysis of Single Pancreatic Islets. *J. Proteome Res.* 2022;21(4):1167–1174. doi: 10.1021/acs.jproteome.2c00047.
28. Trindade F, Amado F, Gomes PS, *et al.* EndoProteoFASP: A Novel FASP Approach to Profile Salivary Peptidome and Disclose Salivary Proteases. *Talanta.* 2015;132:486–493. doi: 10.1016/j.talanta.2014.09.026.
29. Ostasiewicz P, Zielinska DF, Mann M, *et al.* Proteome, Phosphoproteome, and N-Glycoproteome Are Quantitatively Preserved in Formalin-Fixed

Paraffin-Embedded Tissue and Analyzable by High-Resolution Mass Spectrometry. *J. Proteome Res.* 2010;9(7):3688–3700. doi: 10.1021/pr100234w.

30. Ostasiewicz P, Wiśniewski JR. A Protocol for Large-Scale Proteomic Analysis of Microdissected Formalin Fixed and Paraffin Embedded Tissue. In *Methods in Enzymology*. Elsevier, 2017;585:159–176. doi: 10.1016/bs.mie.2016.09.017.
31. Wiśniewski JR. Proteomic Sample Preparation from Formalin Fixed and Paraffin Embedded Tissue. *J. Vis. Exp.* 2013;(79). doi: 10.3791/50589.
32. Tanca A, Abbondio M, Pisanu S, *et al.* Critical Comparison of Sample Preparation Strategies for Shotgun Proteomic Analysis of Formalin-Fixed, Paraffin-Embedded Samples: Insights from Liver Tissue. *Clin. Proteomics.* 2014;11(1):28. doi: 10.1186/1559-0275-11-28.
33. Quesada-Calvo F, Bertrand V, Longuespée R, *et al.* Comparison of Two FFPE Preparation Methods Using Label-Free Shotgun Proteomics: Application to Tissues of Diverticulitis Patients. *J. Proteomics.* 2015;112:250–261. Elsevier Enhanced Reader. doi: 10.1016/j.jprot.2014.08.013.
34. Davalieva K, Kiprijanovska S, Dimovski A, *et al.* Comparative Evaluation of Two Methods for LC-MS/MS Proteomic Analysis of Formalin Fixed and Paraffin Embedded Tissues. *J. Proteomics.* 2021;235:104117. Elsevier Enhanced Reader. doi: 10.1016/j.jprot.2021.104117.
35. Pirog A, Faktor J, Urban-Wojciuk Z, *et al.* Comparison of Different Digestion Methods for Proteomic Analysis of Isolated Cells and FFPE Tissue Samples. *Talanta.* 2021;233:122568. ScienceDirect. https://www.sciencedirect.com/science/article/pii/S0039914021004896?via%3Dihub (accessed 2022-07-21).
36. Huber ML, Sacco R, Parapatics K, *et al.* AbFASP-MS: Affinity-Based Filter-Aided Sample Preparation Mass Spectrometry for Quantitative Analysis of Chemically Labeled Protein Complexes. *J. Proteome Res.* 2014;13(2):1147–1155. doi: 10.1021/pr4009892.
37. Yu Y, Bekele S, Pieper R. Quick 96FASP for High Throughput Quantitative Proteome Analysis. *J. Proteomics.* 2017;166:1–7. Elsevier Enhanced Reader. doi: 10.1016/j.jprot.2017.06.019.
38. Berger ST, Ahmed S, Muntel J, *et al.* MStern Blotting–High Throughput Polyvinylidene Fluoride (PVDF) Membrane-Based Proteomic Sample

Preparation for 96-Well Plates. *Mol. Cell. Proteomics.* 2015;14(10):2814–2823. doi: 10.1074/mcp.O115.049650.

39. McDowell GS, Gaun A, Steen H. IFASP: Combining Isobaric Mass Tagging with Filter-Aided Sample Preparation. *J. Proteome Res.* 2013;12(8):3809–3812. doi: 10.1021/pr400032m.
40. Zougman A, Selby PJ, Banks RE. Suspension Trapping (STrap) Sample Preparation Method for Bottom-Up Proteomics Analysis. *Proteomics.* 2014;14(9):1006–1000. doi: 10.1002/pmic.201300553.
41. Haile Mariam M, Eguez RV, Singh H, *et al.* S-Trap, an Ultrafast Sample-Preparation Approach for Shotgun Proteomics. *J. Proteome Res.* 2018;17(9):2917–2924. doi: 10.1021/acs.jproteome.8b00505.
42. Elinger D, Gabashvili A, Levin Y. Suspension Trapping (S-Trap) Is Compatible with Typical Protein Extraction Buffers and Detergents for Bottom-Up Proteomics. *J. Proteome Res.* 2019;18(3):1441–1445. doi: 10.1021/acs.jproteome.8b00891.
43. Zougman A, Wilson JP, Roberts LD, *et al.* Detergent-Free Simultaneous Sample Preparation Method for Proteomics and Metabolomics. *J. Proteome Res.* 2020;19(7):2838–2844. doi: 10.1021/acs.jproteome.9b00662.
44. Bang G, Lee H, Kim H, *et al.* Comparison of Protein Characterization Using in Solution and S-Trap Digestion Methods for Proteomics. *Biochem. Biophys. Res. Commun.* 2022;589:197–203. ScienceDirect. https://www.sciencedirect.com/science/article/pii/S0006291X21016612?via%3Dihub (accessed 2022-07-22).
45. Kuras M, Woldmar N, Kim Y, *et al.* Proteomic Workflows for High-Quality Quantitative Proteome and Post-Translational Modification Analysis of Clinically Relevant Samples from Formalin-Fixed Paraffin-Embedded Archives. *J. Proteome Res.* 2021;20(1):1027–1039. doi: 10.1021/acs.jproteome.0c00850.
46. Yang Y, Anderson E, Zhang S. Evaluation of Six Sample Preparation Procedures for Qualitative and Quantitative Proteomics Analysis of Milk Fat Globule Membrane. *Electrophoresis.* 2018;39(18):2332–2339. doi: 10.1002/elps.201800042.
47. Wu C, Zhou S, Mitchell MI, *et al.* Coupling Suspension Trapping-Based Sample Preparation and Data-Independent Acquisition Mass

Spectrometry for Sensitive Exosomal Proteomic Analysis. *Anal. Bioanal. Chem.* 2022;414(8):2585–2595. doi: 10.1007/s00216-022-03920-z.

48. Guergues J, Zhang P, Liu B, *et al.* Improved Methodology for Sensitive and Rapid Quantitative Proteomic Analysis of Adult-Derived Mouse Microglia: Application to a Novel In Vitro Mouse Microglial Cell Model. *Proteomics.* 2019;19(11):1800469. doi: 10.1002/pmic.201800469.
49. Costanzo M, Caterino M, Cevenini A, *et al.* Dataset of a Comparative Proteomics Experiment in a Methylmalonyl-CoA Mutase Knockout HEK 293 Cell Model. *Data Brief.* 2020;33:106453. Elsevier Enhanced Reader. doi: 10.1016/j.dib.2020.106453.
50. Cheung JK-W, Bian J, Sze Y-H, *et al.* Human Tear Proteome Dataset in Response to Daily Wear of Water Gradient Contact Lens Using SWATH-MS Approach. *Data Brief.* 2021;36:107120. doi: 10.1016/j.dib.2021.107120.
51. Sze YH, Zhao Q, Cheung JKW, *et al.* High-PH Reversed-Phase Fractionated Neural Retina Proteome of Normal Growing C57BL/6 Mouse. *Sci. Data.* 2021;8(1):27. doi: 10.1038/s41597-021-00813-1.
52. Hughes CS, Moggridge S, Müller T, *et al.* Single-Pot, Solid-Phase-Enhanced Sample Preparation for Proteomics Experiments. *Nat. Protoc.* 2019;14(1):68–85. doi: 10.1038/s41596-018-0082-x.
53. Hayoun K, Gouveia D, Grenga L, *et al.* Evaluation of Sample Preparation Methods for Fast Proteotyping of Microorganisms by Tandem Mass Spectrometry. *Front. Microbiol.* 2019;10. doi: 10.3389/fmicb.2019.01985.
54. Batth TS, Tollenaere MX, Rüther P, *et al.* Protein Aggregation Capture on Microparticles Enables Multipurpose Proteomics Sample Preparation. *Mol. Cell. Proteomics.* 2019;18(5):1027–1035. doi: 10.1074/mcp.TIR118.001270.
55. Johnston HE, Yadav K, Kirkpatrick JM, *et al.* Solvent Precipitation SP3 (SP4) Enhances Recovery for Proteomics Sample Preparation Without Magnetic Beads. *Anal. Chem.* 2022;94(29):10320–10328. doi: 10.1021/acs.analchem.1c04200.
56. Tognetti M, Sklodowski K, Müller S, *et al.* Biomarker Candidates for Tumors Identified from Deep-Profiled Plasma Stem Predominantly

from the Low Abundant Area. *J. Proteome Res.* 2022;21(7):1718–1735. doi: 10.1021/acs.jproteome.2c00122.

57. Franciosa G, Smits JGA, Minuzzo S, *et al.* Proteomics of Resistance to Notch1 Inhibition in Acute Lymphoblastic Leukemia Reveals Targetable Kinase Signatures. *Nat. Commun.* 2021;12(1):2507. doi: 10.1038/s41467-021-22787-9.
58. Marconi S, Santamaria S, Bartolucci M, *et al.* Trastuzumab Modulates the Protein Cargo of Extracellular Vesicles Released by ERBB2+ Breast Cancer Cells. *Membranes.* 2021;11(3):199. doi: 10.3390/membranes11030199.
59. Liu Y, Tempelaars MH, Boeren S, *et al.* Extracellular Vesicle Formation in Lactococcus Lactis Is Stimulated by Prophage-Encoded Holin–Lysin System. *Microb. Biotechnol.* 2022;15(4):1281–1295. doi: 10.1111/1751-7915.13972.
60. Liu C, Boeren S, Rietjens IMCM. Intra- and Inter-Individual Differences in the Human Intestinal Microbial Conversion of (–)-Epicatechin and Bioactivity of Its Major Colonic Metabolite 5-(3′,4′-Dihydroxy-Phenyl)-γ-Valerolactone in Regulating Nrf2-Mediated Gene Expression. *Front. Nutr.* 2022;9.
61. Chirania P, Holwerda EK, Giannone RJ, *et al.* Metaproteomics Reveals Enzymatic Strategies Deployed by Anaerobic Microbiomes to Maintain Lignocellulose Deconstruction at High Solids. *Nat. Commun.* 2022;13(1):3870. doi: 10.1038/s41467-022-31433-x.
62. Nogle R, Nagaraju S, Utturkar SM, *et al.* Clostridium Autoethanogenum Isopropanol Production via Native Plasmid PCA Replicon. *Front. Bioeng. Biotechnol.* 2022;10.
63. Palmer KS, Makarewicz CA, Tishkin AA, *et al.* Comparing the Use of Magnetic Beads with Ultrafiltration for Ancient Dental Calculus Proteomics. *J. Proteome Res.* 2021;20(3):1689–1704. doi: 10.1021/acs.jproteome.0c00862.
64. Rüther PL, Husic IM, Bangsgaard P, *et al.* SPIN Enables High Throughput Species Identification of Archaeological Bone by Proteomics. *Nat. Commun.* 2022;13(1):2458. doi: 10.1038/s41467-022-30097-x.
65. Fang P, Ji Y, Silbern I, *et al.* A Streamlined Pipeline for Multiplexed Quantitative Site-Specific N-Glycoproteomics. *Nat. Commun.* 2020;11(1):5268. doi: 10.1038/s41467-020-19052-w.

66. Duan X, Young R, Straubinger RM, *et al.* A Straightforward and Highly Efficient Precipitation/On-Pellet Digestion Procedure Coupled with a Long Gradient Nano-LC Separation and Orbitrap Mass Spectrometry for Label-Free Expression Profiling of the Swine Heart Mitochondrial Proteome. *J. Proteome Res.* 2009;8(6):2838–2850. doi: 10.1021/pr900001t.
67. Ouyang Z, Furlong MT, Wu S, *et al.* Pellet Digestion: A Simple and Efficient Sample Preparation Technique for LC-MS/MS Quantification of Large Therapeutic Proteins in Plasma. *Bioanalysis.* 2012;4(1):17–28. doi: 10.4155/bio.11.286.
68. Nouri-Nigjeh E, Sukumaran S, Tu C, *et al.* Highly Multiplexed and Reproducible Ion-Current-Based Strategy for Large-Scale Quantitative Proteomics and the Application to Protein Expression Dynamics Induced by Methylprednisolone in 60 Rats. *Anal. Chem.* 2014;86(16): 8149–8157. doi: 10.1021/ac501380s.
69. Tu C, Li J, Jiang X, *et al.* Ion-Current-Based Proteomic Profiling of the Retina in a Rat Model of Smith-Lemli-Opitz Syndrome. *Mol. Cell. Proteomics.* 2013;12(12):3583–3598. doi: 10.1074/mcp.M113.027847.
70. Yuan L, Arnold ME, Aubry A-F, *et al.* Simple and Efficient Digestion of a Monoclonal Antibody in Serum Using Pellet Digestion: Comparison with Traditional Digestion Methods in LC–MS/MS Bioanalysis. *Bioanalysis.* 2012;4(24):2887–2896. doi: 10.4155/bio.12.284.
71. Ma JH, Shen S, Wang JJ, *et al.* Comparative Proteomic Analysis of the Mitochondria-Associated ER Membrane (MAM) in a Long-Term Type 2 Diabetic Rodent Model. *Sci. Rep.* 2017;7(1):2062. doi: 10.1038/s41598-017-02213-1.
72. Klont F, Bras L, Wolters JC, *et al.* Assessment of Sample Preparation Bias in Mass Spectrometry-Based Proteomics. *Anal. Chem.* 2018;90(8):5405–5413. doi: 10.1021/acs.analchem.8b00600.
73. An B, Zhang M, Johnson RW, *et al.* Surfactant-Aided Precipitation/on-Pellet-Digestion (SOD) Procedure Provides Robust and Rapid Sample Preparation for Reproducible, Accurate and Sensitive LC/MS Quantification of Therapeutic Protein in Plasma and Tissues. *Anal. Chem.* 2015;87(7):4023–4029. doi: 10.1021/acs.analchem.5b00350.
74. Shen S, An B, Wang X, *et al.* Surfactant Cocktail-Aided Extraction/Precipitation/On-Pellet Digestion Strategy Enables Efficient and Reproducible Sample Preparation for Large-Scale Quantitative

Proteomics. *Anal. Chem.* 2018;90(17):10350–10359. doi: 10.1021/acs.analchem.8b02172.

75. Ebeling MC, Fisher CR, Kapphahn RJ, *et al.* Inflammasome Activation in Retinal Pigment Epithelium from Human Donors with Age-Related Macular Degeneration. *Cells.* 2022;11(13):2075. doi: 10.3390/cells11132075.
76. Ma F, Liu F, Xu W, *et al.* Surfactant and Chaotropic Agent Assisted Sequential Extraction/On-Pellet Digestion (SCAD) for Enhanced Proteomics. *J. Proteome Res.* 2018;17(8):2744–2754. doi: 10.1021/acs.jproteome.8b00197.
77. Tabang DN, Cui Y, Tremmel DM, *et al.* Analysis of Pancreatic Extracellular Matrix Protein Post-Translational Modifications via Electrostatic Repulsion-Hydrophilic Interaction Chromatography Coupled with Mass Spectrometry. *Mol. Omics.* 2021;17(5):652–664. doi: 10.1039/d1mo00104c.
78. Rodrigues-Ribeiro L, Melo-Braga MN, Kjeldsen F, *et al.* Assessment of Protein Extraction and Digestion Efficiency of Well-Established Shotgun Protocols for Heart Proteomics. *Anal. Biochem.* 2019;578:51–59. doi: 10.1016/j.ab.2019.05.006.
79. Doellinger J, Schneider A, Hoeller M, *et al.* Sample Preparation by Easy Extraction and Digestion (SPEED) — a Universal, Rapid, and Detergent-Free Protocol for Proteomics Based on Acid Extraction. *Mol. Cell. Proteomics.* 2020;19(1):209–222. doi: 10.1074/mcp.TIR119.001616.
80. Treitz C, Gottwald J, Gericke E, *et al.* Quantitative Proteome Profiling Provides Evidence of an Activation of the Complement Cascade in ATTR Amyloidosis. *Amyloid.* 2022;29(2):102–109. doi: 10.1080/13506129.2021.2015316.
81. Guo T, Kouvonen P, Koh CC, *et al.* Rapid Mass Spectrometric Conversion of Tissue Biopsy Samples into Permanent Quantitative Digital Proteome Maps. *Nat. Med.* 2015;21(4):407–413. doi: 10.1038/nm.3807.
82. Powell BS, Lazarev AV, Carlson G, *et al.* Pressure Cycling Technology in Systems Biology. In *Microbial Systems Biology: Methods and Protocols*; Navid A, Ed.; *Methods Mol. Biol.* Totowa: Humana Press, 2012:27–62. doi: 10.1007/978-1-61779-827-6_2.
83. Shao S, Guo T, Gross V, *et al.* Reproducible Tissue Homogenization and Protein Extraction for Quantitative Proteomics Using

MicroPestle-Assisted Pressure-Cycling Technology. *J. Proteome Res.* 2016;15(6):1821–1829. doi: 10.1021/acs.jproteome.5b01136.

84. López-Ferrer D, Petritis K, Hixson KK, *et al.* Application of Pressurized Solvents for Ultrafast Trypsin Hydrolysis in Proteomics: Proteomics on the Fly. *J. Proteome Res.* 2008;7(8):3276–3281. doi: 10.1021/pr7008077.
85. Lucas N, Robinson AB, Marcker Espersen M, *et al.* Accelerated Barocycler Lysis and Extraction Sample Preparation for Clinical Proteomics by Mass Spectrometry. *J. Proteome Res.* 2018;18(1):399–405. doi: 10.1021/acs.jproteome.8b00684.
86. Talib J, Hains PG, Tumanov S, *et al.* Barocycler-Based Concurrent Multiomics Method to Assess Molecular Changes Associated with Atherosclerosis Using Small Amounts of Arterial Tissue from a Single Mouse. *Anal. Chem.* 2019;91(20):12670–12679. doi: 10.1021/acs.analchem.9b01842.
87. Gao H, Zhang F, Liang S, *et al.* Accelerated Lysis and Proteolytic Digestion of Biopsy-Level Fresh-Frozen and FFPE Tissue Samples Using Pressure Cycling Technology. *J. Proteome Res.* 2020;19(5):1982–1990. doi: 10.1021/acs.jproteome.9b00790.
88. Sun R, Lyu M, Liang S, *et al.* A Prostate Cancer Tissue Specific Spectral Library for Targeted Proteomic Analysis. *Proteomics.* 2022;22(7):2100147. doi: 10.1002/pmic.202100147.
89. Sun Y, Li L, Zhou Y, *et al.* Stratification of Follicular Thyroid Tumours Using Data-Independent Acquisition Proteomics and a Comprehensive Thyroid Tissue Spectral Library. *Mol. Oncol.* 2022;16(8):1611–1624. doi: 10.1002/1878-0261.13198.
90. Bao K, Li X, Poveda L, *et al.* Proteome and Microbiome Mapping of Human Gingival Tissue in Health and Disease. *Front. Cell. Infect. Microbiol.* 2020;10.
91. Cheng Y, Chen Y, Yu C. Fast and Efficient Non-Reduced Lys-C Digest Using Pressure Cycling Technology for Antibody Disulfide Mapping by LC-MS. *J. Pharm. Biomed. Anal.* 2016;129:203–209. doi: 10.1016/j.jpba.2016.07.002.
92. Xuan Y, Bateman NW, Gallien S, *et al.* Standardization and Harmonization of Distributed Multi-Center Proteotype Analysis Supporting Precision

Medicine Studies. *Nat. Commun.* 2020;11(1):5248. doi: 10.1038/s41467-020-18904-9.

93. Tarney CM, Wang G, Bateman NW, *et al.* Biomarker Panel for Early Detection of Endometrial Cancer in the Prostate, Lung, Colorectal, and Ovarian Cancer Screening Trial. *Am. J. Obstet. Gynecol.* 2019;221(5):472.e1–472.e10. doi: 10.1016/j.ajog.2019.06.005.
94. Lee S, Zhao L, Rojas C, *et al.* Molecular Analysis of Clinically Defined Subsets of High-Grade Serous Ovarian Cancer. *Cell Rep.* 2020;31(2):107502. doi: 10.1016/j.celrep.2020.03.066.
95. Mitchell D, Hunt AL, Conrads KA, *et al.* Industrialized, Artificial Intelligence-Guided Laser Microdissection for Microscaled Proteomic Analysis of the Tumor Microenvironment. *J. Vis. Exp.* 2022;(184). doi: 10.3791/64171.
96. Betancourt LH, Sanchez A, Pla I, *et al.* Quantitative Assessment of Urea In-Solution Lys-C/Trypsin Digestions Reveals Superior Performance at Room Temperature over Traditional Proteolysis at 37°C. *J. Proteome Res.* 2018;17(7):2556–2561. doi: 10.1021/acs.jproteome.8b00228.
97. Olszowy PP, Burns A, Ciborowski PS. Pressure-Assisted Sample Preparation for Proteomic Analysis. *Anal. Biochem.* 2013;438(1):67–72. doi: 10.1016/j.ab.2013.03.023.
98. Kulak NA, Pichler G, Paron I, *et al.* Minimal, Encapsulated Proteomic-Sample Processing Applied to Copy-Number Estimation in Eukaryotic Cells. *Nat. Methods.* 2014;11(3):319–324. doi: 10.1038/nmeth.2834.

https://doi.org/10.1142/9789811283376_0005

Chapter 5

Amino Acid Sequence Databases and Search Engines — Basic Tools in Proteomic/Peptidomic Research

Aleksandra E. Bogucka[1,3] and Stanisław Ołdziej[2]

[1]*Institute of Biochemistry, Medical Faculty, Justus Liebig University of Giessen, Giessen, Germany*
[2]*Intercollegiate Faculty of Biotechnology, University of Gdańsk and Medical University of Gdańsk, Gdańsk, Poland*
[3]*Department of Physiology, Medical University of Gdańsk, Gdańsk, Poland*

Contents

Abstract

This article describes the basic databases and tools for analyzing data from mass spectrometry measurements in proteomic/peptidomic research. First, we present the basic amino acid sequence database (UniProt) along with its subsequent developmental versions considering, for example, protein isoforms. Next, we also discuss the development of databases collecting amino acid sequence data derived from the transcription of so-called non-coding RNA or so-called small open reading frames. The following paragraphs present the general principles and examples of sequence database search engines with particular consideration of the proper choice of software parameters and their influence on the search space from the potential user's perspective. We also shortly mention alternative strategies for analyzing mass spectrometry data: de novo sequencing and direct analysis of data-independent acquisition measurements.

1. Amino acid sequence databases — introduction

Proteomic or peptidomic studies of individual organisms or specific samples are highly dependent on the amino acid sequence libraries available for a given organism or sample type. Already in the 1960s, thanks to the efforts of Richard V. Eck and Margaret Dayhoff, the first collection of amino acid sequences (Atlas of Protein Sequence and Structure) were created, which were the basis for the development of current databases of amino acid sequences.[1,2] Initially, information on amino acid sequences came only from direct sequencing, but at the end of the 20th century, as a result of advances in DNA sequencing and a deeper understanding of the conversion of genetic information process into amino acid sequences, amino acid sequence information was obtained directly from genetic material.[3] Thanks to many years of work by Amos Bairoch from the University of Geneva and his colleagues, and based on the data collected by Margaret Dayhoff, the SWISS-PROT database of amino acid sequences was created, complemented by the TrEMBL database. The TrEMBL database (Translation of EMBL) contained amino acid sequences read directly from DNA sequences deposited in the European Molecular

Biology Laboratory databases (EMBL). As a result of further unification of the methods of storing and annotating amino acid sequences, in 2004 a single European database of amino acid sequences, UniProt, was created.[4] Of course, in parallel with the creation of European databases of amino acid sequences, the databases maintained by the U.S. National Library of Medicine as part of the National Center for Biotechnology Information (NCBI)[5] have also evolved; however, the basic database on amino acid sequences is the UniProt database, and this database is most often used in proteomic and peptidomic research.[6]

2. UniProt database

Since its creation in 2004, the UniProt database has undergone a number of modifications concerning, among others, the scope of stored information, links to other databases, and organization of resources.[6] From the point of view of proteomic or peptidomic research, the most important resources offered by UniProt are UniProt Knowledgedatabase (KD) and the Proteomes collection.

2.1. *UniProt Knowledgedatabase and Proteomes*

The KD collection consists of approximately 250 million records (amino acid sequences). Each of the records (amino acid sequences) is assigned to one of two sets: SWISS-PROT or TrEMBL. The SWISS-PROT collection is manually supervised and contains only selected amino acid sequences based on a variety of high-quality data [e.g., direct sequencing by mass spectrometry (MS)]. Due to the high-quality criteria set for the sequences contained in the SWISS-PROT collection, their number is small (about 550,000) in relation to the total number of records in the UniProt database. The TrEMBL set contains all other records that are not eligible for the SWISS-PROT part. Most of the sequences collected in the TrEMBL collection are so-called predicted sequences, which are based on the direct conversion of genomic data into amino acid sequences. In the conversion of genomic data, especially in the case of eukaryotic organisms, the homology of genes between closely related species is used. From this very cursory description, the reliability of the sequences from the SWISS-PROT and TrEMBL collections is very different. In most proteomic/peptidomic works, only amino acid sequences from the SWISS-PROT collection are used, which gives higher reliability of the obtained results; on the other hand, however, this approach narrows down the number of identified sequences. For example, information on the human proteome (*Homo sapiens*) deposited in the UniProt database is a collection of about 80,000 amino acid sequences, of which only about 20,000 sequences are in the SWISS-PROT collection.[7] UniProt database records corresponding to individual amino acid sequences, in addition to information about the sequence itself, contain a lot of other information, such as basic information about the function of the protein, the encoding gene, Gene Ontology (GO)[8] classification, post-translational modifications (PTMs), information about the spatial structure, and information about isoforms or naturally occurring variants (see Section 2.3), and references to other databases. Of course, the range of information available for individual records (amino acid sequences) is extremely diverse. Each record of the UniProt database is marked with a unique identifier (protein ID) which is a sequence of letters and numbers. For example, the identifier Q9NLZ4 is assigned to the human HSP70 protein encoded by the HSPBP1 gene. Amino acid sequence identifiers from the UniProt database are commonly used in all available resources (databases and software) for the analysis and processing of proteomic/peptidomic data, especially Gene Ontology resources and tools.[8,9]

2.2. *UniProt proteomes*

The UniProt database also offers access to the collection of amino acid sequences assigned to one species (taxon). At the moment (2023), the resources of the UniProt database are assigned to about 23,000 so-called reference proteomes. Reference proteomes contain collections of amino acid sequences encoded by the genome of a given organism (taxon). Reference proteomes are characterized by a high degree of completeness of proteomic data in relation to genomic data. The degree of completeness is defined by the so-called BUSCO score (Benchmarking Universal Single-Copy Ortholog),[10] which provides intuitive quantitative measures of genomic data completeness in terms of expected gene content to gauge the quality and completeness of genome assemblies and their annotations. Moreover, every reference proteome is described by Complete Proteome Detector (CPD). CPD statistically analyzes each proteome against a group of closely related proteomes to determine its completeness.[11] When the number of available amino acid sequences for a given species is quite poor (often only one or a few sequences), such proteomes are classified as Other proteomes. The number of such datasets in the UniProt database is approximately 140,000.

In the case of bacteria and fungi, it was quite common for many proteomes to be defined within one species, corresponding to different strains or variants. Since 2015, to simplify and reduce the amount of data, each species is described with one proteome of the best-studied strain or variant. Proteomes of all strains or variants of a given species are still available in the UniProt database but are labeled as Redundant proteomes. The number of these proteomes is enormous, over 300,000. The huge number of these datasets is due to the availability and ease of interpretation of genomic data, especially in the case of bacteria. Thus, in the redundant proteomes' collection, we can find, for example, about 1700 proteome variants for *Mycobacterium tuberculosis* bacteria and thousands of proteome variants of popular organisms such as *Escherichia coli* or *Bacillus subtilis*.

2.3. *UniProt — protein isoforms and natural variants*

Single or the combination of various biological processes such as alternative splicing, alternative initiation, ribosomal frameshifting, or alternative promoter could lead to alternative protein sequences (isoforms) originating from the same gene sequence. Due to this, the structure of UniProt database records closely links the amino acid sequence with the gene that encodes this sequence. Therefore, isoforms cause the problem of unambiguous assignment of a gene product (amino acid sequence) to a specific gene. To overcome this problem, the following solutions have been introduced in the UniProt database. First, the definition of a canonical (representative) amino acid sequence is introduced. A canonical sequence should meet at least one of the following criteria: have a specific function, be the most common product of a given gene, be encoded by conserved exons found in orthologous sequences, and have the same sequence as sequences reported in other than UniProt supervised databases.[12,13] In the absence of any information, the longest amino acid sequence produced by a given gene is the canonical sequence. The canonical sequence defines a record in the UniProt database. Thus, if the products of a given gene are several amino acid sequences, then one of these sequences (see rules described above) is designated as the canonical (representative) sequence. For example, for the entry Q9NLZ4 (human HSP70 protein encoded by the HSPBP1 gene), three alternative splicing isoforms have been identified. All three of these amino acid sequences are deposited together in one record but designated as Q9NLZ4-1 (canonical sequence) and Q9NLZ4-2, Q9NLZ4-3 (isoforms), respectively. Such grouping of canonical sequences and isoforms is used when the sequences do not differ significantly in functionality and sequence length. When products of the same gene display different functional features or significantly differ in sequence length, they are classified as separate records. Thus, for example, HSPBP1 gene products are also found in separate records K7EKM6, K7EL16, K7EMM7, K7EN20, K7EQQ0, and KTERT9. To sum up, in the UniProt database, we have a total of nine HSPBP1 gene products, three of these products are in the Q9NLZ4 record, and the remaining six are each in a separate record with a unique identifier. The dual classification of isoforms described above has serious repercussions when used to analyze proteomic data. The UniProt database can be downloaded in two different ways. The first method commonly used in the analysis of proteomic data is to take a standard set of sequences. In this version, only the canonical (representative) sequences from each record will be downloaded, in our example it will be six sequences with identifiers Q9NLZ4, K7EKM6, K7EL16, K7EMM7, K7EN20, K7EQQ0, and KTERT9. Only the canonical sequence will be retrieved from the Q9NLZ4 record, and the other two isoform sequences will be ignored. However, data from the UniProt database can be downloaded in such a way that the sequences of all available isoforms are downloaded (in our example, it will be nine sequences). Unfortunately, a review of the proteomics literature indicates that in most cases, proteomic data are derived only from canonical sequences to the exclusion of a significant number of isoforms.

In the case of many sequences, natural variants are known, which are most often (but not exclusively) point mutations of canonical sequences. Information on natural variants of protein sequences comes mainly from NCBI resources,[14–16] but the UniProt database itself also conducts a literature review and complements data obtained from other databases.[7] In the case of the human body, sequence variants very often affect the functionality of a given sequence, which has a direct impact, for example, on the occurrence of genetic diseases. Sequence variant data are provided in UniProt records in the form of the amino acid position in the canonical sequence affected by the change and the following difference between the canonical sequence and the variant. From the point of view of proteomic research, especially those focused on research related to medical diagnostics, information on natural variants is very valuable. Unfortunately, now, even though the number of known and described variants is very large, there are no appropriate tools to automatically download data from the UniProt database and then process this data in such a way that it can be used to analyze proteomic data. Analyses to identify canonical sequence variants are practically limited to targeted analyses.

3. Other databases — small open reading frame and non-coding RNA

Knowledge about how organisms use genetic information is constantly evolving. The classical understanding of reading genetic information as unambiguous (one gene, one gene product) is completely out of date. As shown in the example of the UniProt database (see Section 2.3), protein sequence isoforms were introduced to it as products of ambiguous reading of genetic material. The development of genetic material sequencing techniques and a deeper understanding of transcription and translation processes have led to the discovery of new types of genes encoded by amino acid sequences.[17–19] Originally, when analyzing genetic material, a rather arbitrary definition of coding RNA (open reading frame — ORF) was adopted as having a classic AUG "start" codon and a length of at least 300 bases.[18,19] However, it turned out that transcripts of genetic material contain a large number of ORFs that are less than 300 nucleotides long (small ORF — sORF) and often contain non-classical start codons (near-homologous start codons).[18] In addition, it was discovered that some RNA sequences considered as non-coding RNA (ncRNA) can be treated as template RNA and be transformed into protein sequences.[20] The discovery of these unusual coding sequences was possible, among others, thanks to the in-depth analysis of high-quality proteomic data.[21]

There are currently several databases or repositories that offer collections of amino acid sequences derived from translations of sORFs or so-called alternative ORFs (see Table 1).

From the point of view of applying this type of data to proteomic research, the OpenProt database seems to be the most interesting.[21]

Table 1. Selected databases or repositories of small or alternative ORF.

Database name	Short description
OpenProt[22] (https://www.openprot.org/)	Database focused on *Homo sapiens* and contains information about isoforms and proteins from alternative ORFs
MetamORF (https://metamorf.hb.univ-amu.fr/)	Database of sORF for *Homo sapiens* and *Mus musculus*
PsORF (http://psorf.whu.edu.cn/#/)	Database of sORF for more than 30 plant species
nORF.org (https://norfs.org/home)	Secondary database combining information: OpenProt database, as well as other repositories
SmProt[23] (http://bigdata.ibp.ac.cn/SmProt/)	Open repository related to sORF from several model organisms
sORF.org[24] (http://www.sorfs.org/database/)	Open repository related to sORF

This database offers combined data from the classic UniProt database with data on small or alternative ORFs. For a user unfamiliar with the methods of converting or combining different types of data, the offer of ready-to-use databases is extremely convenient. So far, the use of the OpenProt database in the analysis of proteomic data is quite limited. On the one hand, this is because the OpenProt database offers only data on the *Homo sapiens* proteome, and on the other hand, it may be due to the conservative attitude of the scientists themselves. Nevertheless, an in-depth exploration of the proteomes/peptidomes of organisms, especially in unusual mechanisms of transcription and translation, can be of great importance, especially in immunology. A significant number of amino acid sequences derived from sORF or ncRNA transcription have strong and targeted immunogenic properties.[25]

4. Sequence database search engines: Basic workflow in shotgun proteomics and examples

In MS–based proteomics and peptidomics, protein sequence databases are widely used in bioinformatic analysis workflows. The complexity of MS measurements increased exponentially with the rapid development of high-resolution instruments, putting an effective end to the manual solving of obtained spectra. Thus, arose an urgent need for computational tools to assist in the analysis of the huge amounts of generated data. The most popularly used are sequence database search engines, which allow the identification of peptides (proteins) in the sample, usually without the need for extensive computational power.[26,27]

Database search engines have been primarily designed for shotgun proteomics.[28–30] The standard shotgun (bottom-up) experiment involves the digestion of proteins to peptides and their analysis by a mass spectrometer in the data-dependent acquisition (DDA) mode.[30,31] There, a determined number of highest-intensity precursor ions are fragmented over time, resulting in a series of MS1 spectra with information on precursor ion *m/z* and linked MS2 spectra with fragmentation patterns. Database search engine performs an in silico digestion of the selected protein sequence database abiding by the rules set by several parameters — mimicking the experimental conditions and compares the actual experimental spectra with the obtained theoretical spectra by an implemented scoring function. After the identification of peptide sequences, they are assigned to their proteins of origin (protein inference), providing relevant biological information.

The first database search algorithm — SEQUEST[28] was released in 1994, followed by its commercial counterpart — Mascot[29] in 1998; since then, many various engines have become available for potential users (see selected examples in Table 2). Many of those are built in more complex software for the analysis of proteomics data, either

Table 2. Selected popular sequence database search algorithms with examples of software suites using those algorithms (commercial and free).

Algorithm	Examples of currently supported software	Free (owned by company)
Andromeda[34]	Galaxy, MaxQuant, SearchGUI	Yes
Byonic[35]	Proteome Discoverer	No (Protein Metrics)
Comet[36]	Galaxy, IP2, OpenMS, SearchGUI, TPP	Yes
Mascot[29]	Galaxy, Mascot Server, ProteinPilot, Proteome Discoverer, TPP	No (Matrix Science)
MS Amanda[37]	Proteome Discoverer, SearchGUI	Yes
MSFragger[38]	FragPipe, Proteome Discoverer	Yes
MS-GF+[39]	Galaxy, SearchGUI, TPP	Yes
MyriMatch[40]	Galaxy, SearchGUI, TPP	Yes
OMSSA[41]	SearchGUI, TPP	Yes
Paragon[42]	ProteinPilot	No (SCIEX)
PEAKS DB[43]	PEAKS Studio	No (Bioinformatics Solutions Inc.)
ProLuCID[44]	IP2	Yes
SEQUEST[28]	Proteome Discoverer, TPP	No (Thermo Fisher Scientific)
Tide[45]	SearchGUI	Yes
X!Tandem[46]	Galaxy, SearchGUI, TPP	Yes

Notes: IP2 — Integrated Proteomic Pipeline; TPP — Trans Proteomic Pipeline.

commercial (e.g., Proteome Discoverer, PEAKS Studio) or freely available (e.g., MaxQuant, Trans Proteomic Pipeline). These programs are often user-friendly to some extent: featuring a graphical user interface and a set of standard settings for a few different experimental setups. The default settings are usually appropriate for basic data, so the user is only required to provide or select a relevant sequence database and check the conditions of the experiment. Some of that software also allow for searching with multiple algorithms at once and subsequent merging of the result (e.g., SearchGUI[32] and PeptideShaker[33] free tools developed by Vaudel *et al.*) or multiple successive rounds of searching with different algorithms or settings (e.g., Proteome Discoverer or PEAKS Studio).

All database search algorithms aim to find and assign peptide-spectrum matches and subsequently score them to assess the probability of them being true positive hits. A variety of scoring algorithms have been adopted for this purpose: for example, SEQUEST uses two tiers of scoring — a preliminary score based on the ion sequence present in the fragment spectra and subsequent cross-correlation of the experimental spectra with the theoretical spectra calculated by the algorithm, whereas Mascot bases its scoring on probability models.[28,29,47] After the initial scoring, in most cases, another procedure — false discovery rate (FDR)[48] analysis — is applied to the results to lower the number of false positives in the result. This step usually involves an additional round of searching and scoring using a decoy database, for example, constructed from reverse or otherwise scrambled sequences of all the records present in the employed database.[49]

5. Parameters and search space considerations: User's perspective

In a specific or more complicated experimental setup, the individual parameters can be adjusted to obtain the best results. However, these changes should be made with careful consideration of the search space generated by the algorithm: a spectral database built in silico to compare against the experimentally obtained spectra. Expanded search space increases the computational burden on the software: time and resources necessary to complete the analysis. Moreover, it influences the scoring of the actual true positive hits and usually decreases the number of correct identifications. The parameters that have the highest effect on the size of the search space are the following: mass error tolerance, specificity of digestion, choice of protein database, and variable modifications.[26]

Several settings depend on the measurement itself and the MS instrument (especially the utilized mass analyzer): mass error tolerance, mass ranges, intensity threshold, precursor ion charge, or the type of fragment ions present in the spectra (depending on the fragmentation: primarily b and y ions for collision-induced dissociation and high energy collisional dissociation, and c and z ions for energy-transfer dissociation). These parameters should be selected according to the MS instrument and rarely need to be adjusted for single measurements. The mass error tolerance should be kept very low for the new generation of high-precision instruments as it can lead to a high false positive rate (while also increasing the search space in a considerable manner[26]).

Another group of parameters is related to the setup of the proteomic experiment, notably proteolytic digestion, PTMs present in the sample, and the choice of the proteomic database.

The rule of digestion defines the exact start- or endpoints of peptide sequences and is usually set simply as an enzyme (i.e., trypsin: cleavage after K or R, not before P). Depending on the efficiency of digestion, the number of missed cleavages can be adjusted (usually set to 2). The length range of resulting peptides can be important in cases where specific short or long peptide sequences are sought for. This number is usually set to start at 7, as shorter sequences are difficult to confidently identify using MS.[50] Digestion mode can be set to be specific, semispecific (not required to follow the digestion rule in case of one terminus), or unspecific. This setting has an enormous influence on the potential size of the search space,[26,51] as the unspecific digestion requires the inclusion of all possible peptides (of specified length) resulting from any digestion of the protein in the created in silico database. However, it is a crucial part of the peptidomic experimental workflow in search of endogenous peptide sequences.

The next parameters are related to PTMs present in the sample; these settings as well as the experimental design itself must be carefully considered as those can also substantially increase the size of the search space or result in incorrect identifications.[26,51,52] Modifications can be set as fixed — present on each designated amino acid residue — or variable — possibly present on some designated residues. Fixed modifications should be set only in cases where the modification reaction undergoes with near total efficiency as unmodified versions of peptides will be ignored; thus, they do not result in greater search space — unmodified version is simply replaced by the modified version of the sequence. Examples are the carbamidomethylation of cysteines in a standard proteomic workflow or introduced labels for label-based quantification workflows, such as iTRAQ[53] or TMT.[54] Variable modifications designate either biologically significant PTMs studied in an experiment (e.g., STY phosphorylation) or modifications introduced during sample storage and handling (e.g., methionine oxidation). The first group usually requires proper experiment design: sample preparation, enrichment for desired PTMs, or adjustment of measurement conditions as the modified peptides often are present in much lower quantities and/or they are not ionized to an efficient degree.[52] In case of the second group, it often makes sense to test the parameters on a sample to establish which modifications can be introduced to the sample to a meaningful degree — to find a proper balance between resources dedicated to the bioinformatic analysis and the loss of biological information. Especially so, that it is the variable modifications setting that increases the size of the search space: sequences containing modifications in various patterns must be included. Some of the available algorithms allow for a comprehensive search for unknown PTMs (MSFragger[38] or SPIDER[55] in PEAKS Studio) without a significant increase in required resources.

From the search engine analysis perspective, protein sequence database should be selected so that it contains all proteins possibly present in the sample (and not too many more). In practice, this usually translates to the choice of the organism protein database and inclusion of a potential contaminant database (e.g., cRAP: https://www.thegpm.org/crap/). The absence of the protein present in the sample from the database as well as the presence of too many similar sequences in the database (e.g., multiple species) can lead to false identifications and errors in scoring.[26] This issue, however, becomes much more complex considering metaproteomic workflows, novel approaches of ncDNA or pseudogene analysis, or simply experiments on an organism without a well-defined sequence database. Various solutions can be applied in case of these situations depending on the research approach and the decisions of the people analyzing data, for example, the choice of a more comprehensive database from related species, use of multiround search with different sequence databases (see Section 4), or relaxation of FDR cutoff. It should be kept in mind that the way that the sequence database search engines are used will be reflected in the obtained results, and the researcher should always consider the limitations of the applied workflow.

6. Summary and last remarks: De novo and discovery DIA

Protein sequence databases and sequence database search engines have been perfected over the last years to make the bioinformatic analysis of proteomic and peptidomic experiments easy, quick, and reliable. However, there are some instances where different kinds of analysis could answer the experimental question better or at the very least complement the obtained solution. De novo sequencing algorithms can be beneficial in cases where it is difficult to properly define the sequence database: unknown protein/peptide sequence, undescribed organism, or multiple unknown organisms (metaproteomics). These tools infer and score the possible amino acid sequences directly from the fragment ion spectra; however, this approach requires more computational resources, especially in complex measurements. A commercial software PEAKS Studio combines the de novo sequencing approach with subsequent database searching,[43] resulting in comprehensive outcomes, especially in case of peptidomic workflows.[51] A free software DeNovoGUI[56] allows for analysis with multiple free sequencing algorithms: DirecTag,[57] Novor,[58] PepNovo+,[59] and pNovo+.[60] Database search engines are used to analyze DDA data (see Section 4). Some novel algorithms based on machine learning techniques can also be applied to DIA (data-independent acquisition) measurements — without the traditionally required spectral library or using a predicted library,[61]:for example, DIA-NN[62] or MaxDIA.[63] Since DIA is superior to DDA in accurate quantification, these solutions allow for the identification and quantification of proteins (peptides) in a single bioinformatic analysis, especially suitable for (but not exclusive to) the label-free approach.

References

1. Eck RV. A Simplified Strategy for Sequence Analysis of Large Proteins. *Nature*. 1962;193(4812):241–243. doi: 10.1038/193241a0.
2. Chang MA, Dayhoff MO, Eck RV, *et al. Atlas of Protein Sequence and Structure*. Silver Spring, MD: National Biomedical Research Foundation, 1965.
3. O'Donovan C, Martin MJ, Gattiker A, *et al.* High-Quality Protein Knowledge Resource: SWISS-PROT and TrEMBL. *Brief. Bioinform.* 2002;3(3):275–284. doi: 10.1093/bib/3.3.275.
4. Apweiler R, Bairoch A, Wu CH, *et al.* UniProt: The Universal Protein Knowledgebase. *Nucleic Acids Res.* 2004;32(suppl_1):D115–D119. doi: 10.1093/nar/gkh131.
5. Sayers EW, Bolton EE, Brister JR, *et al.* Database Resources of the National Center for Biotechnology Information in 2023. *Nucleic Acids Res.* 2023;51(D1):D29–D38. doi: 10.1093/nar/gkac1032.
6. Bowler-Barnett EH, Fan J, Luo J, *et al.* UniProt and Mass Spectrometry-Based Proteomics — a 2-Way Working Relationship. *Mol. Cell. Proteomics.* 2023;100591. doi: 10.1016/j.mcpro.2023.100591.
7. The UniProt Consortium. UniProt: The Universal Protein Knowledgebase in 2023. *Nucleic Acids Res.* 2023;51(D1):D523–D531. doi: 10.1093/nar/gkac1052.
8. The Gene Ontology Consortium, Aleksander SA, Balhoff J, *et al.* The Gene Ontology Knowledgebase in 2023. *Genetics.* 2023;224(1):iyad031. doi: 10.1093/genetics/iyad031.

9. Thomas PD, Ebert D, Muruganujan A, *et al.* PANTHER: Making Genome-Scale Phylogenetics Accessible to All. *Protein Sci.* 2022;31(1): 8–22. doi: 10.1002/pro.4218.
10. Simão FA, Waterhouse RM, Ioannidis P, *et al.* BUSCO: Assessing Genome Assembly and Annotation Completeness with Single-Copy Orthologs. *Bioinformatics.* 2015;31(19):3210–3212. doi: 10.1093/bioinformatics/btv351.
11. Chen C, Natale DA, Finn RD, *et al.* Representative Proteomes: A Stable, Scalable and Unbiased Proteome Set for Sequence Analysis and Functional Annotation. *PLoS One.* 2011;6(4):e18910. doi: 10.1371/journal.pone.0018910.
12. Pujar S, O'Leary NA, Farrell CM, *et al.* Consensus Coding Sequence (CCDS) Database: A Standardized Set of Human and Mouse Protein-Coding Regions Supported by Expert Curation. *Nucleic Acids Res.* 2018;46(D1):D221–D228. doi: 10.1093/nar/gkx1031.
13. Morales J, Pujar S, Loveland JE, *et al.* A Joint NCBI and EMBL-EBI Transcript Set for Clinical Genomics and Research. *Nature.* 2022;604(7905):310–315. doi: 10.1038/s41586-022-04558-8.
14. Sherry ST, Ward M-H, Kholodov M, *et al.* DbSNP: The NCBI Database of Genetic Variation. *Nucleic Acids Res.* 2001;29(1):308–311. doi: 10.1093/nar/29.1.308.
15. Landrum MJ, Chitipiralla S, Brown GR, *et al.* ClinVar: Improvements to Accessing Data. *Nucleic Acids Res.* 2020;48(D1):D835–D844. doi: 10.1093/nar/gkz972.
16. Lappalainen I, Lopez J, Skipper L, *et al.* DbVar and DGVa: Public Archives for Genomic Structural Variation. *Nucleic Acids Res.* 2013;41(D1):D936–D941. doi: 10.1093/nar/gks1213.
17. Albuquerque JP, Tobias-Santos V, Rodrigues AC, *et al.* Small ORFs: A New Class of Essential Genes for Development. *Genet. Mol. Biol.* 2015;38(3):278–283. doi: 10.1590/S1415-475738320150009.
18. Orr MW, Mao Y, Storz G, *et al.* Alternative ORFs and Small ORFs: Shedding Light on the Dark Proteome. *Nucleic Acids Res.* 2020;48(3):1029–1042. doi: 10.1093/nar/gkz734.
19. Couso J-P, Patraquim P. Classification and Function of Small Open Reading Frames. *Nat. Rev. Mol. Cell Biol.* 2017;18(9):575–589. doi: 10.1038/nrm.2017.58.

20. Statello L, Guo C-J, Chen L-L, *et al.* Gene Regulation by Long Non-Coding RNAs and Its Biological Functions. *Nat. Rev. Mol. Cell Biol.* 2021;22(2):96–118. doi: 10.1038/s41580-020-00315-9.
21. Brunet MA, Roucou X. Mass Spectrometry-Based Proteomics Analyses Using the OpenProt Database to Unveil Novel Proteins Translated from Non-Canonical Open Reading Frames. *J. Vis. Exp.* 2019;(146):e59589. doi: 10.3791/59589.
22. Brunet MA, Lucier J-F, Levesque M, *et al.* OpenProt 2021: Deeper Functional Annotation of the Coding Potential of Eukaryotic Genomes. *Nucleic Acids Res.* 2021;49(D1):D380–D388. doi: 10.1093/nar/gkaa1036.
23. Li Y, Zhou H, Chen X, *et al.* SmProt: A Reliable Repository with Comprehensive Annotation of Small Proteins Identified from Ribosome Profiling. *Genom. Proteom. Bioinform.* 2021;19(4):602–610. doi: 10.1016/j.gpb.2021.09.002.
24. Olexiouk V, Van Criekinge W, Menschaert G. An Update on SORFs. Org: A Repository of Small ORFs Identified by Ribosome Profiling. *Nucleic Acids Res.* 2018;46(D1):D497–D502. doi: 10.1093/nar/gkx1130.
25. Kesner JS, Chen Z, Shi P, *et al.* Noncoding Translation Mitigation. *Nature.* 2023;617(7960):395–402. doi: 10.1038/s41586-023-05946-4.
26. Verheggen K, Ræder H, Berven FS, *et al.* Anatomy and Evolution of Database Search Engines — a Central Component of Mass Spectrometry Based Proteomic Workflows. *Mass Spectrom. Rev.* 2020;39(3):292–306. doi: 10.1002/mas.21543.
27. Verheggen K, Martens L, Berven FS, *et al.* Database Search Engines: Paradigms, Challenges and Solutions. In *Modern Proteomics — Sample Preparation, Analysis and Practical Applications*; Mirzaei H, Carrasco M, Eds.; *Adv. Exp. Med. Biol.* Cham: Springer International Publishing, 2016:147–156. doi: 10.1007/978-3-319-41448-5_6.
28. Eng JK, McCormack AL, Yates JR. An Approach to Correlate Tandem Mass Spectral Data of Peptides with Amino Acid Sequences in a Protein Database. *J. Am. Soc. Mass Spectrom.* 1994;5(11):976–989. doi: 10.1016/1044-0305(94)80016-2.
29. Perkins DN, Pappin DJC, Creasy DM, *et al.* Probability-Based Protein Identification by Searching Sequence Databases Using Mass

Spectrometry Data. *Electrophoresis.* 1999;20(18):3551–3567. doi: 10.1002/(SICI)1522-2683(19991201)20:18<3551::AID-ELPS3551>3.0.CO;2-2.

30. Link AJ, Eng J, Schieltz DM, *et al.* Direct Analysis of Protein Complexes Using Mass Spectrometry. *Nat. Biotechnol.* 1999;17(7):676–682. doi: 10.1038/10890.
31. Wolters DA, Washburn MP, Yates JR. An Automated Multidimensional Protein Identification Technology for Shotgun Proteomics. *Anal. Chem.* 2001;73(23):5683–5690. doi: 10.1021/ac010617e.
32. Vaudel M, Barsnes H, Berven FS, *et al.* SearchGUI: An Open-Source Graphical User Interface for Simultaneous OMSSA and X!Tandem Searches. *Proteomics.* 2011;11(5):996–999. doi: 10.1002/pmic.201000595.
33. Vaudel M, Burkhart JM, Zahedi RP, *et al.* PeptideShaker Enables Reanalysis of MS-Derived Proteomics Data Sets. *Nat. Biotechnol.* 2015;33(1):22–24. doi: 10.1038/nbt.3109.
34. Cox J, Neuhauser N, Michalski A, *et al.* Andromeda: A Peptide Search Engine Integrated into the MaxQuant Environment. *J. Proteome Res.* 2011;10(4):1794–1805. doi: 10.1021/pr101065j.
35. Bern M, Kil YJ, Becker C. Byonic: Advanced Peptide and Protein Identification Software. *Curr. Protoc. Bioinform.* 2012;40(1):13.20.1–13.20.14. doi: 10.1002/0471250953.bi1320s40.
36. Eng JK, Jahan TA, Hoopmann MR. Comet: An Open-Source MS/MS Sequence Database Search Tool. *Proteomics.* 2013;13(1):22–24. doi: 10.1002/pmic.201200439.
37. Dorfer V, Pichler P, Stranzl T, *et al.* MS Amanda, a Universal Identification Algorithm Optimized for High Accuracy Tandem Mass Spectra. *J. Proteome Res.* 2014;13(8):3679–3684. doi: 10.1021/pr500202e.
38. Kong AT, Leprevost FV, Avtonomov DM, *et al.* MSFragger: Ultrafast and Comprehensive Peptide Identification in Mass Spectrometry–Based Proteomics. *Nat. Methods.* 2017;14(5):513–520. doi: 10.1038/nmeth.4256.
39. Kim S, Pevzner PA. MS-GF+ Makes Progress Towards a Universal Database Search Tool for Proteomics. *Nat. Commun.* 2014;5(1):5277. doi: 10.1038/ncomms6277.
40. Tabb DL, Fernando CG, Chambers MC. MyriMatch: Highly Accurate Tandem Mass Spectral Peptide Identification by Multivariate

Hypergeometric Analysis. *J. Proteome Res.* 2007;6(2):654–661. doi: 10.1021/pr0604054.
41. Geer LY, Markey SP, Kowalak JA, *et al.* Open Mass Spectrometry Search Algorithm. *J. Proteome Res.* 2004;3(5):958–964. doi: 10.1021/pr0499491.
42. Shilov IV, Seymour SL, Patel AA, *et al.* The Paragon Algorithm, a Next Generation Search Engine that Uses Sequence Temperature Values and Feature Probabilities to Identify Peptides from Tandem Mass Spectra*. *Mol. Cell. Proteomics.* 2007;6(9):1638–1655. doi: 10.1074/mcp.T600050-MCP200.
43. Zhang J, Xin L, Shan B, *et al.* PEAKS DB: De Novo Sequencing Assisted Database Search for Sensitive and Accurate Peptide Identification*. *Mol. Cell. Proteomics.* 2012;11(4):M111.010587. doi: 10.1074/mcp.M111.010587.
44. Xu T, Park SK, Venable JD, *et al.* ProLuCID: An Improved SEQUEST-like Algorithm with Enhanced Sensitivity and Specificity. *J. Proteomics.* 2015;129:16–24. doi: 10.1016/j.jprot.2015.07.001.
45. Diament BJ, Noble WS. Faster SEQUEST Searching for Peptide Identification from Tandem Mass Spectra. *J. Proteome Res.* 2011;10(9):3871–3879. doi: 10.1021/pr101196n.
46. Craig R, Beavis RC. TANDEM: Matching Proteins with Tandem Mass Spectra. *Bioinformatics.* 2004;20(9):1466–1467. doi: 10.1093/bioinformatics/bth092.
47. Sadygov RG, Cociorva D, Yates JR. Large-Scale Database Searching Using Tandem Mass Spectra: Looking up the Answer in the Back of the Book. *Nat. Methods.* 2004;1(3):195–202. doi: 10.1038/nmeth725.
48. Benjamini Y, Hochberg Y. Controlling the False Discovery Rate: A Practical and Powerful Approach to Multiple Testing. *J. R. Stat. Soc. Ser. B Methodol.* 1995;57(1):289–300. doi: 10.1111/j.2517-6161.1995.tb02031.x.
49. Choi H, Nesvizhskii AI. False Discovery Rates and Related Statistical Concepts in Mass Spectrometry-Based Proteomics. *J. Proteome Res.* 2008;7(1):47–50. doi: 10.1021/pr700747q.
50. Swaney DL, Wenger CD, Coon JJ. Value of Using Multiple Proteases for Large-Scale Mass Spectrometry-Based Proteomics. *J. Proteome Res.* 2010;9(3):1323–1329. doi: 10.1021/pr900863u.

51. De La Toba EA, Anapindi KDB, Sweedler JV. Assessment and Comparison of Database Search Engines for Peptidomic Applications. *J. Proteome Res.* 2023. doi: 10.1021/acs.jproteome.2c00307.
52. Parker CE, Mocanu V, Mocanu M, *et al.* Mass Spectrometry for Post-Translational Modifications. In *Neuroproteomics*; Alzate O, Ed.; *Front. Neurosci.* Boca Raton, FL: CRC Press/Taylor & Francis, 2010.
53. Ross PL, Huang YN, Marchese JN, *et al.* Multiplexed Protein Quantitation in Saccharomyces Cerevisiae Using Amine-Reactive Isobaric Tagging Reagents*. *Mol. Cell. Proteomics.* 2004;3(12):1154–1169. doi: 10.1074/mcp.M400129-MCP200.
54. Thompson A, Schäfer J, Kuhn K, *et al.* Tandem Mass Tags: A Novel Quantification Strategy for Comparative Analysis of Complex Protein Mixtures by MS/MS. *Anal. Chem.* 2003;75(8):1895–1904. doi: 10.1021/ac0262560.
55. Han Y, Ma B, Zhang K. Spider: Software for Protein Identification from Sequence Tags with de Novo Sequencing Error. *J. Bioinform. Comput. Biol.* 2005;3(3):697–716. doi: 10.1142/S0219720005001247.
56. Muth T, Weilnböck L, Rapp E, *et al.* DeNovoGUI: An Open Source Graphical User Interface for de Novo Sequencing of Tandem Mass Spectra. *J. Proteome Res.* 2014;13(2):1143–1146. doi: 10.1021/pr4008078.
57. Tabb DL, Ma Z-Q, Martin DB, *et al.* DirecTag: Accurate Sequence Tags from Peptide MS/MS through Statistical Scoring. *J. Proteome Res.* 2008;7(9):3838–3846. doi: 10.1021/pr800154p.
58. Ma B. Novor: Real-Time Peptide de Novo Sequencing Software. *J. Am. Soc. Mass Spectrom.* 2015;26(11):1885–1894. doi: 10.1007/s13361-015-1204-0.
59. Frank A, Pevzner P. PepNovo: De Novo Peptide Sequencing via Probabilistic Network Modeling. *Anal. Chem.* 2005;77(4):964–973. doi: 10.1021/ac048788h.
60. Chi H, Chen H, He K, *et al.* PNovo+: De Novo Peptide Sequencing Using Complementary HCD and ETD Tandem Mass Spectra. *J. Proteome Res.* 2013;12(2):615–625. doi: 10.1021/pr3006843.
61. Tiwary S, Levy R, Gutenbrunner P, *et al.* High-Quality MS/MS Spectrum Prediction for Data-Dependent and Data-Independent

Acquisition Data Analysis. *Nat. Methods.* 2019;16(6):519–525. doi: 10.1038/s41592-019-0427-6.

62. Demichev V, Messner CB, Vernardis SI, *et al.* DIA-NN: Neural Networks and Interference Correction Enable Deep Proteome Coverage in High Throughput. *Nat. Methods.* 2020;17(1):41–44. doi: 10.1038/s41592-019-0638-x.
63. Sinitcyn P, Hamzeiy H, Salinas Soto F, *et al.* MaxDIA Enables Library-Based and Library-Free Data-Independent Acquisition Proteomics. *Nat. Biotechnol.* 2021;39(12):1563–1573. doi: 10.1038/s41587-021-00968-7.

https://doi.org/10.1142/9789811283376_0006

Chapter 6

Chemometrics in Proteomics

Beata Walczak

Institute of Chemistry, University of Silesia, Katowice, Poland

Contents

Abstract

Analysis of proteomic data is a multi-step process, and each step of this process influences the final results and conclusions drawn from the conducted experiment. The first stage of data analysis involves data pre-processing, which aims to eliminate unwanted data variance. This stage is data-dependent and typically involves procedures such as enhancement, warping, normalization, and scaling or transformation. The subsequent steps of data analysis involve exploration and modeling, requiring the application of chemometric methods well-suited for multivariate data. Among these methods are those for analyzing variance of multidimensional data derived from well-designed experiments (Experiment Design). They allow for the decomposition of induced data variance into variance associated with the studied factors and their interactions, while simultaneously addressing issues related to multiple testing of research hypotheses. The principles of these methods are described, and relevant references are provided to support further exploration in the field.

1. Preprocessing of signals

In proteomics, we deal with different types of analytical signals, which depend on an applied analytical technique. Among the most popular techniques, we can specify gel electrophoresis (GEP), mass spectrometry (MS), liquid chromatography with mass spectrometric detection (LC–MS), spectroscopy of the nuclear magnetic resonance (NMR), and so on.[1]

For a set of studied samples, the set of instrumental signals is organized in the matrix form. Further in this text, matrix containing m signals in rows (each signal described by n data points) will be denoted by the bold capital letter, for example, $\mathbf{X}$, and it will have dimensionality $(m \times n)$, where m denotes the number of samples and n denotes the number of variables. The bold lowercase letter, for example, $\mathbf{x}$, will be used to denote the column vector.

Due to the fact that many variables describe the investigated samples, this data type is called multivariate data and it requires special methods of data analysis, well-suited for the multivariate correlated

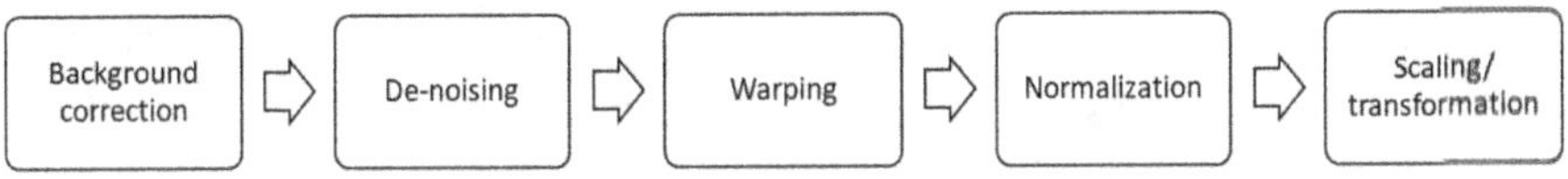

Figure 1. Main steps of data preprocessing.

data sets. Prior to data analysis though, careful preprocessing of instrumental signals is required. Preprocessing can be defined as a multistep process that aims at the elimination of an unwanted data variation. The first two steps focus on signal enhancement, followed by the steps that are necessary to make samples and variables comparable (Figure 1).

1.1. *Signal enhancement*

Each instrumental signal has at least three basic components:

1. background,
2. relevant part containing chemical information, and
3. noise.

Background and noise are, of course, undesired components, and they should be removed to increase signal quality. This process is called signal enhancement, and it is based mainly on frequency differences characteristic of the discussed components. Noise is the high-frequency component, and background is the low-frequency component, whereas the relevant part of the signal characterizes by intermediate frequencies. There are many methods of signal enhancement, but their choice is not trivial and it significantly depends on data characteristics.

1.1.1. *Background estimation and elimination*

Usually, the first step in data preprocessing and analysis is associated with baseline correction. Simple methods (such as polynomial fitting) can be effective in certain cases, but they require user's intervention and are not recommended in low signal-to-noise ratio environments.

One of the most powerful methods of background estimation is the penalized asymmetric least squares (PALS) method developed by Eliers.[2] The objective function of PALS is defined as:

$$Q = \sum_{i=1:n} v_i (x_i - f_i)^2 + \lambda (\Delta^d f_i)^2 \tag{1}$$

where **x** denotes the original signal (1 × n), **f** denotes its smoothed approximation, Δ^d denotes the derivatives of the d-th degree, λ is the penalty coefficient, whereas v denotes the weights.

The first component of Equation (1) describes signal fit, and the second one represents the penalty term introduced to control the smoothness of background approximation. Usually, the first-order derivatives are used to reflect the smoothness of background approximation, whereas the parameter λ has to be optimized for the signals at hand.

PALS is an iterative procedure that starts with all the weights set to 1, and then it weighs data points in an asymmetric way, depending on whether they are above or below an actual approximation of the signal. Points that appear above the signal approximation, **f**, are included with very small weights, whereas all points below the approximation, **f**, are included with weights approaching 1:

$$v_i = \begin{cases} p & \text{if } x_i > f_i \\ 1 - p & \text{if } x_i < f_i \end{cases} \tag{2}$$

where p can vary in the range from 0 to 1, but for the background elimination it is usually set to a very small value, such as 0.001.

PALS at work is presented in Figure 2. It can be used for background correction of both one- and two-dimensional signals.[3]

1.1.2. *Signals de-noising*

All instrumental signals are affected by noise, which reduces the quality of collected data and influences precision, accuracy, and limit of detection of the measurements, in that way influencing data analysis and final conclusions as well. To improve the quality of signals, it is

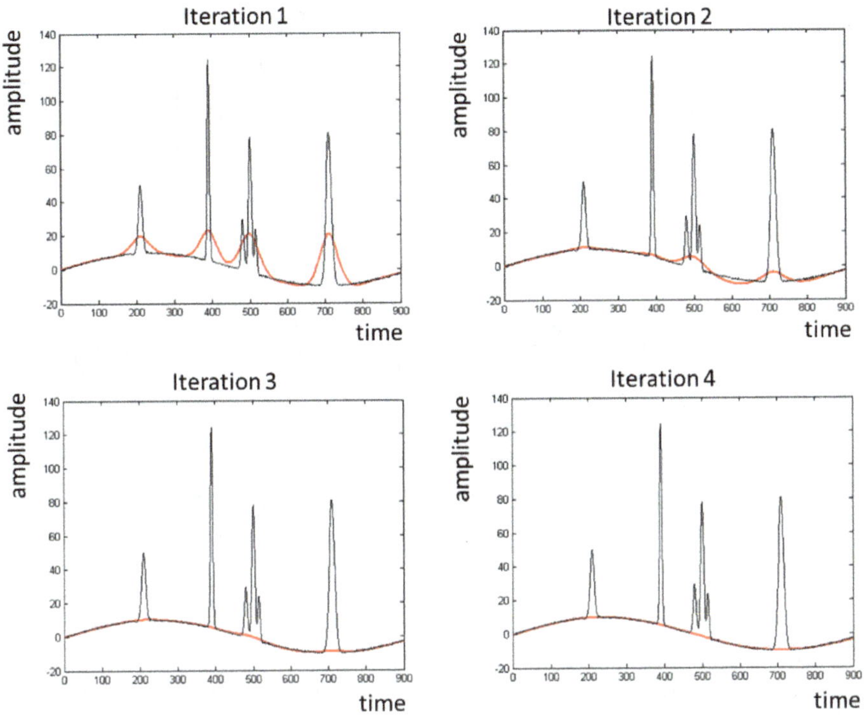

Figure 2. PALS at work: four iterations of the PALS algorithm of background estimation (numerical data from Ref. 4).

necessary to eliminate or minimize noise contribution to the measured signals.

There are many effective noise filtering methods, which can generally be divided into three groups:

1. Classical filters (such as moving window average smoothing, derivative filters, Savitzky–Golay smoothing, Whittaker smoothers, and Kalman filters),[5]
2. Fourier transformation (FT),[6,7] and
3. Wavelet transformations (WT).[7–9]

Choice of the method should be determined by the type of instrumental signal and the type of its noise. If the signal is stationary,

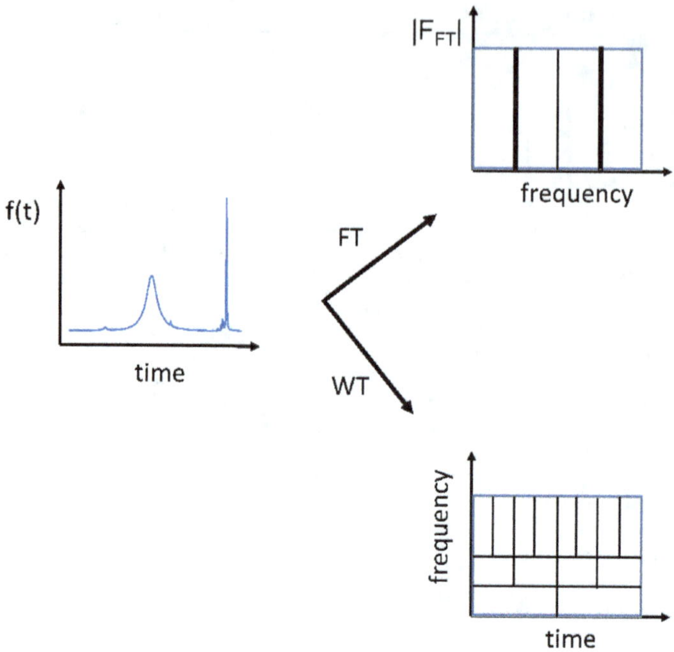

Figure 3. Original signal in the time domain and after FT and WT.

classical filters or FT will do the job. In the case of non-stationary signals, that is, such signals for which the frequency of relevant components varies in the time domain, WT should be applied.

All filters are able to remove a high-frequency component (i.e., noise), but in the case of a non-stationary signal, filtering can remove a high-frequency component of the relevant part of that signal as well. Classical filters are the best-known tools for signal enhancement, and their description can be found in many textbooks. Unfortunately, their applicability is limited to stationary signals which in the case of proteomic studies are rare.

FT and WT allow representing an original signal (i.e., that in the time domain) in different domains (i.e., in the frequency and the time–frequency domains, respectively; see Figure 3), which are better suited for the signal de-noising. New representations are expressed as a combination of a particular set of basis functions multiplied by the transform coefficients. The main difference between FT and WT consists in the type of the applied basis functions. In FT, the sine and cosine functions are applied and each coefficient of the FT corresponds to a single frequency component with no time localization. While sines and cosines provide complete frequency localization, wavelets are well localized in both the time and frequency domain. It means that in the FT basis, functions are orthogonal and global, whereas in the WT basis, they are orthogonal and local. Unique properties of the wavelets allow distinguishing between the relevant high-frequency component and the high-frequency noise (based on the amplitude of the transform coefficients). Due to this unique property, WT is perfectly well suited for de-noising of the non-stationary signals, such as chromatograms, mass spectra, and NMR spectra.

Among different WT, still the most popular ones are discrete wavelet transformation and wavelet package transformation (WPT). An efficient hierarchical algorithm to perform wavelet decomposition is known as a pyramid algorithm, and it can be applied to any discrete-time finite signal (its length being equal to the integer power of 2).[10]

Decomposition of a signal in the wavelet domain requires the selection of wavelet filter, level of decomposition, type of thresholding (universal, sure, visu, etc.), and the type of thresholding policy (soft or hard). In the case of WPT, an additional joint best basis for the set of signals should be established.[11,12] All these steps are important, and they can be time-consuming, yet it is worth devoting a certain amount of time to optimize these parameters because optimization significantly affects the final results of data analysis. Moreover, it should be stressed that the majority of filters are designed to work with white Gaussian noise. In the case of heteroscedastic noise, that is, the noise proportional to the signal, log transformation can be applied prior to signal de-noising. If necessary, the signal after de-noising can be back-transformed to its original form. Heteroscedasticity of noise can easily be revealed by plotting the dependence of the standard deviation of technical replicates versus their mean values. In the wavelet domain, it is also possible to deal with correlated noise (different thresholds at different decomposition levels).

1.2. *Warping*

When dealing with a set of signals, it is necessary to ensure that these signals are well aligned. Many factors can cause peak shifts. For instance in chromatography, small fluctuations in the mobile phase flow rate, pressure, or temperature, the mobile phase composition changes due to poor mixing, stationary phase decomposition, instrumental drift, matrix effect, and so on, can cause peak shifts. If such shifts really occur, signals cannot be organized in the matrix form and further processed.

There are many methods of signal alignment (warping) and among them, we find correlation optimized warping,[13] parametric time warping,[14] automated warping,[15] fuzzy warping,[16] peak Alignment using reduced set mapping,[17] TracMass,[18] peak alignment by genetic algorithm,[19] and so on. Comprehensive surveys of different alignment techniques are given in, for example, Ref. 20. The cost function of these methods is usually defined as the maximization of correlation between the pair of the aligned signals and all methods (except TracMass) assuming that the peak order should not be changed. Still, these methods differ with respect to their flexibility, number of input parameters, and computational time.

Each warping procedure starts with target selection, that is, with the selection of one standard signal, to which all the remaining signals are further aligned. The choice of the target is very important and a good rule is to select a target as a high-quality signal, which has the maximal mean correlation with the remaining signals.[21] If possible, the recommended practice is to construct a reference sample from the mixture of true samples (which then can be used multiple times between the measurements as a quality control sample). To trace the retention time problem, it is enough to examine the sample signal against the reference signal plot.[20]

Once the target is selected, all the remaining signals have to be aligned to it. For each pair of signals (target, T, and signal to be aligned, P), it is necessary to optimize the parameters of the warping algorithm, because different transformation is needed for each pair.

The efficiency of the signal alignment methods strongly depends on the type of peak shifts (linear or non-linear) and on the flexibility of the applied warping function. An idea of exploring the non-linear flexible alignment function is applied, for example, in the semiparametric time warping approach and in automated warping, where the shape of the warping function is approximated by a number of the B-spline functions of a given degree.[21] (The spline basis consists of several polynomial fragments which at certain points of the time axis are joined and usually distributed in a uniform manner along this axis). Spline functions can be replaced with radial functions as well.[22] Warping function $w(t_i)$ can be modeled by using a linear combination of a sufficient number of the spline functions, and an entire iterative procedure aims at the calculation of the spline coefficients to maximize the cost function that describes an overall quality of the alignment (e.g., the correlation coefficient in AA):

$$w(t_i) = \sum_{j=1}^{k} a_j b_j (t_i) \tag{3}$$

where $w(t_i)$ "denotes warping function, j denotes the number of spline functions, a is the vector of the spline coefficients, and b is the j-th spline.

There are, of course, different algorithms that can be used to solve the above optimization problem, which consists of calculation of an optimal set of coefficients in Equation (3).[22]

1.3. *Normalization*

Normalization should be applied to make all studied signals comparable, which is needed in these cases when an overall sample concentration differs or other types of systematic effects are observed. This procedure relies on the estimation of a scaling factor for each row of the data matrix. Unfortunately, the majority of the normalization methods (such as total sum normalization and signal normal variate) introduce the so-called closure[23,24] (i.e., spurious correlation of variables[25]), which can obviously lead to wrong identification of biomarkers in the systems studied.

In the close feature vector, one of its elements can always be expressed as a function of the remaining set of parameters:

$$x_i = 1 - \sum_{j \neq i} x_j \tag{4}$$

Closing data set can only be justified, once the different samples have a joint relation differing by the sample-specific multiplication factor, that is, when this multiplication factor is proportional to the sum of individual variables for all samples. With chemical signals, this is usually not the case, and then the introduced closure effect can lead to wrong identification of biomarkers in the systems studied.

To avoid spurious correlation of variables due to an induced closure, probabilistic quotient normalization (PQN) is recommended.[26] Its idea can be presented in the following way: Let us assume that vector **mx** denotes the mean (or median) signal of the data set. For the i-th signal, it is necessary to calculate the ratios of its individual parameters and the corresponding parameters of **mx**, and the median of these ratios is the scaling factor for the i-th signal:

$$s_i = \text{median}\left(\frac{x_{i1}}{mx_1}, \frac{x_{i2}}{mx_2}, \ldots, \frac{x_{in}}{mx_n}\right) \tag{5}$$

It ought to be noticed that the PQN normalization can be applied for the set of signals for which the majority, that is, 51% features, do not differ. If this is not the case, that is, if the majority of features for the studied sample classes are different, then the SVN method (variable sorting for normalization) is recommended.[27]

In the case of a limited number of variables, the pair-wise log-ratio approach[28] can be considered as an alternative to PQN. If the majority of variables do not differ for the studied classes of samples, it is possible to identify individual biomarkers in terms of individual variables. In a different case, their ratios can only be identified as significant features.

When choosing the normalization methods, we should always carefully maintain correlation patterns between the variables and eliminate the risk of introducing spurious correlation.

1.4. *Scaling/transformations*

Once all studied samples are comparable with each other, it is necessary to make sure that all studied variables (parameters, features) are comparable as well. By comparable features, we understand features with comparable variance for an entire data set.

With analytical instrumental signals from chemical measurements, the variance of individual variables quite often depends on their mean values, which means that the standard deviation versus the mean values' plot can simply be treated as a good diagnostic tool to cope with the heteroscedasticity problem. In this case, power transformation, or log transformation, will do a job.

A list of the most popular scaling/transformation methods is presented in Table 1. Their choice depends on both the data characteristics and the goal of data analysis.

Data centering is usually combined with other methods of data scaling/transform. Its goal is to focus on the relevant part of data variation, that is, on variation around zero (to adjust for differences in the offset of variables).

With the procedure of scaling, we aim to adjust fold differences observed for individual variables by dividing each variable by a suitable

Table 1. Popular scaling/transformation methods.[29]

Method	Equation
Centering	$x_{ij} = x_{ij} - \bar{x}_j$
Autoscaling	$x_{ij} = (x_{ij} - \bar{x}_j)/s_j$
Pareto scaling	$x_{ij} = (x_{ij} - \bar{x}_j)/\sqrt{x_j}$
Range scaling	$x_{ij} = (x_{ij} - \bar{x}_j)/(x_{j(max)} - x_{j(min)})$
Level scaling	$x_{ij} = (x_{ij} - \bar{x}_j)/\bar{x}_j$
Power transform	$x_{ij} = \sqrt[k]{x_{ij}}$
Log transform	$x_{ij} = \log(x_{ij})$

scaling factor. There are two well-known scaling subclasses[29]: the first subclass is based on the data dispersion measure (e.g., on the standard deviation), and the second subclass is based on the size measure (e.g., on the mean value). Among the scaling methods that belong to the first subclass, we find autoscaling, Pareto scaling, range scaling, and vast scaling. The second subclass belongs level scaling, and its aim is to convert changes in the parameter magnitude into changes relative to the average magnitude. Level scaling uses the mean value of the variable as the scaling factor, so that the level-scaled variable represents the changes in percentage, compared with the variable mean.

The log and power transformations represent non-linear modifications of the data, and they are usually applied to correct for data heteroscedasticity and to convert multiplicative effects into the additive ones (e.g., see Ref. 30). They also have a pseudo-scaling effect, as they reduce large variable values more effectively than small ones. In many cases, these transforms significantly reduce the magnitude differences of the data, but if these differences still remain too big, proper scaling after non-linear transformation can be applied. The impact of scaling methods on data analysis are well illustrated in Ref. 31.

2. Data analysis

Once the data set is properly preprocessed, it can further be analyzed using the available chemometric tools. Whatever the final goal of data analysis is, this part of the study usually starts with data exploration, which aims at the examination of data structure, evaluation of their complexity, identification of outliers, and so on. For this particular purpose, one can use methods, such as principal component analysis (PCA),[32–34] independent component analysis,[35–37] and all other types of linear and non-linear projection methods.[38–40] Certain insights into data structure can be gained based on the clustering method as well.

In proteomic studies, the exploration step (limited to studying the **X** data, i.e., the set of independent variables) is usually followed by the supervised data analysis. Its goal is to represent the dependent variable(s) **y** (or **Y**) as a function of the set of independent variables **X**. Depending on the nature of the variable **y**, we deal with calibration (continuous **y**) or discrimination (discrete **y**, describing belongingness of samples to the groups studied).

Discrimination methods are applied in the majority of proteomic studies. They allow the construction of models that reveal differences among the studied classes. Discriminant partial least squares,[41,42] classification and regression tree,[43] and random forest[44] count among the popular discrimination methods.

Each supervised modeling method requires representative data for both model construction and validation. Under the representative data set, we understand that this set contains all possible sources of data variance. To divide a given data set into the model and the validation set, we can apply the Duplex algorithm.[45]

When studying complex samples or systems, the application of robust versions of the available methods is recommended to make sure that further data analysis is not influenced by a few outlying samples. Robust approaches (e.g., robust PCA[47]) allow the construction of a model which well describes the data majority (more than 50% data samples) and identifies the outlying/unique objects. The outlying samples (leverages) should be re-examined and eventually removed from the data set. It should be emphasized that samples outlying in the X-space not necessarily are the outlying samples in regression. To distinguish between bad and good leverage points in regression, we need to apply the robust regression methods [e.g., the robust partial least squares (PLS) regression or discrimination[48]].

When the goal of the research study is the evaluation of a complex system influenced by multiple factors, the application of the design of experiments approach (also known as experimental design) is necessary.[49,50] Its main objective is to carefully plan and conduct the experiments in such a way that the experimental domain is systematically studied with as few experiments, as only possible. The resulting and well-balanced data set with an unbiased sample selection ensures scientifically valid conclusions.

2.1. *Experimental design and analysis of variance*

The aim of experimental design is to understand and use statistical approaches to evaluate the controlled factors (variables) that influence a studied system. In a properly designed experiment, the maximum amount of useful information is obtained by performing a small number of trials.

Data obtained are referred to as the response variable, or simply, the response. Nowadays, there are many different types of experimental design and their choice depends on the objective of the study and the number of factors involved. If we assume that there are k factors to be considered, the relationship between the response (x) and these factors (factor 1, factor 2, .., factor k) can be expressed in a general form as:

$$x = f\,(\text{factor } 1, \text{factor } 2, \ldots, \text{factor } k) \tag{6}$$

The values (or settings) attributed to each factor are called factor levels.

The method that is usually applied to the analysis of the univariate response measured at different design points is known as the analysis of variance (ANOVA), and it allows the decomposition of an overall variability of the response variable according to the assumed linear model.

To be more specific, let us focus on a fully crossed factorial design with two factors, each of them appearing at two levels. This design is described by four design points, representing all possible combinations of each level of the first factor with each level of the second factor. To have an estimate of the size of random error, two replicates at each design point should be considered, leading to eight experiments altogether. A linear model that describes the design considered can be presented in the vector–matrix notation as:

$$\begin{aligned}\mathbf{x} &= \hat{\mathbf{x}} + \mathbf{e} = \hat{\mathbf{x}}_{\mu} + \hat{\mathbf{x}}_{\alpha} + \hat{\mathbf{x}}_{\beta} + \hat{\mathbf{x}}_{\alpha\beta} + \mathbf{e} \\ &= \mathbf{D}_{\mu}\mathbf{b}_{\mu} + \mathbf{D}_{\alpha}\mathbf{b}_{\alpha} + \mathbf{D}_{\beta}\mathbf{b}_{\beta} + \mathbf{D}_{\alpha\beta}\mathbf{b}_{\alpha\beta} + \mathbf{e}\end{aligned} \tag{7}$$

where $\mathbf{x}_k$ denotes the effect of the k-th factor and $\mathbf{e}$ denotes statistical noise. Then each effect can be described as $\mathbf{D}_k\mathbf{b}_k$, where matrix $\mathbf{D}$ contains coded levels of the k-th factor and vector $\mathbf{b}$ denotes vector of regression coefficients associated with the corresponding factor levels.

A more compact form of Equation (7) is given as:

$$\mathbf{x} = \mathbf{Db} + \mathbf{e} \tag{8}$$

where $\mathbf{D}$ and $\mathbf{b}$ are defined as:

$$\mathbf{D} = [\mathbf{D}_\mu \; \mathbf{D}_\alpha \; \mathbf{D}_\beta \; \mathbf{D}_{\alpha\beta}] \tag{9}$$

$$\mathbf{b} = [\mathbf{b}_\mu \; \mathbf{b}_\alpha \; \mathbf{b}_\beta \; \mathbf{b}_{\alpha\beta}]. \tag{10}$$

For the discussed two factors at two levels, the full factorial design of the coded factors is:

$$\begin{matrix}
1 & 1 & 1 & 1 \\
1 & 1 & 1 & 1 \\
1 & 1 & 1 & 1 \\
1 & 1 & 1 & 1 \\
1 & 1 & 1 & 1 \\
1 & 1 & -1 & -1 \\
1 & 1 & -1 & -1 \\
1 & 1 & -1 & -1 \\
1 & 1 & -1 & -1 \\
1 & 1 & -1 & -1 \\
1 & -1 & 1 & -1 \\
1 & -1 & 1 & -1 \\
1 & -1 & 1 & -1 \\
1 & -1 & 1 & -1 \\
1 & -1 & 1 & -1 \\
1 & -1 & -1 & 1 \\
1 & -1 & -1 & 1 \\
1 & -1 & -1 & 1 \\
1 & -1 & -1 & 1 \\
1 & -1 & -1 & 1
\end{matrix}$$

2.2. *ANOVA of the multivariate data*

Owing to the considerable development of instrumental analytical techniques, nowadays not one but many responses are measured for each studied sample. In the case of proteomic studies, techniques such as GEP, LC-MS, and NMR have become the standard working tools. They allow the collection of analytical signals that contain thousands (or even millions) of data points. The analysis of this type of multivariate collinear data requires special methods for their analysis.[51,52] Among the techniques well-suited for this purpose, we have at our disposal ANOVA-simultaneous component analysis (ASCA),[53,54] ANOVA-PCA,[55] ANOVA-ComDim,[56] ANOVA-multiblock OPLS,[57] regularized MANOVA,[58] and ANOVA-target projection (ANOVA-TP).[59–61] Now we are going to focus on ASCA and ANOVA-TP.

2.2.1. *Principles of data decomposition*

ASCA[53,54] and ANOVA-TP[59–61] both consist of two steps. At step one, the centered data matrix, **X**, is decomposed in the same way as it is in the multiple ANOVA analysis, that is, into the sum of matrices associated with the studied factors (and their interactions), according to the principles of ANOVA. This decomposition is determined by the structure of the applied experiment design (factors can be crossed or nested), the nature of the studied factors (factors can be fixed or random), and restrictions introduced for mixed effects, which means that for a given design, specific linear model of experimental data is assumed. For the sake of an example, let us consider an experiment with three factors (A, B, and C) and repeated measurements within one factor (C). Assuming that a_1, a_2, …, a_p denote levels of factor A; b_1, b_2, ..., b_g denote levels of factor B; c_1, c_2, …, c_p denote levels of factor C, and n holds for the number of samples in each group, we can organize the experimental data in the form presented in Table 2. In the case of the univariate response, measurement is represented by the scalar x_{ijkz}, whereas with multiple responses, the results of individual measurement are represented by the vector $\mathbf{x}_{ijkz}$ (e.g., the NMR profile). This type of experiment requires that at least one factor, A or

Table 2. Scheme of data organization in the experiment with three factors A, B, and C, with the repeated measurements within factor C. Factor B* is assumed as random, and factors A and C are assumed as fixed.

			Factor C		
Factor A	**Factor B***	**Sample**	c_1	……	c_r
a_1	b_1	1	x_{1111}	……	x_{11r1}
		.			
		.			
		n	x_{111n}		x_{11rn}
	……	……	……	……	……
	b_q	1	x_{1q11}	……	x_{1qr1}
		.			
		.			
		n	x_{1q1n}		x_{1qrn}
	….	……	……	……	
a_p	b_1	1	X_{p111}	……	X_{p1r1}
		.			
		.			
		n	X_{p11n}		X_{p1rn}
	….	….	……	……	……
	b_q	1	X_{pq11}	……	x_{pqr1}
		.			
		.			
		n	X_{pq1n}		x_{pqrn}

B, is randomized, and factor C is usually fixed (but it can also be randomized).

Value of the response variable (for k-th sample) has the following linear structure:

$$y_{ijkz} = \mu + \alpha_i + \alpha_i\beta_j + \pi_{k(ij)} + \gamma_z + \alpha_i\gamma_z + \beta_j\gamma_z + \alpha_i\beta_j\gamma_z + \gamma_z\pi_{k(ji)} + \in_{k(ijz)} \tag{11}$$

where μ is the overall mean; α_i is the effect of the i-th level of factor A; β_j is the effect of the j-th level of factor B; $\alpha_i\ \beta_j$ is the effect of interaction of the i-th level of A and the j-th level of B; $\Pi_{k(ij)}$ is a constant associated with the k-th sample, for which factor A has level a_i and factor B has level b_j; γ_z is the effect of the z-th level of factor C; $\alpha_i \gamma_z$ is the effect of interactions of the i-th level of factor A and the z-th level of factor C; $\beta_j \gamma_z$ is the effect of interaction of the j-th level of factor B and the z-th level of factor C; $\alpha_i \beta_j \gamma_z$ is the effect of interactions of the i-the level of factor A, the j-th level of factor B, and the z-th level of factor C; $\gamma_z \pi_{k(ij)}$ is a component associated with the response of the k-th sample, for which factor A is at the i-th level, factor B is at the j-th level, and measurement is performed for the z-th time; and $\in_{k(ijz)}$ is the experimental error.

In the discussed example of experiment design, the following null hypothesis can be tested:

$$H_0\colon \alpha_i = 0 \quad \text{for all i},$$
$$H_0\colon \beta_j = 0 \quad \text{for all j},$$
$$H_0\colon \gamma_z = 0 \quad \text{for all z},$$
$$H_0\colon \alpha_i \beta_j = 0 \quad \text{for all ij},$$
$$H_0\colon \alpha_i \gamma_z = 0 \quad \text{for all iz},$$
$$H_0\colon \beta_j \gamma_z = 0 \quad \text{for all jz},$$
$$H_0\colon \alpha_i \beta_j \gamma_z = 0 \quad \text{for all ijz},$$

against the corresponding alternative hypothesis:

$$H_0\colon \alpha_i \neq 0 \quad \text{for at least one i},$$
$$H_0\colon \beta_j \neq 0 \quad \text{for at least one j},$$
$$H_0\colon \gamma_z \neq 0 \quad \text{for at least one z},$$
$$H_0\colon \alpha_i \beta_j \neq 0 \quad \text{for at least one ij},$$
$$H_0\colon \alpha_i \gamma_z \neq 0 \quad \text{for at least one iz},$$
$$H_0\colon \beta_j \gamma_z \neq 0 \quad \text{for at least one jz},$$
$$H_0\colon \alpha_i \beta_j \gamma_z \neq 0 \quad \text{for at least one ijz}.$$

Decomposition of the total variation (expressed as the sum of squares, SS) on its components and the corresponding degree of freedom are presented in Figure 4.

It is noteworthy that in the discussed example, two types of errors are observed and the corresponding SS are SS_s and SS_{Cs}. This fact has to be taken into account when testing the significance of individual factors. In the standard approach, the SS are divided by the corresponding degrees of freedom to calculate the means of squares, MS, and the statistics, F:

$$F_A = \frac{MS_A}{MS_S} \tag{12}$$

$$F_B = \frac{MS_B}{MS_S} \tag{13}$$

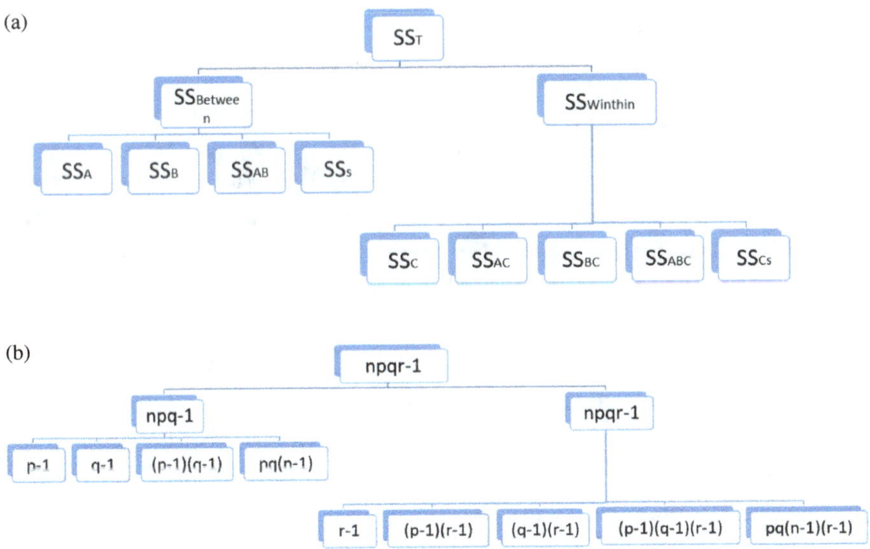

Figure 4. (a) Decomposition of the total sum of squares (SS_T) in the experiment design with three factors A, B, and C, and the repeated measurements within factor C; index s denotes the SS between the samples within groups and (b) the corresponding degrees of freedom.

$$F_{AB} = \frac{MS_{AB}}{MS_S} \tag{14}$$

$$F_C = \frac{MS_C}{MS_{CS}} \tag{15}$$

$$F_{AC} = \frac{MS_{AC}}{MS_{CS}} \tag{16}$$

$$F_{BC} = \frac{MS_{BC}}{MS_{CS}} \tag{17}$$

$$F_{ABC} = \frac{MS_{ABC}}{MS_{CS}} \tag{18}$$

Instead of applying standard statistics, F, we can use as an alternative the randomization test to assess the significance of the factors (interactions) studied.

2.2.2. *Randomization tests*

Contrary to the traditional parametric ANOVA, randomization tests are gaining an increasing attention due to the fact that they do not require an assumption as to the normally distributed errors. These tests are particularly relevant to the multivariate data, for which this assumption is rarely fulfilled.

Randomization test allows the calculation of null distribution of the relevant test statistics by calculating all possible values of that statistics under all possible data arrangements (or under a large random subset of such arrangements). The goal is to determine whether the observed difference between the observed SS is at a certain significance level large enough to be rejected. The p-value is calculated as proportion of the statistics values obtained under permutation that are equal to or higher than the observed value.

A detailed description of possible permutation strategies, such as restricted permutations, permutations of the whole groups of units, and permutations of certain forms, or combinations of residuals, is given in Ref. 62. The results of simulations presented in that study show that permutations of residuals under the reduced model usually overperform the remaining strategies.

It should be emphasized that even if permutation tests do not require an assumption as to the normality of errors, they assume the exchangeability of the relevant units under the null hypothesis. The good news is that for each particular term, an information about the exchangeable units is provided by the F-ratio, because these units are identified in the mean square of its F-ratio denominator. The exchangeable units have the same distribution under the null hypothesis.

A very helpful tool to identify these components of variation, which contribute to variability in the tested term, is Hasse diagrams that illustrate the studied model and allow finding the test denominator.[63]

2.3. *Principal component analysis*

PCA[32–34] probably is the most popular method of multivariate data exploration. It allows data compression, visualization, and interpretation. It aims to represent the centered data matrix **X**, which contains m samples and n variables as multiplication of the two matrices **T**(m, f) and **P** (n, f), where f<< n, which contain coordinates of samples and variables, respectively, on new variables (principal components) that maximize description of data variance. The PCA model can be presented as:

$$\mathbf{X}_{(m,n)} = \mathbf{T}_{(m,f)}\mathbf{P}^{T}_{(f,n)} + \mathbf{E}_{(m,n)} \tag{19}$$

where **E** denotes the matrix of residuals from the PCA model with f factors.

Score matrix, **T**, contains information about the studied samples, whereas loading matrix, **P**, contains information about data variables. Columns of score matrix, **T**, are orthogonal, whereas columns of the loading matrix, **P**, are orthonormal (orthogonal and their length equals to 1). The number of significant principal components, f, that is, the number of the columns of the score (**T**) matrix and the loading (**P**) matrix is determined by the chemical rank of the data. Usually, the f value is much lower than the mathematical rank defined as the minimum value of (m, n). We can estimate an optimal number of significant components based on the cross-validation (CV) method, but for data exploration, the number of significant components usually is evaluated based on the bar plot, which represents the cumulative percentage of the explained variance versus model complexity (the number of factors). The efficiency of data compression is determined by the data correlation structure. The first principal component describes the highest percentage of the total data variance, and each consecutive component describes it in a stepwise declining degree.

Both matrices, **T** and **P**, are used to visualize data structure and relationships among the variables. The score matrix is used to construct score plots, and the loading matrix is used to construct loading plots (Figure 5).

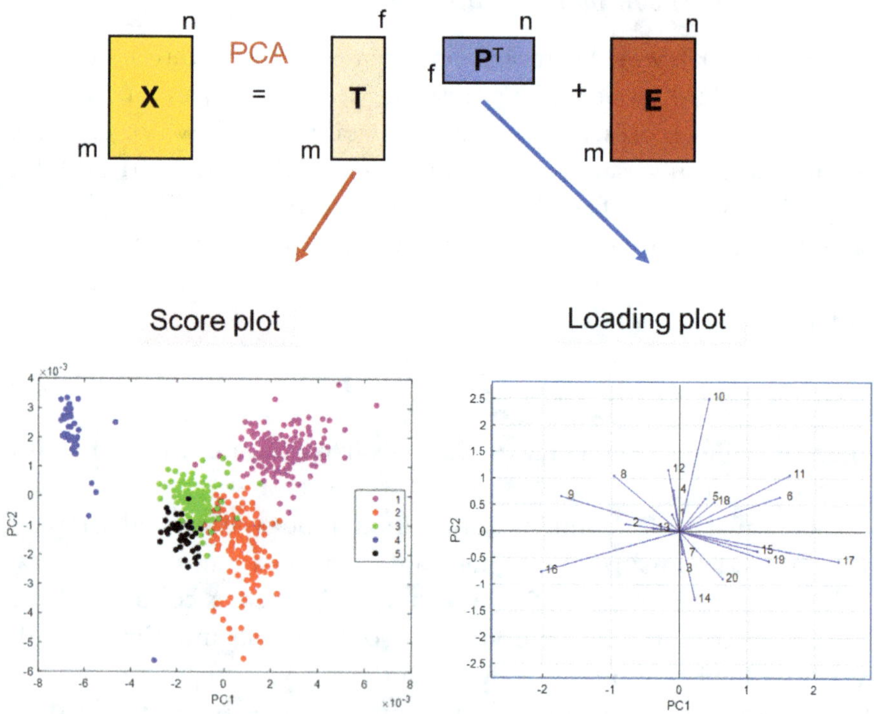

Figure 5. Schematic illustration of PCA.

Interpretation of the score plot is based on distances among the samples in the space of PCs. Samples are usually projected on the plane described by PC1 and PC2, then by PC1 and PC3, and so on. Distribution of data points (representing samples) provides an information on data structure, its clustering tendency, the number of subgroups, the outlying samples, and so on.

Relationship among the variables manifests itself in the loading plots, and its interpretation is based on the angle between the vectors drawn from the data origin to the points representing individual variables. The cosine of the angle between the two non-zero vectors equals to the dot product of the vectors divided by the product of their lengths. For the centered data set (i.e., for the centered variables), it is equivalent to the Pearson correlation coefficient. In view of this statement, the angle between two vectors close to 0° indicates a high and positive correlation of variables, the angle close to 180° reveals their high yet negative correlation, and the angle close to 90° indicates their independence. Joint interpretation of the corresponding score and loading plots reveals which variables are responsible for the observed pattern of samples. Thus, PCA can be used for data exploration, but it can also serve as a building block of the below-described ASCA method.[53,54]

2.4. *ASCA: ANOVA combined with simultaneous component analysis (SCA)*

Let us return to the simple linear model describing data that originate from the two crossed factors design. In the case of the multivariate response, data matrix **X** is decomposed as:

$$\mathbf{X} = \mathbf{X}_{\mu} + \mathbf{X}_{\alpha} + \mathbf{X}_{\beta} + \mathbf{X}_{\alpha\beta} + \mathbf{E} \tag{20}$$

where matrix $\mathbf{X}_k$ represents the k-th effect studied (i.e., k ∈ α, β, αβ), and matrix **E** represents the residuals.

Once the data matrix **X** is decomposed according to the underlying experiment design, PCA can be applied to reduce the dimensionality of individual components. If, for example, we intended to study the effect of factor A in the above-described example, we should focus on matrix $\mathbf{X}_A$ (m × n), which represents this effect. Matrix $\mathbf{X}_A$ contains average profiles of all levels. The PCA model of matrix $\mathbf{X}_A$ can be presented as:

$$\mathbf{X}_A = \mathbf{S}_A \mathbf{L}_A' \quad (21)$$

where $\mathbf{S}_A$ denotes the score matrix (m × (a − 1)), and $\mathbf{L}_A'$ ((a − 1) × n) denotes the transposed loading matrix.

It is noteworthy that for factor A with two levels, data matrix $\mathbf{X}_A$ is compressed to one score vector and one loading vector and then with three levels, it is compressed to two score vectors and two loading vectors, and so on. The resulting score and loading matrices can be used to visualize and interpret the effect of factor A. Score plot reveals variation among the levels of factor A, but the natural variation of samples within the levels cannot directly be observed. Certain insight into the variation of samples within the levels of factor A can be gained by the projection of matrix $\mathbf{X}_A$ and the residual matrix $\mathbf{R}$ onto loadings $\mathbf{L}_A$ of the PCA model, given by Equation (21):

$$\mathbf{S}_A^* = (\mathbf{X}_A + \mathbf{R})\, \mathbf{L}_A \quad (22)$$

As variation in $\mathbf{S}_A^*$ is not maximized (like it was in the case of $\mathbf{S}_A$), the resulting projection not always allows a good estimation of the significance of factor A.

To evaluate the statistical significance of factor A, a permutation test is applied. If we work with residuals of the reduced model, it is necessary to remove the component which represents the effect of factor B, that is, matrix $\mathbf{X}_B$, from the data matrix $\mathbf{X}$ and to perform permutations of samples of the resulting matrix ($\mathbf{X} - \mathbf{X}_B$), followed by the PCA model. Each permutation leads to the SS (SS^p) of $\mathbf{T}_A{}^p$. Distribution of a large number of SS^p defines the null distribution of no-effect of the factor A hypothesis. The SS observed for the experimental data, SS^{obs}, is then compared with SS^p, and the probability p is defined as the number of permutations for which SS^p is higher than SS^{obs}, divided by the total number of the permutations performed. If the p value is higher than a predefined probability value, the no-effect of the factor A hypothesis is rejected, and an alternative hypothesis is accepted. Analogous calculations are performed for factor B and interactions AB.

Once the statistical significance of factor A is evaluated, and the hypothesis of its effect is accepted, the loading matrix can be examined to identify the variables that differentiate its levels. This task can be performed with the use of bootstrapping to calculate confidence intervals of individual loadings.[64,65] Bootstrapping belongs to an extensive class of resampling methods. It uses random sampling with replacement (a number of resamples with replacement of the experimental data, each resample of equal size as the experimental data) to assess properties of an estimator by measuring them when sampling from the empirical distribution function of the experimental data. Confidence interval for an individual loading can be constructed based on empirical quantiles from the bootstrap distribution of that loading.

2.5. *ANOVA-TP: ANOVA combined with target projection (TP)*

Anova-TP[59–61] is based on the PLS method[66,67] and on its modification called target projection (TP).[68,69] PLS is a well-known multivariate regression method suitable for the correlated data. It performs bilinear decomposition of the independent and dependent variable blocks to maximize the covariance of the resulting scores. For a univariate-dependent variable **y**, the PLS model with the optimal number of latent variables, f^{opt}, can be given as:

$$\mathbf{X} = \mathbf{TP}' + \mathbf{E} \tag{23}$$

$$\mathbf{y} = \mathbf{Tq} + \mathbf{r} \tag{24}$$

where **X** (m × n) represents the centered independent data matrix, **y** describes the centered dependent variable, **T** (m × f^{opt}) and **P** (n × f^{opt}) denote X-scores and X-loadings, respectively, and **q** (f^{opt} × 1) represents the vector of the y-loadings. Matrix **E** and vector **r** denote the X-residuals matrix and the vector of y-residuals, respectively.

Usually, an optimal number of latent variables, f^{opt}, is calculated based on leave k-out CV.[66] The number of left-out samples depends on the total number of samples in the data set studied, but 10 validation groups can be considered as a standard approach. Model with the minimal root mean square error of cross-validation (RMSCV), that is, model with the best predictive power of the left-out samples, is considered as an optimal one.

$$\mathrm{RMSCV}(f) = \sqrt{\frac{\sum (y - \hat{y}(f))^2}{m}} \tag{25}$$

Final PLS model can also be presented in the following form:

$$\mathbf{y} = \mathbf{X}\mathbf{b}_{PLS} + \mathbf{r} \tag{26}$$

where $\mathbf{b}_{PLS}$ denotes the vector (n × 1) of the regression coefficients.

As proposed by Kvalheim and Karstang, the PLS model can be replaced by the model which contains a single latent variable (the target-projected component) constructed by projecting data matrix onto the regression vector, $\mathbf{b}_{PLS,}$ of the optimal PLS model.[69] It means that the target-projected scores, $\mathbf{t}_{TP}$, and the target-projected loadings, $\mathbf{p}_{TP,}$ in TP are defined as:

$$\mathbf{t}_{TP} = \mathbf{X}\mathbf{b}_{PLS}/||\mathbf{b}_{PLS}|| = \mathbf{y}^{predicted}/||\mathbf{b}_{PLS}|| \tag{27}$$

$$\mathbf{p}'_{TP} = \mathbf{t}'_{TP}\,\mathbf{X}/(\mathbf{t}'_{TP}\,\mathbf{t}_{TP}) = \mathbf{b}'_{PLS}/||\mathbf{b}_{PLS}|| * (\mathbf{X}'\mathbf{X})/||\mathbf{t}_{TP}||^2 \tag{28}$$

Thus, the data matrix **X** can be presented as:

$$\mathbf{X} = \mathbf{t}_{TP}\mathbf{p}'_{TP} + \mathbf{E}_{TP} = \mathbf{X}_{TP} + \mathbf{E}_{TP} \tag{29}$$

and the target-projected scores are proportional to the vector of the predicted response, that is, to $\mathbf{y}^{predicted}$. Moreover, the target-projected loadings, $\mathbf{p}_{TP}$, are the product of the normalized regression coefficients of the optimal PLS model ($\mathbf{b}'_{PLS}/||\mathbf{b}_{PLS}||$) and the covariance matrix of the data **X** scaled by the inverse of the variance of the target-projected scores ($(\mathbf{X}'\mathbf{X})/||\mathbf{t}_{TP}||^2$). In that way, the target-projected loading vector combines the predictive power of the variables with their explanatory power and it can be used for identification of biomarkers.

In cases with more than one dependent variable, the same concepts can be generalized and as many target projected components as the **Y** matrix columns can be extracted.

Evaluation of the statistical significance of the studied factors and their interaction is performed in ANOVA-TP in an analogous way, as it is done in ASCA, that is, using the randomization test for the reduced model residuals. Confidence intervals of the target-projected loadings are estimated based on the bootstrapping sampling method.

The main difference between ASCA and ANOVA-TP is that ANOVA-TP allows the analysis of all possible designs and not only of the crossed-factor fixed-effect models, and it provides a better estimate of the within-level variability than ASCA.

3. Conclusions

A well-formulated research hypothesis and a well-designed and carefully performed experiment are the basic building blocks of a successful scientific story. However, even high-quality experimental data can lead to doubtful or even wrong conclusions if improper methods of data preprocessing and/or analysis are used. Especially, vulnerable step is data preprocessing and once it is not properly performed, an irreversible loss of relevant chemical information can occur, leading to the wrong interpretation of the final results. The same effort which is undertaken to standardize different experimental procedures should also be invested in the standardization of data pretreatment and analysis, leaving space though for their improvement and new approaches.

References

1. Aslam B, Basit M, Nisar MA, *et al.* Proteomics: Technologies and Their Applications. *J. Chromatogr. Sci.* 2017;55(2):182–196.
2. Eilers PHC, Marx BD. Flexible Smoothing with B-Splines and Penalties. *Stat. Sci.* 1996;11(2):89–121.
3. Kaczmarek K, Walczak B, de Jong S, *et al.* Baseline Reduction in Two Dimensional Gel Electrophoresis Images. *Acta Chromatogr.* 2005;15:82–96.

4. Nielsen NPV, Carstensen JM, Smedsgaard J. Aligning of Single and Multiple Wavelength Chromatographic Profiles for Chemometric Data Analysis Using Correlation Optimised Warping. *J. Chromatogr. A.* 1998;805(1–2):17–35.
5. Thekkudan DF, Rutan SC. Denoising and Signal-to-Noise Ratio Enhancement: Classical Filtering. In *Comprehensive Chemometrics*; Brown S, Tauler R, Walczak B, Eds. Amsterdam: Elsevier, 2009.
6. Reis MS, Saraiva PM, Bakshi BR. Denoising and Dignal-to-Noise Ratio Enhancement: Wavelet Transform and Fourier Transform. In *Comprehensive Chemometrics*; Brown S, Tauler R, Walczak B, Eds. Amsterdam: Elsevier, 2009.
7. Hubbard BB. *The World According to Wavelets.* Wellesley, MA: A.K. Peters, 1998.
8. Walczak B, Ed. *Wavelets in Chemistry.* Amsterdam: Elsevier, 2000.
9. Walczak B, Massart DL. Noise Supression and Signal Compression Using Wavelet Packet Transform. *Chemom. Intell. Lab. Syst.* 1997;36:81–94.
10. Mallat SG. A Theory for Multiresolution Signal Decomposition: The Wavelet Representation. *IEEE Trans. Pattern Anal. Mach. Intell.* 1989;11(7):674–693.
11. Pasti L, Walczak B, Massart DL, *et al.* Optimization of Signal Denoising in Wavelet Transform (DWT). *Chemom. Intell. Lab. Syst.* 1999;48:21–34.
12. Coifman RR, Wickerhauser MV. Entropy-Based Algorithms for Best Basis Selection. *IEEE Trans. Inform. Theory.* 1992;38(2):713–718.
13. Tomasi G, van den Berg F, Andersson C. Correlation Optimized Warping and Dynamic Time Warping as Preprocessing Methods for Chromatographic Data. *J. Chemom.* 2004;18(5):231–241.
14. Eilers PHC. Parametric Time Warping. *Anal. Chem.* 2004;76(2):404–411.
15. Daszykowski M, Vander Heyden Y, Boucon C, *et al.* Automated Alignment of One-Dimensional Chromatographic Fingerprints. *J. Chromatogr. A.* 2010;1217:6127–6133.
16. Walczak B, Wu W. Fuzzy Warping of Chromatograms. *Chemom. Intell. Lab. Syst.* 2005;77:173–180.
17. Torgrip RJO, Aberg KM, Karlberg B, *et al.* Peak Alignment Using Reduced Set Mapping. *J. Chemom.* 2003;17(11):573–582.

18. Aberg KM, Torgrip RJO, Kolmert J, *et al.* Feature Detection and Alignment of Hyphenated Chromatographic–Mass Spectrometric Data: Extraction of Pure Ion Chromatograms Using Kalman Tracking. *J. Chromatogr. A.* 2008;1192(1):139–146.
19. Forshed J, Schuppe-Koistinen I, Jacobsson SP. Peak Alignment of NMR Signals by Means of a Genetic Algorithm. *Anal. Chim. Acta.* 2003;487(2):189–199.
20. Jellema RH. Variable Shift and Alignment. In *Comprehensive Chemometrics*; Brown S, Tauler R, Walczak B, Eds. Amsterdam: Elsevier, 2009, 85–107.
21. Daszykowski M, Walczak B. Target Selection for Alignment of Chromatographic Signals Obtained Using Monochannel Detectors. *J. Chromatogr. A.* 2007;1176:1–11.
22. Marini F, Walczak B. Particle Swarm Optimization (PSO): A Tutorial. *Chemom. Intell. Lab. Syst.* 2015;149:153–165.
23. Bylesjo M, Cloarec O, Rantalainen M. Normalization and Closure. In *Comprehensive Chemometrics*; Brown S, Tauler R, Walczak B, Eds. Amsterdam: Elsevier, 2009, 109–127.
24. Filzmoser P, Walczak B. What Can Go Wrong at the Data Normalization Step for Identification of Biomarkers? *J. Chromatogr. A.* 2014; 1362:194–205.
25. Pearson K. On a Form of Spurious Correlation Which May Arise When Indices are Used in the Measurement of Organs. *Proc. R. Soc. Lond.* 1897;60:489–498.
26. Dieterle F, Ross A, Schlotterbeck G, *et al.* Probabilistic Quotient Normalization as Robust Method to Account for Dilution of Complex Biological Mixtures. Application in 1H NMR Metabonomics. *Anal. Chem.* 2006;78(13):4281–4290.
27. Rabatel G, Marini F, Walczak B, *et al.* VSN: Variable Sorting for Normalization. *J. Chemom.* 2020;34:e3164.
28. Malyjurek Z, de Beer D, Joubert E, *et al.* Working with Log-Ratios. *Anal. Chim. Acta.* 2019;1059:16–27.
29. van den Berg RA, Hoefsloot HC, Westerhuis JA, *et al.* Centering, Scaling, and Transformations: Improving the Biological Information Content of Metabolomics Data. *BMC Genomics.* 2006;7:142.

30. Kvalheim OM, Brakstad F, Liang Y. Preprocessing of Analytical Profiles in the Presence of Homoscedastic or Heteroscedastic Noise. *Anal. Chem.* 1994;66:43–51.
31. Sanccenti E. Correlation Patterns in Experimental Data are Affected by Normalization Procedures: Consequences for Data Analysis and Network Inference. *J. Proteome Res.* 2017;16:619–634.
32. Jolliffe IT. *Springer Series in Statistics: Principal Component Analysis*, 2nd ed. New York: Springer-Verlag, 2002.
33. Mardia KV, Kent JT, Bibby JM. *Multivariate Data Analysis.* London: Academic Press, 1979.
34. Massart DL, Vandeginste BGM, Deming S, *et al. Chemometrics: A Textbook.* Amsterdam: Elsevier, 1988.
35. Hyvärinen A, Oja E. Independent Component Analysis: Algorithms and Applications. *Neural Netw.* 2000;13(4–5):411–430.
36. Hyvarinen A, Karhunen J, Oja E. *Independent Component Analysis.* New York: John Wiley & Sons, 2001.
37. Stone JV. *Independent Component Analysis: A Tutorial Introduction.* Cambridge, MA: MIT Press, 2004.
38. Belkin M, Niyogi P. Laplacian Eigenmaps and Spectral Techniques for Embedding and Clustering. *Adv. Neural Inf. Process. Syst.* 2001;14:586–691. MIT Press.
39. Donoho D, Grimes C. Hessian Eigenmaps: Locally Linear Embedding Techniques for High-Dimensional Data. *Proc. Natl. Acad. Sci. U.S.A.* 2003;100(10):5591–5596.
40. Daszykowski M, Walczak B, Massart DL. Projection Methods in Chemistry. *Chemom. Intell. Lab. Syst.* 2003;65:97–112.
41. Ståhle L, Wold S. Partial Least Squares Analysis with Cross-Validation for the Two-Class Problem: A Monte Carlo Study. *J. Chemom.* 1987;1(3):185–96.
42. Barker M, Rayens W. Partial Least Squares for Discrimination. *J. Chemom.* 2003;17(3):166–173.
43. Breiman L, Friedman JH, Olshen RA, *et al. Classification and Regression Trees.* Pacific Grove, CA: Wadsworth, 1984.
44. Breiman L. Random Forests. *Mach. Learn.* 2001;45(1):5–32.
45. Snee RD. Validation of Regression Models: Methods and Examples. *Technometrics.* 1977;19:415–428.

46. Daszykowski M, Walczak B, Massart DL. Representative Subset Selection. *Anal. Chim. Acta.* 2002;468:91–103.
47. Verboven S, Hubert M. LIBRA: A MATLAB Library for Robust Analysis. *Chemom. Intell. Lab. Syst.* 2005;75(2):127–136.
48. Serneels S, Croux C, Filzmoser P, *et al.* Partial Robust M-Regression. *Chemom. Intell. Lab. Syst.* 2005;79:55–64.
49. Ohlert GW. *A First Course in Design and Analysis of Experiments.* New York: W.H. Freeman, 2000.
50. Montgomery DC. *Design and Analysis of Experiments*, 8th ed. New York: John Wiley & Sons, 2019.
51. Ståhle L, Wold S. Multivariate Analysis of Variance (MANOVA). *Chemom. Intell. Lab. Syst.* 1990;9:127–141.
52. Pillai KCS. Multivariate Analysis of Variance (MANOVA). In *Encyclopedia of Statistical Sciences*; Kotz S, Balakrishnan N, Read CB, *et al.* Eds. New York: Wiley, 2005, 5161–5170.
53. Smilde AK, Jansen JJ, Hoefsloot HCJ, *et al.* ANOVA-Simultaneous Component Analysis (ASCA): A New Tool for Analyzing Designed Metabolomic Data. *Bioinformatics.* 2005;21:3043–3048.
54. Jansen JJ, Hoefsloot HCJ, van der Greef J, *et al.* ASCA: Analysis of Multivariate Data Obtained from an Experimental Design. *J. Chemom.* 2005;19:469–481.
55. Harrington PB, Vieira NE, Espinoza J, *et al.* Analysis of Variance — Principal Component Analysis: A Soft Tool for Proteome Discovery. *Anal. Chim. Acta.* 2005;544:118–127.
56. Bouveresse DJR, Pinto RC, Schmidtke LM, *et al.* Identification of Significant Factors by an Extension of ANOVA–PCA Based on Multi-Block, Analysis. *Chemom. Intell. Lab. Syst.* 2011;106:173–182.
57. Boccard J, Rudaz S. Exploring Omics Data from Designed Experiments Using Analysis of Variance Multiblock Orthogonal Partial Least Squares. *Anal. Chim. Acta.* 2016;920:18–28.
58. Engel J, Blanchet L, Bloemen B, *et al.* Regularized MANOVA (rMANOVA) in Untargeted Metabolomics. *Anal. Chim. Acta.* 2015;899:1–12.
59. Marini F, de Beer D, Joubert E, *et al.* Analysis of Variance of Designed Chromatographic Data Sets: The Analysis of Variance — Target Projection Approach. *J. Chromatogr. A.* 2015;1405:94–102.

60. Marini F, de Beer D, Walters NA, *et al.* Multivariate Analysis of Variance of Designed Chromatographic Cata. A Case Study Involving Fermentation of Rooibos Tea. *J. Chromatogr. A.* 2017;1489:115–125.
61. Marini F, Walczak B. ANOVA-Target Projection, In *Comprehensive Chemometrics*; Brown S, Tauler R, Walczak B, Eds. Amsterdam: Elsevier, 2020.
62. Anderson MJ, ter Braak CJF. Permutation Tests for Multi-Factorial Analysis of Variance. *J. Stat. Comput. Simul.* 2003;73:85–113.
63. Lohr SL. Hasse Diagrams in Statistical Consulting and Teaching. *Am. Stat.* 1995;49:376–381.
64 Efron B. Bootstrap Methods: Another Look at the Jackknife. *Ann. Stat.* 1979;1:1–26.
65. Efron B, Tibshirani RJ. *An Introduction to the Bootstrap.* New York: Chapman & Hall, 1993, 141–152.
66. Martens H, Naes T. *Multivariate Calibration.* Chichester: John Wiley, 1989.
67. Geladi P, Kowalski BP. Partial Least-Squares Regression: A Tutorial. *Anal. Chim. Acta.* 1986;185(1):1–17.
68. Kvalheim OM, Karstang TV. Interpretation of Latent-Variable Regression Models. *Chemom. Intell. Lab. Syst.* 1989;7:39–51.
69. Kvalheim OM. Interpretation of Partial Least Squares Regression Models by Means of Target Projection and Selectivity Ratio Plot. *J. Chemom.* 2010;24:496–504.

https://doi.org/10.1142/9789811283376_0007

Chapter 7

Basics of MS-Based Glycomic and Glycoproteomic Analysis of Glycoproteins

Yehia Mechref,[1,*] Wenjing Peng,[1] Cristian D. Gutierrez Reyes,[1] Andrew I. Bennett,[1] Firas Kobeissy,[2] and Stefania Mondello[3]

[1]*Department of Chemistry and Biochemistry, Texas Tech University, Lubbock, TX*
[2]*Program for Neurotrauma, Neuroproteomics & Biomarkers Research, Department of Emergency Medicine, University of Florida, Gainesville, FL*
[3]*Department of Biomedical and Dental Sciences and Morphofunctional Imaging, University of Messina, Messina, Italy*

Contents

*Corresponding author.
Department of Chemistry and Biochemistry, Texas Tech University, Lubbock, TX 79409-1061, Email: yehia.mechref@ttu.edu, Tel: 806-742-3059, Fax: 806-742-1289.

Abstract

Protein glycosylation is a post-translational modification in mammalian cells that plays an essential role in many biological functions. Additionally, aberrant glycosylation is implicated in many disease pathways, thus suggesting the important role glycoproteins play in the development and progression of many human diseases. This chapter describes the analytical methods employed to qualitatively and quantitatively characterize glycans and glycoproteins, including the most common digestion, purification, and derivatization protocols applied before separation and detection. The applications of capillary electrophoresis and ion mobility for separating glycans and glycopeptides are also described and discussed. As gas-phase dissociation of the glycopeptides is critical for their identification in mass spectrometry (MS) analysis, valuable information regarding the dissociation parameters utilized to prompt efficient fragmentations of glycans and glycoproteins is also included. Metabolic labeling techniques, which focus on modification of the protein's glycan and amino acid moieties, are presented. Such a labeling tool eliminates the need for additional sample treatment and increases MS detection sensitivity. Finally, bioinformatics tools commonly employed to facilitate data processing, analyzing, and interpreting MS-based glycomic and glycoproteomic data are also described and discussed.

1. Introduction

Protein glycosylation is one of the most common post-translational modifications in mammalian cells, with more than half of the membrane proteins being glycosylated.[1,2] Glycosylation has been known to mediate a variety of biological functions such as cell adhesion, cell–cell communication and signaling, cell trafficking, and protein stabilization.[3–9] In past decades, aberrant glycosylation has been linked to multiple diseases such as neurodegenerative diseases,[10–12] diabetes,[13,14] inflammation,[15] infectious diseases,[16,17] and cancers.[18–23] Recently, isomeric glycosylation alterations were observed in biological processes,[8,24–27] implicating the essential role that isomeric glycosylation plays in biofunctions and diseases.

Glycosylation initiates in the endoplasmic reticulum (ER) and completes in the Golgi apparatus. Certain monosaccharides are added to core structures by the regulation of a series of glycosidases and glycotransferases following glycosylation biosynthesis pathways.[28] There are two major types of glycosylation: *N*- and *O*-glycosylation. An *N*-glycan attaches on asparagine with a known motif, asparagine-X-serine/threonine, where X can be any amino acid but proline, while an *O*-glycan attaches on serine/threonine without a consistent motif.[28] There are hundreds of monosaccharides that are listed in databases[29]; however, only a few are involved in building glycan structures. For example, human glycans contain nine monosaccharides: *N*-acetylglucosamine (GlcNAc), *N*-acetylgalactosamine (GalNAc), mannose (Man), galactose (Gal), glucose (Glc), fucose (Fuc), sialic acid (Sia, Neu5Ac), xylose (Xyl), and glucuronic acid (GlcA).[29] However, glycan structures are more complex than linear proteins or genes. The different combinations of monosaccharides and corresponding linkages and branches, with the existence of α- and β-anomers of each monosaccharide, introduce a huge microheterogeneity of glycan structures. In addition, the different structural occupancies of multiple glycosylation sites on a single protein induce complex heterogeneity of glycoproteins. This diversity of glycosylation results in a major challenge in glycomic and glycoproteomic studies, which demands advanced technologies for a comprehensive analysis.

MS is the most powerful tool in glycomic and glycoproteomic analysis due to its high resolution, high sensitivity, and ability to acquire abundant structural information.[30–34] Although nuclear magnetic resonance or fluorescent detection can also be used in glycomic analysis, the fact that glycans are low abundant in nature along with a lack of standard has hindered the application of these techniques in glycosylation analysis of biological samples. However, MS analysis for glycomics and glycoproteomics itself remains a challenge due to the competitive ionization and poor ionization efficiency of glycopeptides relative to peptides. Therefore, enrichment approaches and separation techniques are commonly employed prior to MS to enhance the sensitivity and identification of glycans and glycopeptides. Meanwhile, chemical and metabolic labeling strategies and different derivatization methods have been developed to improve the isomeric separation, ionization efficiency, sensitivity, quantitation accuracy, and throughput of glycan and glycoprotein analysis. To acquire more structural information, especially for isomeric identifications, a variety of dissociation methods have been introduced to enhance the cross-ring fragmentation of glycans and increase the sequence coverage of peptide backbones.

Together, advanced sample preparation, separation, identification, and quantitation strategies in glycomic and glycoproteomic analysis enable in-depth investigation of protein glycosylation in different types of samples. Here, we summarize the recent MS-based approaches and strategies in glycomic and glycoproteomic analysis to guide future research toward a better understanding of the role glycosylation plays in biological processes.

2. Glycan release

The characterization of glycoconjugates released from glycoproteins is commonly completed by MS analysis using enzymatically or chemically released glycans from the protein backbone. *N*-Linked glycans are commonly released via enzymatic cleavage using the peptide *N*-glycosidase F (PNGase F) enzyme.[35] Other enzymes used to release *N*-linked glycans are the peptide *N*-glycosidase A (PNGase A)[36] and the peptide N^4-(*N*-acetyl-β-glucosaminyl) asparagine amidases (PNGase H^+) that works in acidic pHs.[37] Additionally, recent studies have demonstrated that this type of glycan can be chemically released using aqueous ammonia with sodium cyanoborohydride as the reaction medium.[38] For the release of *O*-linked glycans, single enzymatic methods are less efficient and only release certain glycan types.[39] An example is the *endo*-a-*N*-acetylglucosaminidase that releases only core 1 and core 3 *O*-glycans under certain conditions.[40] The addition of a protease mixture (pronase) is another strategy commonly used to release *O*-glycans.[41] Finally, the use of chemical methods such as β-elimination[42] and non-reductive β-elimination such as hydrazinolysis[43,44] is often employed for the release of *O*-linked glycoconjugates.

2.1. *Enzymatic release of glycans*

The *N*-glycan release process with PNGase F is usually completed in a long incubation time; therefore, many recent attempts focus on decreasing the digestion time. GlycoWorks (Waters) utilizes a surfactant in high concentration to expose the glycoprotein surface, facilitating the release of the *N*-glycans in 10 m with an incubation temperature of 50°C.[45] Gly-X (Prozyme) utilizes an enzymatic digestion of 5 min and a total sample preparation of 1 h. Rapid PNGase F (New England BioLabs) enables the complete and rapid deglycosylation of antibodies and glycoproteins in 10 min with an incubation temperature of 50°C.[46] Although rapid digestion methods decrease the incubation time during glycomic analysis, the glycan release is facilitated by the addition of surfactants that need to be removed before liquid chromatography (LC)–MS analysis. Additional methods that improve the release of *N*-glycans using PNGase F were recently reported by Mechref and coworkers. The addition of sodium deoxycholate SDC to the PNGase F digestion made the solubilization of cell membrane glycoproteins more efficient and therefore improved the release of their *N*-glycans.[47] Additionally, the method allowed an easy separation of the SDC by acid precipitation. The same research group compared PNGase F performance in three different approaches that evaluated the *N*-glycan release from standard cerebrospinal fluid (CSF) glycoproteins[35]: two methods had different sample concentrations, buffer dilutions, and an additional SPE–C18 purification step, and one method used filter-aided *N*-glycan separation (FANGS) as the glycan purification system. The results demonstrated the lowest glycan abundance when the FANGS method was used for glycan release. Otherwise, the method with a large buffer volume, a lower sample amount (15 μL of CSF and 85 μL of ammonium bicarbonate buffer), and solid phase extraction (SPE)–C18 purification showed the highest glycan abundance after MS analysis. The diminution of the CSF starting sample is of high importance due to the inherent complexity of acquiring this type of human fluid. Zhang *et al.*[48]

recently reported an efficient *N*-glycan release procedure that immobilized PNGase F on magnetic particles. The oriented immobilization of PNGase F was completed using the affinity of its coexpressed histidine-tag toward iminodiacetic acid/nickel modified on the magnetic particle. In this approach, the digestion process was applied in conjunction with a microwave irradiation system, and the reaction time was reduced to 5 min. The immobilized PNGase F was found to have tolerance for a high temperature (52°C) and strong activity even after using the same enzyme solution more than five times. Figure 1a show the MALDI-TOF analysis of glycans released from fetuin and human serum by immobilized PNGase F assisted with microwave irradiation (*ii* and *iv*) and by free His-PNGase F (*i* and *iii*). The *endo*-a-*N*-acetylglucosaminidase (*O*-glycosidase) is used to cleave core 1 and core 3 *O*-glycans containing *N*-acetylgalactosamine with β1–3 linkage to galactose "core 1" or via β1–3 linkage to *N*-acetylglucosamine "core 3."[39] This strategy is limited to *O*-glycans containing no-additional saccharide units. In a recent approach to mapping the *O*-glycosylation sites, the company Genovis offers OpeRATOR, a specific protease that digests mucin-type *O*-glycoproteins *N*-terminally of the serine and threonine glycosylation sites, generating glycopeptides carrying *O*-glycans that enable their identification. Nordgren *et al.*[49] used the OpeRATOR enzyme to investigate the *O*-glycosylation of Zika virus proteins, human kidney tissues, T-cells, and serum.

2.2. *Chemical release of glycans*

Chemical methods are often employed for the release of *O*-glycans,[44,50–52] and some interesting attempts have studied the ability of these methods to release *N*- and *O*-glycans in a single procedure. Song *et al.*[53] introduced the oxidative release of natural glycans (ORNG) method, which utilizes sodium hypochlorite (NaOCl) at room temperature to release glycoconjugates. Although the utility of ORNG was proved from different biological sources, the optimal conditions for small-scale analysis still need to be developed. Fischler *et al.*[54] compared the chemical and enzymatic release of *N*-glycans from fetuin samples using the ORNG and PNGase F protocols. Further targeted LC–MRM–MS analysis indicated that the relative composition of released *N*-glycans from both protocols was similar, but their absolute recovery was significantly lower when the chemical procedure was used. The results showed a recovery of *N*-glycans that was 20 times lower in the NaOCl protocol in comparison to PNGase F when 300 μg of fetuin was used as a starting sample. A strategy developed by Wang *et al.*[38] demonstrated that both *N*- and *O*-glycans can be chemically released using aqueous ammonia with sodium cyanoborohydride ($NaBH_3CN$) as the reaction medium. This protocol releases *N*-glycans as 1-amino-alditols that can be subsequently labeled with fluorescent cationic molecules to facilitate the LC–MS analysis. Additionally, no degradation of core α-1,3-fucosylated *N*-glycans (peeling) or deacetylation reactions were observed in this approach. Figure 1b shows the *electrospray ionization* (ESI)–MS of the Fmoc-derivatized 1-amino-alditols of *N*-glycans released using aqueous ammonia containing $NaBH_3CN$ from chicken ovalbumin (*i*), fetuin bovine serum (*ii*), and Moench pollen (*iii*). Additionally, some studies have proposed the automatization of glycomic analysis to optimize the throughput.[55–57]

3. Purification of released glycans

Matrix sample complexity and the low abundance of glycans complicate glycomic MS-based analysis. Glycan molecule signals can be suppressed for unwanted sample components such as salts, lipids, and remaining protein residues, making sample purification of glycomic and glycoproteomic samples prior to MS-based analysis essential. Desalting and enrichment procedures are the common techniques applied for the purification of glycans. During the past decade, several studies reported the use of hydrophilic interaction chromatography (HILIC),[58] alkyl phases (C18),[35] porous graphitized carbon (PGC),[59] and lectin[60]-based materials that facilitate the identification and quantitation of glycans. Recently, the development of carbon-based nanoparticles and metal–organic frameworks (MOFs) emerged as alternatives for glycan purification and enrichment. Sun *et al.*[61] designed and synthesized magnetic graphene–mesoporous silica functionalized with polydopamine "C-Mag G@

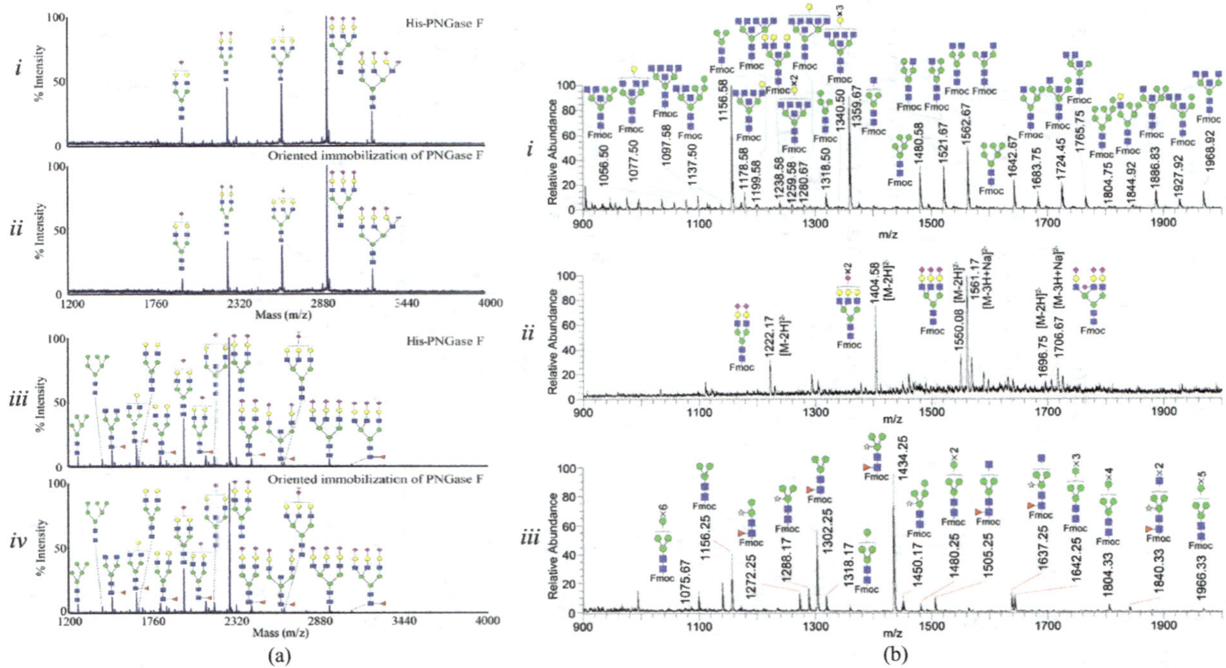

Figure 1. (a) MALDI-TOF analysis of *N*-glycans released from fetuin and human serum by immobilized PNGase F assisted with microwave irradiation (*ii* and *iv*) and by free His-PNGase F (*i* and *iii*). Reprinted with permission from Ref. 48. (b) ESI-MS of the Fmoc-derivatized 1-amino-alditols of *N*-glycans released using aqueous ammonia containing $NaBH_3CN$ from chicken ovalbumin (*i*), fetuin bovine serum (*ii*), and Moench pollen (*iii*). Reprinted with permission from Ref. 38. Glycan symbols: fucose (▼); *N*-acetyl glucosamine (■); mannose (●); galactose (●), and sialic acid (◆).

$mSiO_2$" for the selective enrichment of *N*-glycans facilitated by the glycan–amine interactions. Qin *et al.*[62] prepared the silica–carbon composite nanoparticles "NP-MCM-C" and utilized them to selectively extract *N*-linked glycans from ovalbumin and human serum; the functionality of the material was tested using glycans released from standard glycoproteins. Shu *et al.*[63] synthesized a hybrid carbon material as a platform for glycan enrichment using MOFs as a carbon precursor. These novel materials were successfully used to enrich *N*-glycans from tryptic-digested standard glycoproteins and human serum. Although the aforementioned approaches showed the utility of MOF materials to selectively enrich glycans or glycopeptides from complex biological samples, the number of detected analytes is low compared with non-treated samples. Salih and coworkers developed an HILIC-based method that used poly(amidoamine)–PANAM–dendrimer-coated magnetic nanoparticles,[64] silk-packed micropipette tips,[65] and titania-based material packed into pipette tips.[66] All their approaches were tested for the enrichment of glycans and glycopeptides derived from the tryptic-digested standard glycoproteins immunoglobulin G (IgG), fetuin, transferrin, and human plasma. The results indicated a low number of identified glycans or glycopeptides after enrichment compared to the common methods. Sha *et al.*[67] developed a single-step purification method utilizing cellulose microsphere–filled pipette tips (CM-PTs). To prove the functionality of the CM-PTs for glycan purification, pipette tips were filled with the CM-PTs, zwitterionic HILIC, microcrystalline cellulose, and polyamide resin. The CM-PTs showed better glycan purification efficiency of *N*-glycans from IgG and horseradish peroxide than the other tested materials.

SPE and FANGS are other common methods for glycan purification.[68] SPE is an extractive technique where target compounds are separated from a mixture using their physical and chemical properties and their interactions with a solid phase.[69,70] In the case of FANGS, the method intends to facilitate both enzymatic digestion and glycan purification in a simple workflow.[71] Cho *et al.*[35] compared the efficacy of the two methods; the results demonstrated that the SPE–C18 purification generated a more abundant CSF glycan profile when

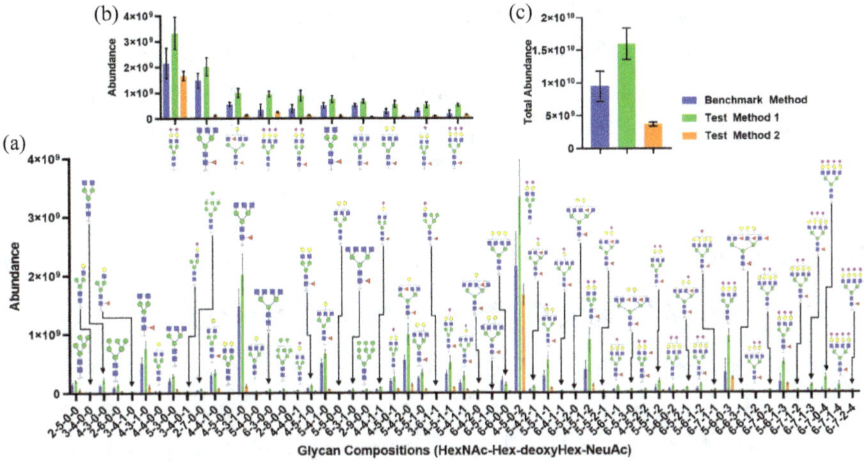

Figure 2. Quantitative analysis comparing three different purification methods of *N*-glycans released from CSF. (a) Absolute abundance of the detected glycans; (b) absolute abundance of the top 10 most abundant glycans; (c) sum of the total glycan abundances from the studied methods. Reprinted with permission from Ref. 35. Glycan symbols: fucose (▼); *N*-acetyl glucosamine (■); mannose (●); galactose (●), and sialic acid (◆).

compared with the FANGS protocol. Figure 2a–2c show the quantitative comparison of the three *N*-glycan purification methods investigated for LC-MS glycomic analysis of CSF. Additional approaches with high glycan selectivity have been developed and tested for desalting and enrichment steps before glycomic analysis. Rehulka *et al.*[72] implemented a gradient method for the purification of permethylated glycans. In this approach, a reverse-phase column was attached to a gastight syringe through which the mobile phase gradient was delivered. The sample purification was completed in three steps: separation of polar contaminants, glycan recovery, and the removal of large-sized remaining protein contaminants. The application of this approach allowed the detection of large tri- and tetra-antennary *N*-glycans during MALDI–MS analysis. Wang *et al.*[73] combined an acetone precipitation protein with the filter-aided sample preparation (FASP) method to remove human milk oligosaccharides (HMOs) and purify the *N*-glycans. The method used acetone at −80°C for protein precipitation and the remotion of non-protein components and HMOs. Thereafter, the remaining HMOs and small molecules were removed by using size exclusion spin ultrafiltration. Using this strategy, the *N*-glycosylation of the bovine and human milk glycoproteins was completed by MALDI-TOF–MS analysis.

4. Chemical labeling of glycans

4.1. *Chemical labeling with hydrazide chemistry*

Hydrazides are a class of organic molecules with the formula R-$NHNH_2$. The R group can be either an acyl, sulfonyl, or phosphoryl substituent. Hydrazides are bound to a reducing glycan with an acid catalyst present. The product of this reaction is hydrazone. One of the benefits of chemical labeling with hydrazide chemistry is that it does not require additional clean-up steps or purification like many other glycan derivatization steps.[74,75] Improvements by way of a positively charged reagent such as Girard's reagent T (GT), Girard's reagent P (GP), and similar compounds have been used to improve the sensitivity of glycans in MS analysis.[74,76–79] Gao *et al.*[80] used a two-step derivatization of sialylated human milk oligosaccharides (SHMOs), which includes the selective amidation or lactone reaction of sialic acid at the non-reducing end and fluorescent labeling of GP at the reducing end (Figure 3a). Recent work conducted by Zang *et al.*[79] demonstrated the use of GT and GP to significantly enhance signal intensity for MS analysis. When using GT, signal intensities were improved more than sevenfold for lactose, sialylated, and neutral *N*-glycans. However, the reaction efficiency for sialylated glycans was insufficient. In the investigation of GP and its effects on derivatized *N*-glycan analysis, formalin-fixed paraffin-embedded (FFPE) tissue was used in MALDI–MS imaging.[81] This method provided a noteworthy improvement in glycan signal intensity. The researchers also developed a new hydrazide tag, 4-(hydrazinecarbonyl)-*N*,*N*,*N*-trimethylbenzenaminium, and showed that complete derivatization was accomplished using mildly acidic conditions at room temperature.[82]

4.2. *Chemical labeling with carbamate*

Carbamate derivatization is a glycan derivatization method that uses chromophores or fluorophores to label glycans, which can then be detected by fluorescence (FLR) or ultraviolet (UV) detectors. The target for the derivatization is glycosylamine of the *N*-acetylglucosamine. It should be noted that glycosylamine is easily hydrolyzed within hours after deglycosylation with PNGase F, so it is important to complete the derivation reaction quickly to ensure proper labeling. Several methods have been developed to address this, including approaches that use *N*-hydroxysuccinimide-activated carbamates and esters[83,84] and 6-aminoquinolyl-*N*-hydroxysuccinimidyl carbamate (AQC).[85,86] The AQC method did not require additional clean-up steps. Wu *et al.*[87] optimized an AQC approach for carbamate derivatization of *N*-glycans in only 50 min. The FLR intensities for high mannose structures increased 10-fold compared to the traditional 2-aminobenzoic acid labeling method (Figure 3d). Additionally, commercial kits RapiFluor-MS (RFMS) and InstantPC are available for enhanced rapid glycan labeling[45,88] (Figure 3b).

4.3. *Chemical labeling with methyl iodide*

Permethylation has become a keystone for glycan analysis as it improves the sensitivity of glycans in traditional reversed-phase liquid chromatography by about 100×. There are two main ways that this method can be carried out: liquid and solid phase. Mechref and coworkers[89] recently reported a solid phase eight-plex sample analysis, using CH_2DI, CHD_2I, CD_3I, $^{13}CH_3I$, $^{13}CH_2DI$, $^{13}CHD_2I$, $^{13}CD_3I$, and CH_3I (Figure 3c).

4.4. *Chemical labeling of sialic acid moieties that prompt isomeric differentiation*

Sialic acid linkages are primarily either α2,6 or α2,3 in nature, and differences in the distribution of these linkages have been reported in several diseases, warranting attention from the scientific community to

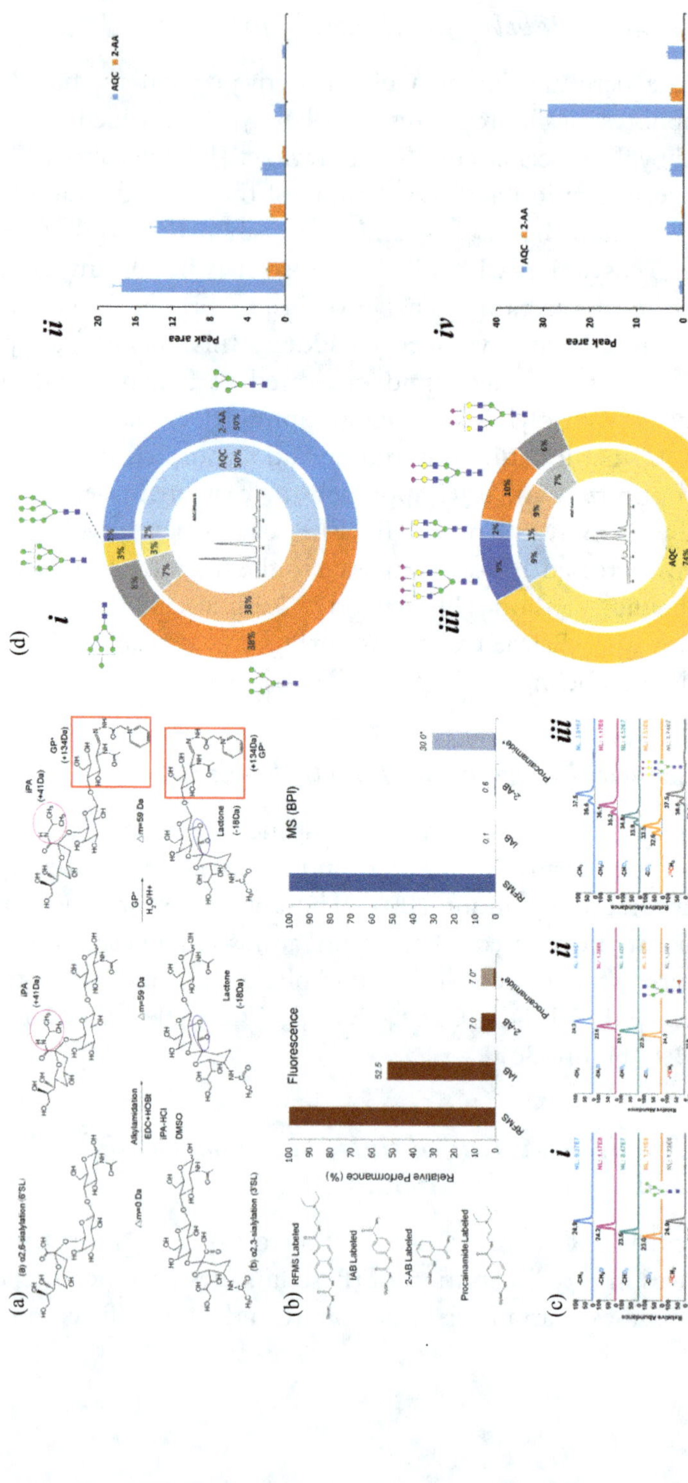

(a)
(b)
(c)
(d)
RFMS Labeled
IAB Labeled
2-AB Labeled
Procainamide Labeled
Fluorescence
MS (BPI)
Relative Performance (%)
Peak area
AQC
2-AA

Figure 3. (a) Reaction scheme of the two-step derivatization of SHMOs. Reprinted with permission from Ref. 80. (b) Relative (%) performance of glycan labels. Response factors are shown as percentages versus the fluorescence and MS response factors of RFMS-labeled *N*-glycans. Reprinted with permission from Ref. 45. (c) EICs of eight-plex permethylated (*i*) Man 5; (*ii*) fucosylated glycan; and (*iii*) trisialylated triantennary glycan. Reprinted with permission from Ref. 89. (d) Doughnut plot showing the relative percentage peak areas of RNase B (*i*) and fetuin (*iii*) *N*-glycans labeled with 2-AA "outer circle" and AQC "inner circle." The bar graphs show the *N*-glycan profile of RNase B (*ii*) and fetuin (*iv*) labeled with AQC and 2-AA, respectively. Reprinted with permission from Ref. 87. Glycan symbols: fucose (▼); *N*-acetyl glucosamine (■); mannose (●); galactose ( ), and sialic acid (◆).

develop methods to accurately investigate these differences. Lageveen-Kammeijer *et al.*[90] report a one-pot reaction with a simple clean-up that increased the sensitivity of the sialic acid glycans (Figure 4a). The method relies on the ethyl esterification of α2,6-linked sialic acid and amidation of α2,3 in a two-step reaction. They also demonstrated this method to be useful for analyzing biofluids and tissues with a capillary electrophoresis (CE)–ESI–MS approach. Cheng *et al.*[91] developed a linkage-specific method using a microchip CE–MS that targets α2,3-linked sialic acids (Figure 4b). This is done by a selective amidation of the α2,3-linked sialic acid with *N,N*-dimethylethylenediamine (DMEN). The positively charged DMEN addition aided in separation and reduced interference from other glycans. An improvement to aminolysis sialic acid linkage-specific alkylamidation has been reported for glycosphingolipid glycan analysis.[92] The new method used methylamine (MA) instead of carboxylic acid activators for amidation, which reduced the reaction time from 2 h to 1 h. Additional optimization of derivatization efficiency was investigated by Pongracz *et al.*,[93] where it was found that three catalysts resulted in high derivatization efficiency, namely hydroxybenzotriazole (HoBt), 6-chloro-1-hydroxybenzotriazole, and 1-hydroxy-6-(trifluoromethyl) benzotriazole.

4.5. *Isotopic labeling of glycans*

To overcome the ionization differences and chromatographic challenges of esters and lactones of the derivatized α2,6- and α2,3-linked sialic acids,[94] Peng *et al.*[95] developed a method utilizing isotope-coded d0- and d3-methylamine hydrochloride for high derivatization selectivity.

5. Chromatographic separation of glycans

Separation techniques prior to MS analysis provide resolution to competing ionization issues and the ability to identify different isomers that have the same *m/z* and MS^2 patterns, thus enhancing sensitivity and isomeric characterization in glycomics. Among a variety of separation techniques, LC is the most commonly used in glycomic analysis

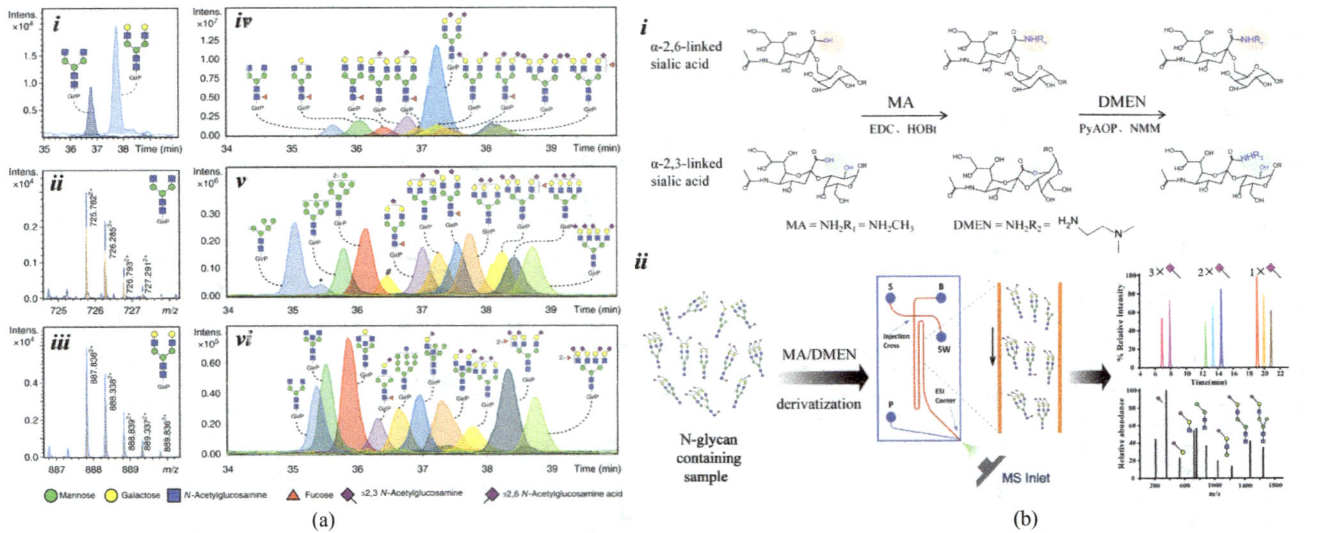

Figure 4. (a) Extracted ion electropherograms of derivatized and labeled *N*-glycans. (*i*) Extracted ion electropherograms of standard *N*-glycans H3N4 and H5N4, and their respective isotopic pattern (*ii* and *iii*). Extracted ion electropherograms of 10 abundant >2% *N*-glycans (*iv*) present in TPNG derivatized with Girard's reagent P, (*v*) 10 with an abundance between 0.5% and 1.0%, and (*vi*) 10 with an abundance <0.25%. Reprinted with permission from Ref. 90. (b) Linkage-specific derivatization of sialylated *N*-glycans (*i*) and workflow of microchip CE–MS analysis of α2,3-sialylated *N*-glycans (*ii*). Reprinted with permission from Ref. 91. Glycan symbols: fucose (▼); *N*-acetyl glucosamine (■); mannose (●); galactose (●), and sialic acid (◆).

due to its promising resolution, robustness, and efficient interface to MS. Since native glycans are not ionized efficiently, the aforementioned labeling methods are usually employed in glycomic studies to enhance sensitivity and separation. In the last decade, increasing cases of isomeric glycan expression alterations were observed in different diseases, implicating the necessity of a comprehensive investigation of glycan isomers. Therefore, multiple separation approaches have been developed to enhance the process. This section summarizes the common LC approaches and their potential capacities for glycan isomeric separation.

5.1. *HILIC separation*

HILIC separates analytes based on hydrophilic interactions, which is efficient in resolving polar molecules. HILIC belongs to normal phase chromatography, where the chromatography of carbohydrates on HILIC is driven by a partitioning mechanism.[96] As multiple hydroxyl groups in glycan structures provide much hydrophilicity, the use of HILIC columns is one of the best choices to separate native and reducing end-labeled glycans and has been utilized for decades.[97–101] The HILIC column uses polar stationary phases such as amine-, amide-, hydroxy-, or zwitterionic particles with an elution gradient from high content of organic (e.g., 90% ACN, 10% water) to high content of aqueous solvent (e.g., 50% water, 50% ACN, 20 mM ammonium formate).[102] Some widely used HILIC column types in glycomic research include zwitterionic (ZIC)-HILIC,[98] HALO columns,[103] and amide/amine columns.[104] For years, HILIC–MS has been utilized to investigate glycan expression changes in several diseases.[105–109] In a recent comparison, three HILIC materials were compared: HALO penta-HILIC, glycan ethylene bridged hybrid (BEH) amide, and ZIC-HILIC.[110]

In addition to glycan profiling, HILIC is also able to separate glycan isomers.[98,111] The different glycan isomeric structures provide differences in their surface hydrophilicities, thus allowing the isomeric separation on HILIC columns. Many recent studies have demonstrated isomeric glycomic analysis for both *N*- and *O*-glycans using

HILIC–MS. Examples include, but are not limited to, the isomeric separations of oligosaccharides,[112] pharmaceutical glycoproteins,[113,114] and diseases such as congenital disorders of glycosylation[115] and Parkinson's disease.[116] In a recent research, Smith *et al.*[117] utilized two-dimensional HILIC–MS to characterize isomeric *N*-glycans from human serum. A weak anion exchange (WAX) was employed for fractionation (four fractions, S1–S4), followed by an HILIC–LC–MS analysis. Figure 5 depicts examples of baseline separation of sialylated linkage isomers from S1 fraction and S2 fraction, respectively.

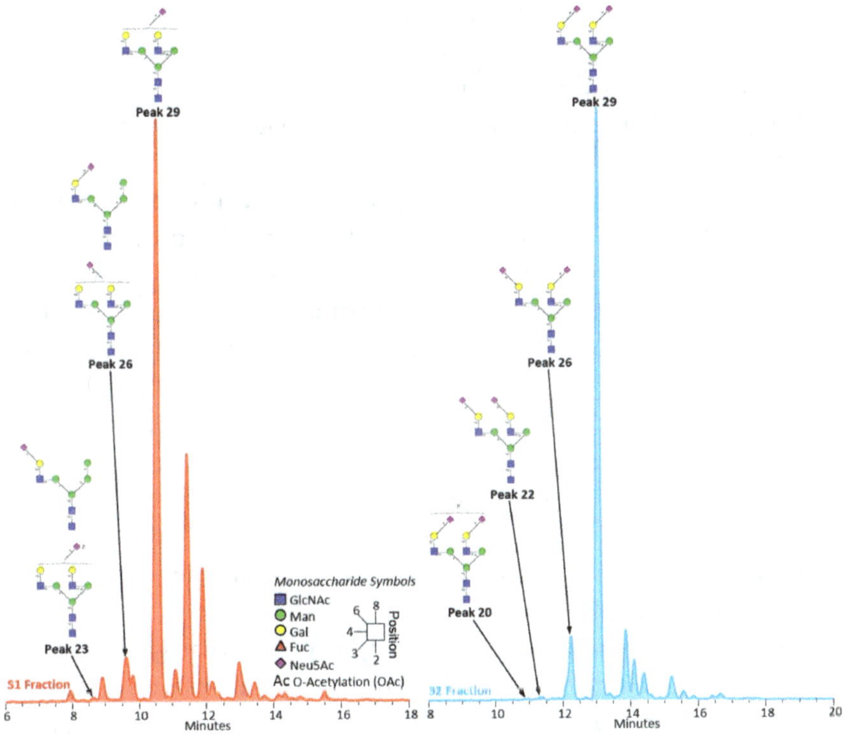

Figure 5. As illustrated in both the S1 (left) and S2 (right) WAX fractions applied to HILIC, *N*-glycans containing α2,3-linked sialic acids elute earlier than their counterparts containing α2,6-linked sialic acids. *Note: Intensities of the S1 and S2 HILIC profiles are not to scale.* Reprinted with permission from Ref. 117. Glycan symbols: fucose (▼); *N*-acetyl glucosamine (■); mannose (●); galactose (●), and sialic acid (◆).

Additionally, α2,3-linked sialylated glycans were proven to be eluted earlier than α2,6-linked sialylated glycans on HILIC columns. The linkages were confirmed by a sialic acid linkage–specific derivatization method that was applied to each eluted peak in the post-HILIC analysis. Although HILIC has demonstrated its potential in the isomeric separation of glycans, the relatively low resolution remains a challenge when analyzing complex biological samples.

5.2. *PGC separation*

PGC has been utilized as an efficient stationary phase to separate glycan isomers. PGC separation is driven by the combination of dispersive interactions and the polar retention effect.[118] These bifunctional interactions of PGC make it an appropriate material to separate both polar and non-polar analytes, and it is thus effective in the separation of both native and permethylated glycans. The unique chromatographic behavior of PGC allows its efficient separation of glycan isomers. At an early stage, the isomeric separation of native glycans on PGC was observed using model glycoproteins.[119,120] Then, the PGC–MS sensitivity was improved by using a nanoPGC column[121] and applied to biological samples.[122] To date, the common PGC columns include analytical and nano Hypercarb PGC columns (Thermo) and the nanoPGC chip (Agilent). The isomeric separation on native and reducing end-labeled glycans has been demonstrated in many studies for both *N*- and *O*-glycan isomers.[123–128] However, Thermo has discontinued the nanoPGC column, creating a challenge for sensitive glycan isomeric separation.

The isomeric separation of permethylated glycans was demonstrated later than native glycans[129] but proved an efficient isomeric separation at a high temperature (75°C).[130,131] The modeling of glycan structures had demonstrated that the three-dimensional (3D) structural difference of fucosylated glycan isomers prompted the isomeric separation on the PGC column.[130] This high-temperature nanoPGC–LC–MS method has been successfully utilized to investigate the isomeric alterations of *N*-glycans in several diseases, including hepatocellular carcinoma and cirrhotic patients,[132] primary restless legs syndrome,[133] idiopathic rapid eye movement sleep behavior disorder,[134] and breast cancer.[24] In addition to *N*-glycans, efficient *O*-glycan isomeric separation has been achieved for both model glycoproteins and complex biological samples.[135,136] The most recent research investigated the increased peak capacity of PGC–MS at an ultrahigh temperature (190°C), which could resolve Lewis X and Lewis A isomers.[137] However, no sialylated glycan isomers were investigated in this study.

One challenge of nanoPGC–LC–MS analysis is the retention time shifts when studying biological samples with a complex matrix. To overcome this issue, a glucose unit index (GUI) was utilized to normalize and correct the retention times between different LC–MS runs by converting retention times to corresponding GU values. This approach has been highlighted in both native[138] and permethylated glycan analysis[139] with high reproducibility. Although nanoPGC–LC–MS enables efficient isomeric separation of glycans, the relatively low stability and reproducibility, the need to regenerate the column,[140,141] and the discontinuation of nanoPGC columns from Thermo Fisher Scientific demand a better alternative to this material.

Most recently, an in-house packed mesoporous graphitized carbon (MGC) column was utilized to replace the nanoPGC column with a better resolution and reproducibility.[141] MGC has the same graphite sheet structure as PGC but is much smaller in particle size (238 nm), allowing it to achieve efficient isomeric separation of permethylated glycans on a 1 cm short column. Figure 6 depicts the isomeric separation of glycans from model glycoproteins and biological samples. The baseline separation was achieved for both linkage isomers (Figure 6a) and positional isomers (Figure 6b). The comparison between MGC and PGC demonstrates a better reproducibility of the MGC column over 3 months (Figure 6a). The potential applications of this material for isomeric glycomic analysis of complex biological samples are also demonstrated in Figure 6b. Although MGC–LC–MS has exhibited promising potential for isomeric glycomic analysis, a commercialized MGC column is not currently available, which hinders the wide application of this method.

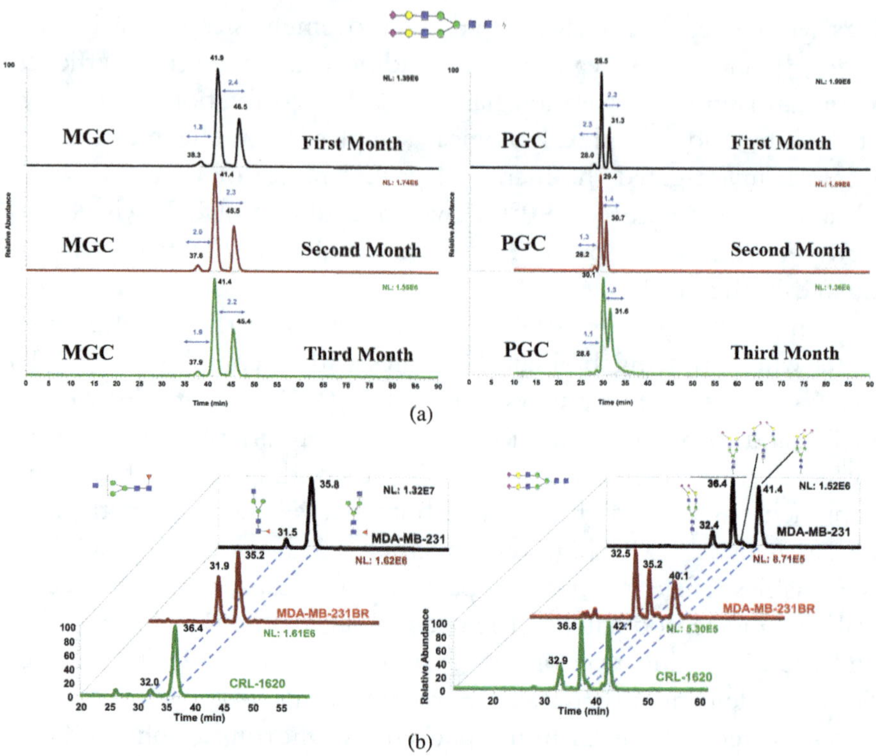

Figure 6. (a) Comparison of the stabilities of the MGC column versus the PGC column over a period of 3 months, demonstrated by the isomeric separation of permethylated biantennary bisialylated glycan derived from bovine fetuin. MGC was tested using 2% ACN in water with 0.1% DFA as MPA and 50% IPA in ACN with 0.1% DFA as MPB. PGC was tested using 2% ACN in water with 0.1% FA as MPA and 100% ACN with 0.1% FA as MPB. Resolution values are depicted as blue numbers, in between the peaks. (b) Examples of isomeric separation of permethylated glycans derived from cell lines MDA-MB231, MDA-MB-231BR, and CRL-1620 on the MGC column. Reproduced with permission from Ref. 141. Glycan symbols: fucose (▼); *N*-acetyl glucosamine (■); mannose (●); galactose (●), and sialic acid (◆).

5.3. *C18 separation*

As one of the most common reverse-phase materials, C18 columns are widely used in analytical chemistry with separation based on hydrophobic interactions.[142] Therefore, the hydrophilicity of native glycans makes them difficult to separate on a C18 column. However, permethylation provides a way to analyze these glycans with C18 columns; after permethylation, hydrophilic hydroxyl and amine groups will change to hydrophobic methyl groups. The C18–LC–MS approach has been utilized for permethylated glycomic profiling for many years.[143–146] However, the isomeric separation of a C18 column is not as efficient as PGC, resulting in partial or no isomeric separation.[8] Since partial isomeric separation for *N*-glycans[147,148] and isomeric separation for smaller glycans such as oligosaccharide and *O*-glycans[149] have been observed, extending the length of the C18 column (e.g., 50 cm) may be one way to achieve efficient isomeric separation of permethylated *N*-glycans on a C18 column.

This hypothesis has been proven by a novel C18 technique utilizing a micro pillar array column (μPAC). Unlike conventional C18 columns, μPAC coats C18 material on the surface of many micro pillars in a microfluidic channel to reduce the backpressure at a high flow rate and enhance the robustness of the column.[150] Recently, the performance of 50 cm and 200 cm μPACs was investigated for glycomic analysis.[151] Figure 7 depicts the efficient separation of *N*-glycan isomers derived from standards and different glycoproteins. The separations of positional isomers and linkage isomers were achieved. Not only *N*-glycans but also *O*-glycan isomeric separation was demonstrated on both 50 cm and 200 cm μPACs.[151] This method was further applied to analyze glycomes derived from different coronavirus spike proteins.[152] Although isomeric separation of permethylated glycans has been observed on C18 columns, it is not the most common method to analyze glycan isomers due to its relatively low resolution and efficiency in isomeric separation. However, it is reliable for permethylated glycome profiling.

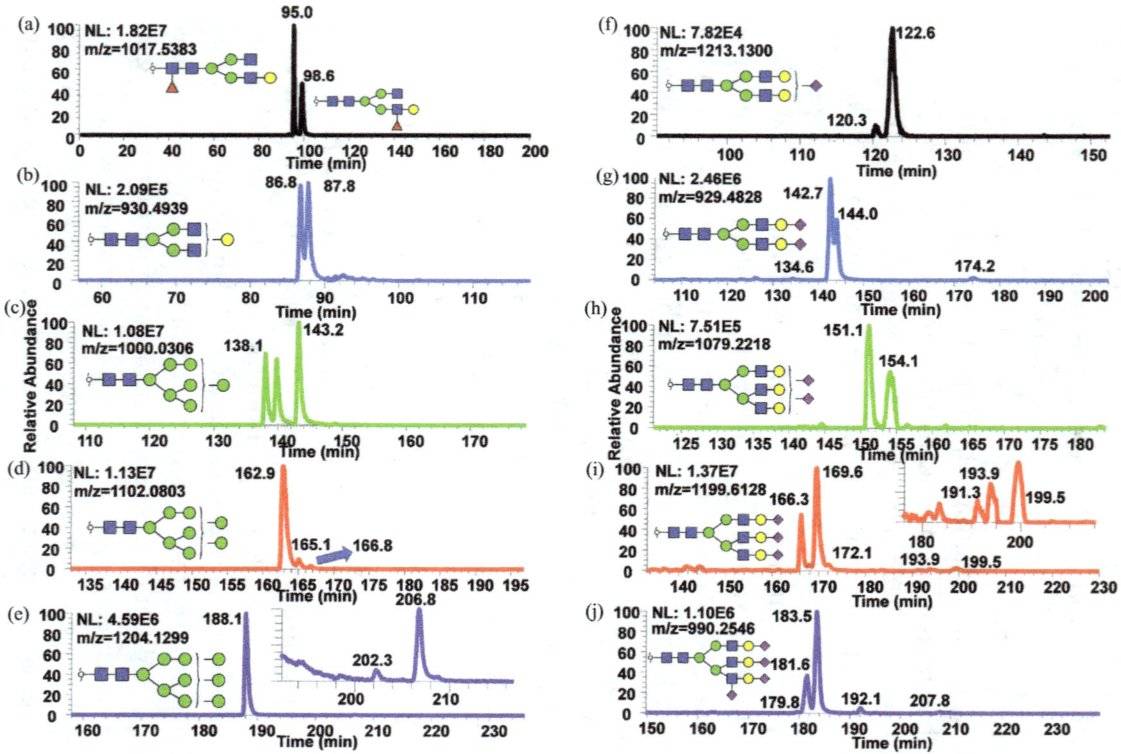

Figure 7. Isomeric separation of permethylated two *N*-glycan standards (a) and *N*-glycans derived from model glycoproteins using 200 cm μPAC. EIC of permethylated *N*-glycans derived from mouse immunoglobulin G (b), ribonuclease B (c–e), and bovine fetuin (f–j) using μPAC 200-cm column. Reprinted with permission from Ref. 151. Glycan symbols: fucose (▼); *N*-acetyl glucosamine (■); mannose (●); galactose (●), and sialic acid (◆).

6. Capillary electrophoretic separation

Capillary electrophoresis (CE) is a high-resolution separation technique that uses voltages to separate analytes by their charge and size. CE coupled with MS leverages the strength of both approaches: the high resolution of CE and the sensitivity of MS. This leads to a powerful technique for the analysis of complex samples. In the analysis of glycopeptides by CE-MS, the size and charge of the glycan portion of the analyte add an additional level of complexity compared to CE-proteomics or CE-glycomics as the glycan portion of the glycopeptide also influences the separation of the analytes.

6.1. *Glycans and glycopeptides*

CE was successfully coupled to drift tube ion-mobility MS for the first time by Jooβ *et al.*[153] They investigated 8-aminopyrene-1,3,6-trisulfonic acid–labeled glycans as well as sialylated native *N*-glycans. Isomers were successfully separated with this method (Figure 8c).

A glycomic and glycoproteomic approach to study the effect of sialic acids on the movement of glycopeptides in the capillary was developed by Khatri *et al.*[154] with a novel CE to MS interface. Kammeijer *et al.*[90] developed a CE–ESI–MS technique to identify the isomers of α2,3- and α2,6-sialylated glycopeptides. This approach was then used to investigate the differences in sialic acid isomers in urine samples between prostate cancer patients and healthy controls.

7. Ion mobility separation

Ion mobility separation (IMS) is a separation technique where ions are moved through a drift cell, and separation is achieved by the differences in the collision cross-section (CCS) areas of the analytes. Smaller molecules travel faster than the larger species creating the separation. The analytes are then passed to an MS analyzer. Because isomers have different 3D spaces, they have different CCS areas and can theoretically be separated by IMS. However, isomeric separation by IMS can be limited.

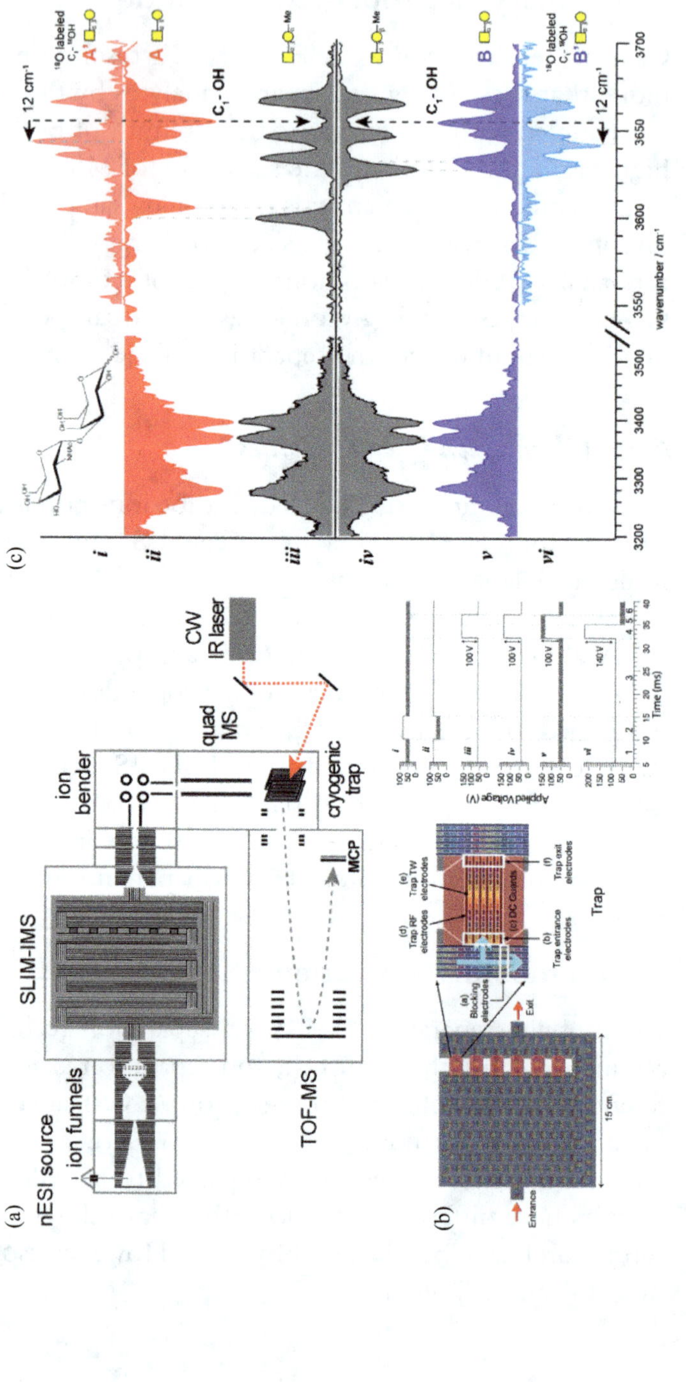
(a)
nESI source
ion funnels
SLIM-IMS
ion bender
quad MS
CW IR laser
cryogenic trap
TOF-MS
MCP
(b)
Entrance
Exit
15 cm
Trap
Applied Voltage (V)
Time (ms)
100 V
140 V
(c)
i
ii
iii
iv
v
vi
12 cm⁻¹
C₁ - OH
A'
A
B
B'
¹⁸O labeled
wavenumber / cm⁻¹

Figure 8. (a) Schematic overview of the instrument. (b) Layout of SLIM–IMS–CID (collision-induced dissociation) device. Left: an overview of SLIM board. Center: details of on-board traps used for CID. Right: potentials used for isolating, trapping, and performing CID in the trap section of the SLIM board. Periods for mobility separation (*i*), trap filling (*ii*), ion trapping (*iii*), bias increase (*iv*), ion release and fragmentation (*v*), and further separation or exiting the board (*vi*) are indicated. Reprinted with permission from Ref. 158. (c) IR spectra of mobility-selected species (*i* and *ii*) of sodiated GalNAc-α(1–3)Gal (*ii* and *v*) and their C_1-^{18}OH-labeled equivalents (*i* and *vi*) and the corresponding C_1–O-methylated species with the C_1 locked in α (*iii*) or β (*iv*) configuration. The C_1–OH stretch vibration at 3655 cm^{-1} is clearly identified by the ^{18}O labeling experiment. Reprinted with permission from Ref. 157.

7.1. *IMS for Glycans*

Recently, a technique known as structures for lossless ion manipulations (SLIM) serpentine ultralong path with extended routing was developed by Smith and coworkers.[155] They reported the successful separation of di-/tri-/tetrasaccharide isomers and validated this method using human milk. Using a variation of SLIM, Rizzo and coworkers[156–159] added an additional level of analysis in the form of cryogenic infrared (IR) spectroscopy to analyze the isomers GalNAcα (1–3)Gal and GlcNAcα(1–3)Gal (Figure 8). Campbell and coworkers[160] used differential mobility spectrometry (DMS) which is also known as field asymmetric ion mobility spectrometry (FAIMS) to successfully separate sialic acid linkage isomers. By incorporating metal ions into the IMS analysis of glycans to create glycan–metal ion complexes, improvements in IMS separation can be seen[161] A combination of techniques by Nguyen-Khuong and coworkers[162] using RFMS-labeled *N*-glycans in an LC–FLR–IMS–MS workflow delivered separation for branched glycopeptide isomers in positive mode for the first time. Isomeric *N*-glycans were able to be identified using intact IMS or fragment-based IMS glycan sequencing.

Lin and coworkers[163] recently incorporated gated-trapped ion mobility spectrometry (G-TIMS) with online electronic excitation dissociation (EED) MS/MS to improve the separation and identification of glycan α-/β-anomers. Boons *et al.*[164] measured arrival time distributions using drift-tube IMS to determine conformer distributions (CDs), which facilitate the identification of *N*-glycan isomers. Additionally, a database of CD fingerprints was compiled to assist in building a library of collision cross-section distributions (CCDs) making structure identification more feasible.

7.2. *IMS for Glycopeptides*

Currently, the analysis of isomeric glycopeptides by IMS is a complementary feature to LC–MS as it provides an additional level of isomeric separation. CCS plays a crucial role in glycopeptide IMS analysis.[155,165–167] Dodds and coworkers[168] investigated the effect that glycan size, composition, and charge had on the separation efficiency.

Pallister *et al.* demonstrated the ability to separate sialic acid linkage isomers[162] using a T-Wave IMS (Synapt G2-Si, Waters). They also showed that positional glycopeptide isomers of IgG can be separated using CCS distribution calculations for both the isomers as well as their fragments. Trapped ion mobility spectrometry (TIMS) is a flavor of IMS where an electric field is applied in the opposite direction of the gas flow and ions are sent to be detected or fragmented as the electric field decreases. This method has been shown to be effective in analyzing peptides in biological samples.[169] A parallel accumulation-serial fragmentation method was developed and utilized for the efficient analysis of peptides.[170–172] A recent method coupling TIMS and electromagnetostatic (EMS) electron-capture dissociation using a TIMS–quadrupole-EMS–time-of-flight–MS achieved the characterization of isomeric acetylated and methylated histone tails.[173] However, additional work using TIMS to investigate glycopeptides needs to be conducted. Cyclic ion mobility is a technique that was developed to improve the performance of the T-Wave IMS.[174] This method provides an additional level of tunability as the ions can be sent around in the circular device several times to deliver optimal separation. This method was used by Sanda *et al.*[175] to analyze the *O*-glycopeptides from the SARS-CoV-2 (severe acute respiratory syndrome coronavirus 2) spike protein. Waveform FAIMS is also known as DMS. A recent study using a custom room-temperature FAIMS unit and LTQ XL ion trap with the energy-transfer dissociation (ETD) option (FAIMS–LTQ–ETD–MS) demonstrated the ability to separate *O*-glycopeptide isomers derived from mucins,[176] with the effective separation of the α- and β-GalNAc anomers. Wu *et al.*[177] demonstrated the ability of a home-built CaptiveSpray DMS device to separate synthetic glycopeptide isomers with either glucose or mannose.

8. Gas phase dissociation of glycans

The applications of MS methods have demonstrated the capacity to assign glycan structural features independent of standards due to the predictability of fragmentation patterns. The complete elucidation of glycans involves the description of their variated isomeric forms, which can be achieved by the cross-ring fragmentation patterns observed during the glycan dissociation.[178] However, the lability of the glycan glycosidic bonds and the relatively large amount of sample needed to trigger MS^n analysis complicate this application for low abundant analytes such as glycans. To address this difficulty, numerous dissociation techniques such as collision-induced dissociation/collisionally activated dissociation (CID/CAD), electronic excitation dissociation (EED), ultraviolet photodissociation (UVPD), and infrared multiphoton dissociation (IRMPD) have been employed in MS to enhance the structural identification of glycans. Additionally, glycan derivatization strategies such as permethylation enhance the cross-ring fragmentation during dissociation, enabling comprehensive characterization of glycans.

8.1. *Collision-induced dissociation/collisionally activated dissociation*

CID (also known as CAD) and is an MS technique that induces fragmentation of the selected ions in the gas phase.[179] The ions are accelerated by applying an electrical potential to increase their kinetic energy, and then they are allowed to collide with neutral gas molecules such as helium or nitrogen.[179,180] MS^n analysis can solve the glycan structural details in the mass analyzers of the MS systems because of the different monomer bonding energies that provide stepwise disassembly to smaller molecules on CID. This dissociation technique produces mostly Y and B glycan ions.[178,181] The formation of Y ions depends on charge localization and originates from a single bond rupture to the C_1-carbon, providing monomer mass intervals. However, this type of fragmentation rarely has linking connectivity.[181] Therefore, glycan derivatization strategies such as permethylation are commonly applied to produce cross-ring fragmentation during CID, enabling comprehensive characterization of glycans during MS^n procedures. This type of fragmentation is dependent on the bonding position of adjacent sugars, which provides evidence of their linkage.[182] Kurz *et al.*[183] used an LC-MS^n approach to describe the sialic acid linkages of isomeric permethylated *N*-glycans derived from human immunodeficiency virus gp120. Data-dependent MS^2 fragmentation of the top three to five most intense peaks was generated by CID with a collision energy of 40%. For target MS, intelligent fragmentation methods were generated using neutral loss function and product-dependent MS^{3-5} acquisition to determine the sialic acid linkage of sialylated *N*-glycans. Reinhold and coworkers[178] used MS^n to characterize the permethylated and 2-amino-*N*-(2-aminoethyl)-benzamide (AEAB) derivatized glycans enzymatically released from human milk, as well as the breast cancer cell lines MCF-7 and MDA-MB-231. The MS^2 fragmentation spectra were acquired by CID with a collision energy of 35%. Further, the fucosylated di-LacNAc motifs of the AEAB-derivatized glycans were utilized for linkage identification. The characterization of the isomeric permethylated *N*- and *O*-glycans was based on the MS^n fragmentation of the MS^2 product ions from the LacNAc and Gal-GalNAc glycan moieties.

8.2. *Electronic excitation dissociation (EED)*

EED is a high-energy dissociation method developed by Lin and coworkers that employs electron energy larger than 9 eV.[184] In the mechanism, an electron detachment on an oxygen atom is initiated, forming a distonic atom. Subsequently, the electron is recaptured, forming a di-radical which facilitates glycan fragmentation including the cross-ring fragments. The EED method was successfully employed in the analysis of isomeric Sialyl Lewis *a* and Sialyl Lewis *x*,[184] the isomeric oligosaccharides lacto-*N*-fucopentaose (LNFP) I, II, III, IV, and V,[185] and the reducing end–derivatized LNFP I, II, III, IV, and V.[186] Figure 9 shows the EED 12 eV MS/MS spectra of Me-PRAGS-labeled LNFP I, II, III, V, and VI. In a more recent approach, Lin and coworkers[187] investigated EED on permethylated *N*-glycans derived from ribonuclease B using a PGC–LC–MS analytical system. This strategy allowed sufficient cross-ring fragmentation to describe mannose linkage on high-mannose isomeric glycan structures. Later, the same research group implemented EED in a gated-trapped ion mobility spectrometer "G-TIMS–EED–MS" to differentiate glycan isomers from conformers.[165]

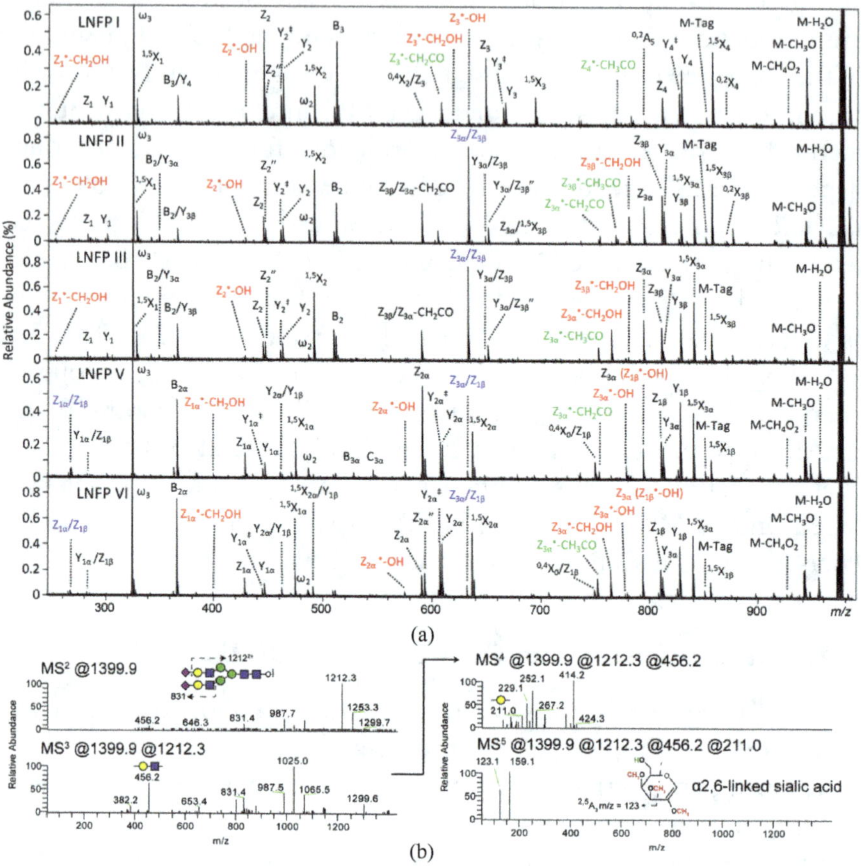

Figure 9. (a) EED 12 eV MS/MS spectra of Me-PRAGS-labeled: LNFP I, LNFP II, LNFP III, LNFP V, and LNFP VI. The secondary fragments useful for linkage analysis at hexose and GlcNAc sites are labeled in red and green, respectively. The internal fragment ions are labeled in blue. Reprinted with permission from Ref. 186. (b) LC–NSI–MS^n analysis of the permethylated *N*-glycan $HexNAc_4$, Hex_5, $Neu5Ac_2$ released from recombinant gp120. The application of MS^{2-4} allows the differentiation of α2,3 and α2,6 sialic acid linkages. Reprinted with permission from Ref. 183. Glycan symbols: fucose (▼); *N*-acetyl glucosamine (■); mannose (●); galactose (●), and sialic acid (◆).

8.3. *Ultraviolet photodissociation*

Photodissociation (PD) was first coupled to MS by Brodbelt and coworkers.[188] To implement UVPD in a mass spectrometer, a UV laser source can be coupled either in a linear ion trap instrument or in the high energy collisional dissociation (HCD) cell of an Orbitrap instrument. During the PD process, the incident photons interact with the selected glycan ions to produce glycosidic and cross-ring cleavages that provide sequential and linkage information, respectively. Ropartz *et al.*[189] used synchrotron radiation as a laser source of UVPD to characterize the isomeric oligosaccharides released from homogalacturonans extracted from citrus fruit. This strategy enables the evaluation of the photon activation over a wide range of energy, from 5 to 24 eV. The spectral data presented significant cross-ring cleavages and its interpretation was found to be easier than in CID due to the absence of neutral losses and internal cleavages. Later, the same technique was used to characterize oligosaccharides derived from red algae *Porphyra umbilicalis.*[190,191] Riggs *et al.*[192] explored the radical direct dissociation (RDD), which generates a radical on UV laser PD inside the mass spectrometer; the generated radical undergoes a subsequent radical migration via hydrogen abstraction highly sensitive to the polysaccharide structure.[192] Compared to the conventional 213-nm UVPD, the 266-nm fourth harmonic pulses induced RDD MS^3 spectra provided more difference in relative abundances of several fragments, which prompted the ability to distinguish between glycan isomers. Brodbelt and coworkers[193] implemented UVPD on an Orbitrap to investigate the PD of glycosaminoglycans (GAGs). The study compared the GAG dissociation using UVPD, HCD, EDD, and NETD and found that the UVPD method exhibited extensive fragmentation ability for both glycosidic and cross-ring products. Later, the same research group combined UVPD with CID in a denominated activated electron photodetachment (aEPD) method that provided more glycosidic and linkage information for in-depth glycan structural identification.[194] Thereafter, the aEPD method was utilized to analyze lipopolysaccharides from *Escherichia coli*. A 193-nm UVPD was utilized for MS^2 followed by collisional activation of charge-reduced precursor ions. This technique provided more abundant linkage information than CID alone and less congested spectra than UVPD alone, which enhanced the characterization of the branching pattern of core oligosaccharides.[195] Although UVPD-based dissociation techniques have provided better identification of glycan isomers, the aforementioned methods still need to be validated with more complex glycan structures and complex biological samples.

8.4. *Infrared multiphoton dissociation (IRMPD)*

IRMPD uses IR radiation that can be absorbed by the glycans by resonant cross sections to perform PD. In the process, the energy is accumulated via the adsorption of hundreds of low-energy photons, and low-energy pathways thus dominated the fragmentation patterns.[188] IRMPD shares similar fragmentation patterns with CID but produces ions with less degradation in high vacuum and fewer ion losses.[196] Additionally, IRMPD can be employed together with MS analysis to improve the glycan characterization. Tan *et al.*[197] implemented a variable wavelength IRMPD–MS approach to evaluate the glycosidic bond linkage in trisaccharides. A stepwise IRMPD process via a tunable CO_2 laser was utilized to differentiate the variants of glucose fragments. After the deprotonated trisaccharides were trapped and fragmented using an FTICR cell, a further variable wavelength IR of the carbon two ion produced a dissociation pattern similar to the dissociation patterns of disaccharides with identical glycosidic bonds. Conversely, the dissociation patterns of different glycosidic linkages exhibited considerable spectral differences. Depraz Depland *et al.*[198] coupled an LC–MS with an IRMPD approach to study the isomeric linkage of sialylated *N*-glycans derived from human milk. The α2,3 and α2,6 linkages of the glycans 3′-sialyllactose sodium salt, 6′-sialyllactose sodium salt, 3′-sialyl-*N*-acetyllactosamine sodium salt, and the 6′-sialyl-*N*-acetyllactosamine sodium salt were easily identified due to the spectral fingerprint observed for both linkages using the representative fragment ions. Renois-Predelus *et al.*[199] reported an IRMPD–MS approach for the identification of GAGs. The results demonstrated the ability of the IRMPD–MS method to distinguish sulfated positional isomers of GAGs. The isomers were irradiated by an IR beam within 2700–3300 cm^{-1} spectral range, and the photofragmentation yields were plotted to generate IRMPD patterns. This technique was used in an LC–MS–IRMPD and a CZE–MD–IRMPD approach to differentiate between isomeric and anomeric sugars.[200,201] The potential IRMPD applications in glycomics still need to be explored in more complex samples.

9. Glycopeptide enrichment

Glycoprotein enrichment is challenging yet necessary to acquire information regarding site-specific glycosylation. Non-glycosylated peptides ionize more efficiently than glycopeptides, so without enrichment, the glycopeptide signal is suppressed. Understanding site-specific variations is critical for a deeper understanding of biological pathways and processes. There are a variety of methods that are used for the enrichment of glycoprotein.

9.1. *Lectin*

Lectins are a class of proteins that selectively bind carbohydrates, specifically glycans and glycoconjugates. This binding is enabled by hydrogen bonding, as well as hydrophobic and van der Waals interactions. Lectins recognize sialic acid and mannose glycans. In biological systems, one of their primary functions is recognition, from cells to individual proteins. They are also found in many foods, such as grains and legumes. Ricin, one of the most well-known toxic lectins, is found in the beans of the castor oil plant *Ricinus communis*. The selective binding of carbohydrates and lectins provides a unique property that can be used for affinity chromatography. Additionally, stationary lectins can be used in microarrays, with magnetic beads or resin materials to provide the infrastructure needed for glycopeptide enrichment. Typically, a variety of lectins are used in separation science to capture as many different types of glycans as possible. Recently, Zhu *et al.*[202] investigated whether certain types of lectins had higher affinities for specific types of glycans. They found that wheat germ agglutinin showed the highest affinity for sialylated complex glycans, whereas concanavalin A (Con A lectin) showed a high affinity for mannose glycans (Figure 10a).

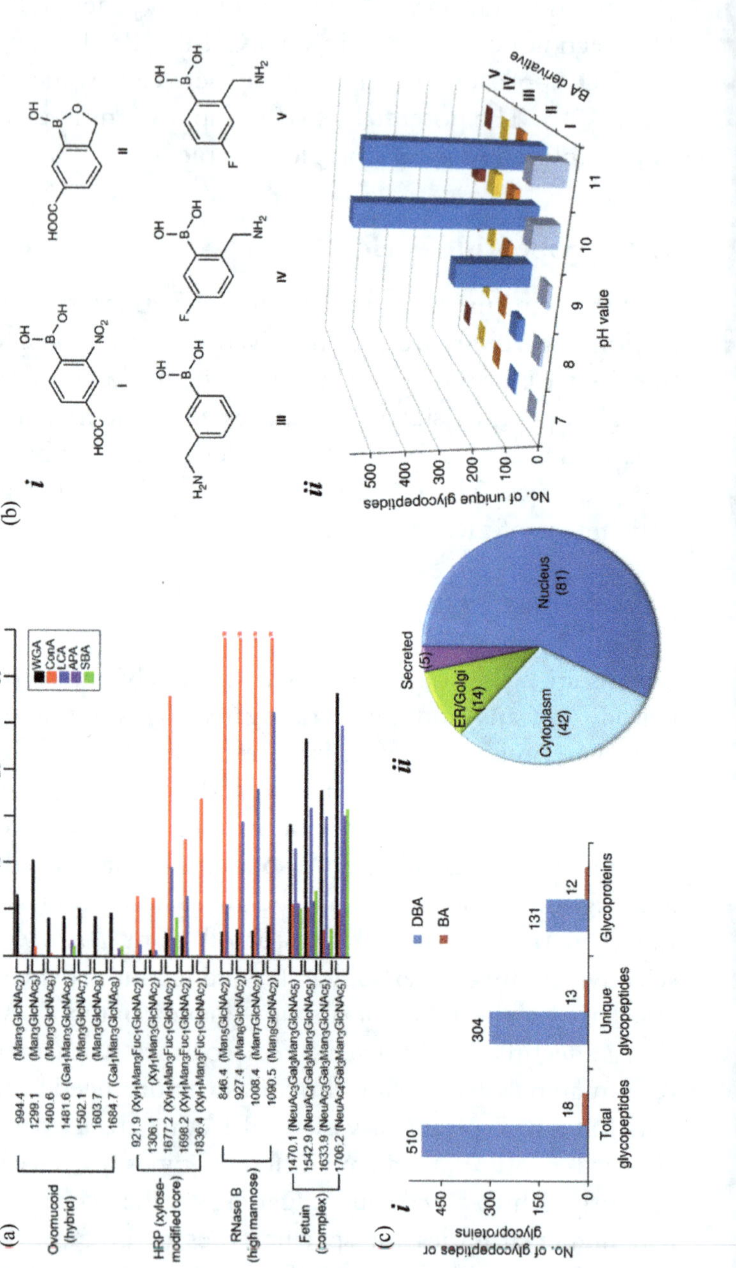
(a)
Adjusted Retention Time (min)
WGA
ConA
LCA
APA
SBA
Ovomucoid (hybrid)
HRP (xylose-modified core)
RNase B (high mannose)
Fetuin (complex)
(b)
pH value
BA derivative
No. of unique glycopeptides
(c)
DBA
BA
Total glycopeptides
Unique glycopeptides
Glycoproteins
No. of glycopeptides or glycoproteins
Nucleus (81)
Cytoplasm (42)
ER/Golgi (14)
Secreted (5)

Figure 10. (a) The adjusted retention times of the glycopeptides on different lectin-coated POROS columns at the flow rate of 0.2 mL/min. These glycopeptides are from four proteins that contain distinct types of glycan structures: high-mannose glycans from RNase B, xylose- (and fucose) modified core pentasaccharide from HRP, hybrid glycans from ovomucoid, and complex glycans from fetuin. Glycopeptides are shown with m/z labels and associated glycan structures. The red star (*) symbol for the high-mannose glycopeptides from RNase B indicates that the retention times of these glycopeptides on the Con A-coated column exceeded the time course of the experiment. Reprinted with permission from Ref. 202. (b) Structures of boronic acid derivatives and experimental results using different derivatives. (*i*) Structures of boronic acid derivatives tested in this work. (*ii*) The number of unique *N*-glycopeptides identified with each BA derivative at varying pH values from the parallel experiments. Reprinted with permission from Ref. 214. (c) Comparison of glycoproteins with one HexNAc identified with BA and DBA, which clearly shows that the results from DBA are substantially better (*i*). Distribution of *O*-glycoproteins modified with HexNAc(1) identified in HEK 293T cells based on the cellular compartment (*ii*). Reprinted with permission from Ref. 214.

9.2. *Electrostatic repulsion hydrophilic chromatography*

ERLIC is based on interactions between a positively charged polyethyleneimine group, which is covalently bound to a modified silica bead stationary phase and negatively charged glycans, primarily sialic acid groups. This separation technique was first used by Alpert *et al.*[203] A strong anion exchange-ERLIC variation was developed by Bermudez and coworkers[204] who showed results comparable to traditional HILIC for *N*-glycopeptide and sialylated *O*-glycopeptide enrichment. Counterion-ERLIC developed by Cui and coworkers[205] leverages the positive charge of a magnesium ion (Mg^{2+}) to interact with the negative charges of glycan and phosphopeptide functional groups, increasing their retention and separation.

9.3. *Hydrophilic interaction chromatography*

HILIC is a type of normal-phase chromatography where the stationary phase is hydrophilic in nature. General benefits of this technique include excellent reproducibility and wide-ranging specificity for glycans. Recent work with other hydrophilic materials including MOFs,[206,207] polymers,[208,209] and magnetic substances[210,211] is being conducted to improve the separation/enrichment of glycopeptides. An automated enrichment method to analyze site-specific *N*-glycoproteomes was developed by Liu *et al.*,[212] where a total of 2081 *N*-glycopeptides from 190 glycoproteins were identified.

9.4. *Hydrazide and boronic acid*

Hydrazides are a class of organic molecules with the formula $RNHNH_2$. The R group can be either an acyl, sulfonyl, or phosphoryl substituent. This variability leads to a wide array of potential interactions that can be engineered to facilitate the effective separation of glycopeptides. A key step in this technique is to oxidate the diol groups on the glycans, forming aldehyde groups which can then be linked to hydrazide groups. This reaction is pH reversible as well, creating the appropriate condition where pH chromatography can be carried out. Bai *et al.*[213] identified 329 *N*-glycosylation sites from 180 *N*-glycoproteins in human plasma using poly-NIPAM-co-hydrazide for *N*-glycoprotein/glycopeptide enrichment.

Separation using boronic acid enrichment is achieved as the stationary phase with boronic acid reacts with the glycan diol groups and retains them in the column. The glycopeptide portions of the sample are eluted with sugars that have a higher affinity for the boronic acid, such as glucose.

Benzoboroxole was used by Xiao and coworkers[214] for glycoprotein enrichment. *N*-glycosylation sites were analyzed in mouse brain tissue (4195), human cells (4691), and yeast (1000). Additionally, *O*-mannosylated proteins were analyzed in yeast and human cells, with 324 sites found in both samples (Figure 10b).

10. Derivatization of glycopeptides

Glycoproteomic analysis is challenging due to factors such as low abundance and low ionization efficiency of glycopeptides, as well as glycan heterogeneity. To address these issues, the chemical derivatization of glycopeptides provides a variety of alternatives to improve their quantitation in complex biological samples. Moreover, derivatization approaches offer the possibility to identify isomeric glycopeptides due to the improvement of the MS^n fragmentation patterns, along with the advantage of identifying the sialic acid linkages of sialylated glycan isomers.[182,215] Isobaric or isotopic labeling approaches also increase the precision and accuracy of glycoproteomic analysis.[216,217]

10.1. *Amidation*

The determination of the sialic acid modifications plays a significant role in understanding glycoprotein-related human diseases. Derivatization methods modify the carboxylic acid residue of the sialic acid using amidation reactions. Additionally, the differences in selectivity of these reactions allow the identification of the sialic acid linkage of isomeric sialylated glycopeptides. The α2,3-linked sialic acids undergo the formation of lactones with the adjacent sugars of the glycan residue. The α2,6-linked sialic acids undergo carboxylic acid derivatization into amides or esters.[218–222] Wang *et al.*[218,223] developed a two-step amidation approach for glycopeptide derivatization. In this method, the carboxylic acids of the α2,6-linked sialic acid moieties of the sialylated glycopeptides were subjected to dimethylamine amidation. Thereafter, the remaining carboxylic acids with α2,3 linkage were subjected to ammonia amidation. The approach was performed in a MALDI–MS system using tryptic-digested glycopeptides from prostate-specific antigen (PSA). Figure 11 shows the representative MALDI–FTICR–MS spectrum of SP PSA glycopeptides and their quantitation. Annotation of the 20 most abundant glycoforms (a) and the 24 low abundant glycoforms (b). Peaks with gray asterisks are non-identified analytes, and the peaks with red asterisks are insource decay products. (c) Relative abundance of the 20 most abundant glycopeptides showing the intra- and interday repeatability of the method. Mechref and coworkers[224] reported the derivatization of sialylated glycopeptides (DOSGs), a strategy that allowed the differentiation of the sialic acid linkages. The carboxyl groups on the α2,6-linked sialic acids and the peptide backbones were selectively derivatized with isopropylamine. In a second reaction, the unreacted carboxylic acid groups on the α2,3-linked sialic acids were derivatized with methylamine.

10.2. *Isobaric labeling*

Isobaric labeling is a quantitative proteomics strategy where the peptides are labeled with chemical groups with identical masses but vary in the distribution of isotopic species around their structure.[225] This technique allows for simultaneous analysis of samples and provides an additional increase in sensitivity due to the molecules with strong positive ions that are incorporated in the peptide structure. Thus, the isobaric labeling approach was adopted for glycoproteomics in an attempt to enhance the glycopeptide quantitation. Tandem mass tags (TMTs), isobaric tags for relative and absolute quantitation, isobaric *N*, *N*-dimethyl leucine (DiLeu) tags, and methylamine stable isotope

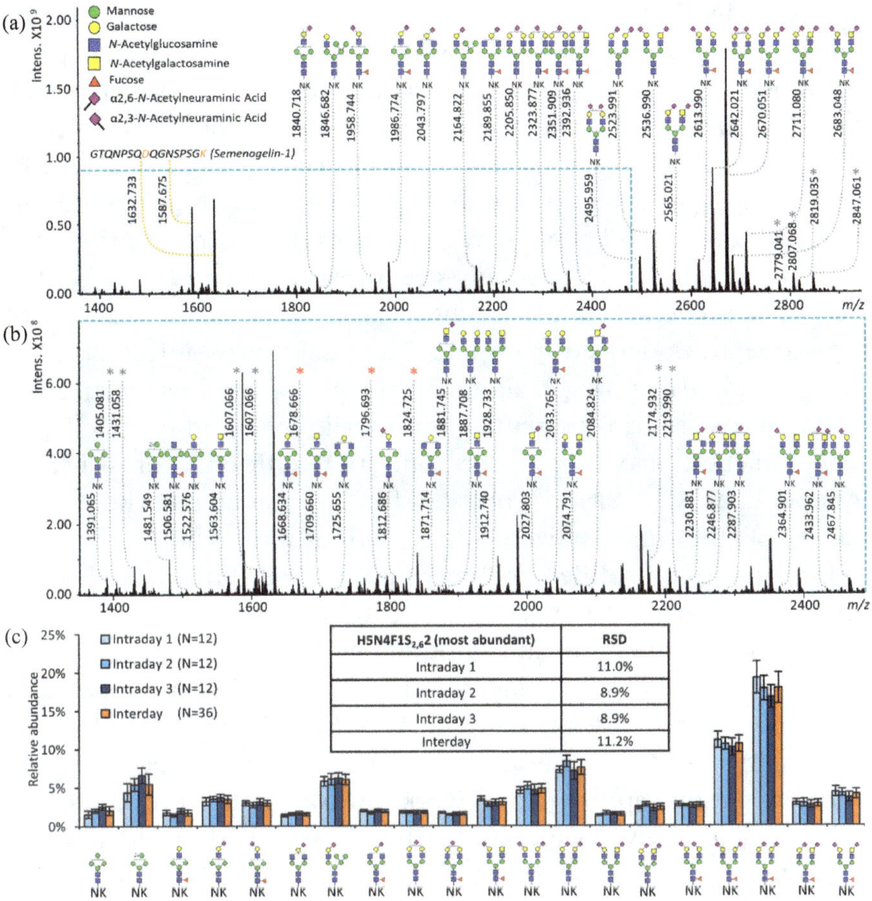

Figure 11. Representative MALDI–FTICR–MS spectrum of SP PSA glycopeptides and their robust quantitation. Annotation of 20 most abundant glycoforms (a) and 24 low abundant glycoforms (b). Peaks with gray asterisks are non-identified analytes, and the peaks with red asterisk are insource decay products. (c) Relative abundance of the 20 most abundant glycopeptides showing the intra- and interday repeatability of the method. Reprinted with permission from Ref. 223. Glycan symbols: fucose (▼); *N*-acetyl glucosamine (■); mannose (●); galactose (●), and sialic acid (◆).

labeling (MeSIL) are the most common approaches.[216,217,226–229] To address glycoproteomic analysis using these methods, a glycopeptide enrichment step should be incorporated into the protocol. Chen *et al.*[228] utilized the DiLeu tags to study the *N*-glycopeptides derived from PANC1 cells. The labeled glycopeptides provided abundant reporter ions, which provided an accurate quantitation, and abundant oxonium, Y, and peptide fragment ions that prompted the identification of 1067 glycopeptide structures. Sun *et al.*[230] used the MeSIL approach for the identification and quantification of *N*-glycopeptides derived from Huh7 and Zika virus–infected Hub7 cells. The MeSIL approach allows for the glycopeptide derivatization of the carboxylic acids presented in the sialic acids and peptide backbone moieties, maximizing the ionization efficiency of sialylated glycopeptides. Cho *et al.*[231] utilized TMT to investigate the performance of the immobilized metal ion affinity chromatography (IMAC) phosphopeptide enrichment protocol at different pHs from a previously enriched sample of intact glycopeptides. The combination of the glycopeptide and phosphopeptide enrichment with the TMT approach allowed the identification of 17,582 phosphopeptides and 3468 glycopeptides from luminal and basal-like types of breast cancer patient–derived xenografts. Fang *et al.*[232] developed the protein aggregation capture (PAC) workflow for the fast preparation of TMT-labeled *N*-glycopeptides. In this approach, the cell lysis was optimized by the addition of 4% SDS, then used PAC bead types and ZIC-HILIC enrichment to achieve a short handling time and good recovery of *N*-glycopeptides. The TMT labeling was then incorporated into the workflow to allow multiplexed relative quantitation. Moreover, the characterization of the MS data was completed through synchronous precursor selection-MS^3. The results showed an identification and quantification of more than 5000 unique glycoforms in Burkitt's lymphoma cells.

11. Gas phase dissociation of glycopeptides

The applications of MS^n methods have been demonstrated as trustworthy tools to assign structural glycopeptide features due to the predictability of their fragmentation patterns. The LC-MS system has a variety of available dissociation techniques that focus on glycopeptide fragmentation, such as CID, HCD, ETD, UVPD, and IRMPD.[182,201,215,233–236] These techniques were initially applied as single strategies to achieve glycopeptide dissociation. Later, research in the field demonstrated that the combination of more than one dissociation technique significantly improved glycopeptide fragmentation. For example, electron-transfer/higher-energy collision dissociation (EThcD) combined with ETD produces abundant fragmentation of the glycopeptide backbone, and HCD produces abundant fragmentation of the glycan moiety.[237]

For glycopeptides, the rupture of glycosidic bonds from the glycan occurs prior to the peptide backbone, leading to the generation of abundant glycan fragments and less information from the peptide backbone. Escobar *et al.*[238] used an HCD–UVPD strategy to address the formation of multiple peptide fragment ions that enabled precise localization of *O*-GlcNAc sites in the *O*-glycopeptides. Using this approach, four positional isoforms of the *O*-glycopeptide VYPVSSVPSSAQSTSK + GlcNAc were identified with the information provided by the fragment ions. Figure 12 shows the applicability of this combined technique to provide extensive peptide fragmentation and the preservation of the *O*-GlcNAc fragments for the *O*-glycopeptides TAB1-*O*-GlcNAc and HCB-*O*-GlcNAc. Riley *et al.*[237] explored the advantages and disadvantages of HCD, stepped HCD, ETD, and EThcD for *N*- and *O*-glycopeptide analysis. Although EThcD produced spectra with superior quality, HCD and stepped HCD provided rapid scan acquisition rates, which improved the *N*-glycopeptide identification. Among the last two dissociation techniques, the findings indicated that stepped HCD produced slightly better spectral quality for *N*-glycopeptides than the standard HCD. Conversely, EThcD provided a better option for the dissociation of *O*-glycopeptides due to the challenges in localizing the *O*-glycosites. Lee *et al.*[239] combined CID and HCD as a fragmentation strategy to increase the glycopeptide oxonium and B/Y ions. The method was applied for the characterization of site-specific *N*-glycopeptide isoforms of α-1-acid glycoprotein (AGP). A total of 165 *N*-glycopeptides

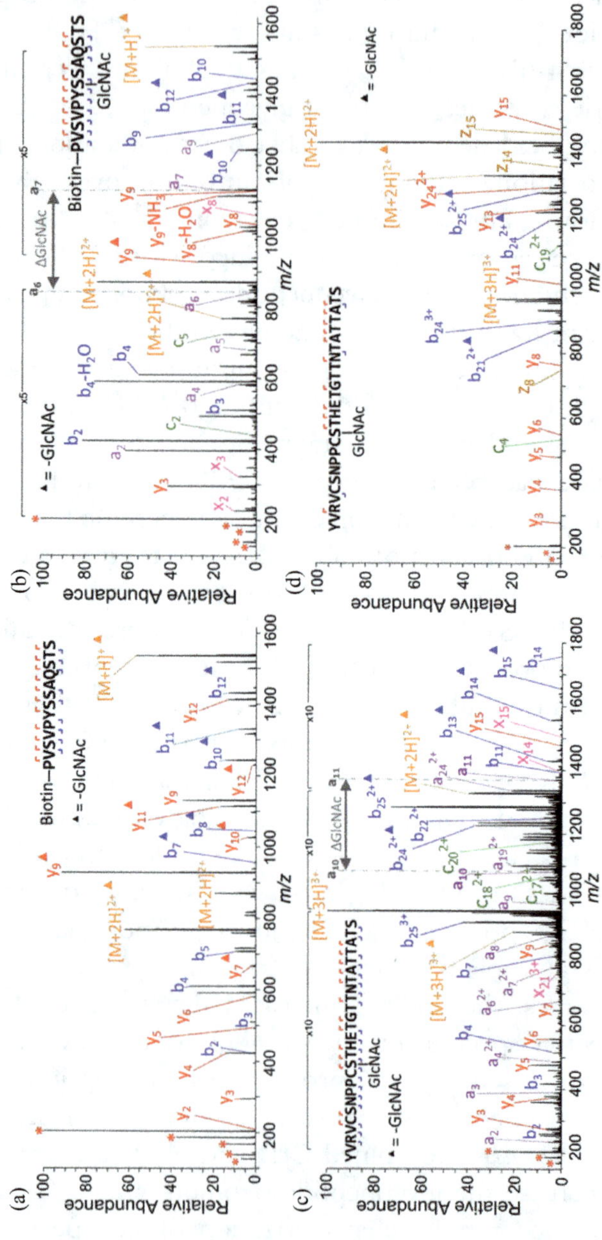

(a)
Biotin—PVSVPYSSAQSTS
▲ = -GlcNAc
(b)
Biotin—PVSVPYSSAQSTS
GlcNAc
▲ = -GlcNAc
ΔGlcNAc
(c)
YVRVCSNPPCSTHETGTTNTATTATS
GlcNAc
▲ = -GlcNAc
ΔGlcNAc
(d)
YVRVCSNPPCSTHETGTTNTATTATS
GlcNAc
▲ = -GlcNAc
Relative Abundance
m/z

Figure 12. HCD oxocarbenium ion-triggered UVPD of *O*-GlcNAcylated peptides generates a rich array of glycosylated sequence ions for confident mapping. (a) HCD mass spectrum "NCE 28" of TAB1-*O*-GlcNAc peptide (NCE 28, $[M+2H]^{+2}$ = 869.90 *m/z*) showing a distinctive oxocarbenium ion pattern, labeled with red asterisks. (b) 193-nm UVPD mass spectrum of TAB1-*O*-GlcNAc peptide, $[M+2H]^{+2}$ = 869.90 *m/z*. (c) 193-nm UVPD mass spectrum of HCB-*O*-GlcNAc, $[M+2H]^{+3}$ = 967.76 *m/z*. (d) EThcD mass spectrum of the peptide HCF-*O*-GlcNAc $[M+2H]^{+3}$ = 967.76 *m/z* using a 50 ms of reaction time for ETD with NCE 25%. The double-headed arrow and ΔGlcNAc indicate the mass shift loss of the *O*-GlcNAc-modified amino acid between fragment ions (labeled in gray at each end of the double-headed arrow) that bracket the site of modification. Oxonium ions are labeled with red asterisks. Triangles denote fragment ions that have lost the GlcNAc modification. Backbone cleavages leading to *N*-terminal and *C*-terminal fragment ions are marked in blue and red, respectively, on the peptide sequence shown in each spectrum. Reprinted with permission from Ref. 238.

were identified and used to depict the microheterogeneity of AGP. Ye *et al.*[240] used a hybrid CID–ETD–HCD approach to characterize TMT-glycopeptides tryptic digested from bovine fetuin. Initially, ETD was used to cleave the peptide backbone without carbohydrate fragmentation. Thereafter, the glycan characterization was accomplished with CID, and the TMT-glycopeptide quantitation was performed based on the HCD spectra. Additional important achievements for the characterization of glycopeptides are described in subsequent approaches: stepped HCD,[241,242] CID/ETD,[243] HCD/ETD,[244] and EThcD.[228,245–247]

12. Metabolic labeling of glycopeptides

Accurate and sensitive quantitation is always a challenge in glycomics and glycoproteomics. Chemical labeling provides enhanced sensitivity and quantitation and is thus successfully applied in many research projects. However, chemical labeling happens after several steps of sample preparation (e.g., after cell/tissue lysis or protein digestion), which increases deviations introduced during these steps. In addition, chemical labeling is affected by the variations introduced by the differences in individual biological samples.[248] To reduce the variation, metabolic labeling strategies are commonly used for the analysis of living systems such as cell cultures or animals.

12.1. *Stable isotope labeling by amino acids in cell culture*

SILAC is the most common metabolic labeling method in proteomics. It was first introduced in 2002[249] and was then applied to glycoproteomics.[250–252] In the early stages, only deuterated leucine (Leu-D3) was used for labeling, but it was impossible to ensure all peptides were labeled after digestion. To solve this problem, ^{13}C-^{15}N-lysine and ^{13}C-^{15}N-arginine were used instead of Leu-D3,[253] enabling the complete labeling of all peptides after tryptic digestion where peptides always have either lysine or arginine at the *C*-termini. The basic protocol of SILAC is to culture one cell line in a "heavy medium" containing isotope-labeled amino acids, with another cell line in a "light medium" containing normal amino acids. The "heavy" amino acids will incorporate into proteins, introducing 8–10 Da mass shift of each peptide in labeled samples as a result. Then, the "heavy" and "light" cell lines can be mixed in a 1 to 1 ratio immediately after the collection of the cells, and then go through sample preparation steps together to eliminate any deviations introduced in the sample preparation steps and instrument runs. Therefore, the SILAC method has been widely used in glycoproteomics for different types of cell lines.[251,254]

12.2. *Azido sugar*

Besides SILAC, the incorporation of an azido group into glycan structures is another metabolic labeling approach to artificially modify glycoproteins. It was first introduced by Bertozzi and coworkers[255] using unnatural *N*-levulinoyl mannosamine in the cell culture, which could finally be converted to ketone-sialic acid. This metabolic labeling is based on the fact that the enzymes that participate in sialic acid and glycan biosynthesis pathways can recognize these unnatural sugars and use them as substrates to build up glycan structures.[255] Azide was considered as a better group than ketone for this metabolic labeling method, prompting the use of ManNAz to incorporate the azido group into sialic acids.[256] In addition to ManNAz, other azido sugar analogs include GalNAz,[257,258] FucNAz,[259,260] and GlcNAz.[261,262] One early limitation of these azido sugar analogs was the low cell membrane permeability that reduced the labeling efficiency of azido-glycans. However, this has been improved by peracetylation with acetyl groups,[263–265] resulting in the most common azido sugar *N*-azidoacetylmannosamine-tetraacylated (AC_4ManNAz).

12.3. *Click chemistry*

One of the most important goals of azido-based metabolic labeling is the subsequent biorthogonal enrichment of glycoproteins through a specific reaction with azido groups.[266] The first functional partner of the azido group was Staudinger ligation[264] before the click chemistry was introduced using a copper-catalyzed cycloaddition reaction.[267] Click chemistry is also well known as copper-catalyzed azide–alkyne 1,3-dipolar cycloaddition (CuAAC), and it allows a rapid and reliable connection between azides and alkynes. The CuAAC method has been used for both *N*- and *O*-glycopeptide analysis.[268,269] However, the use of copper has been proven to be toxic for both bacteria and mammalian cells.[256,270] To overcome the toxicity issue, a copper-free click chemistry was developed using a biotin-conjugated cyclooctyne reagent as "Strain-Promoted [3 + 2] Azide–Alkyne Cycloaddition."[271] This reaction is performed under a physiological condition without copper and has demonstrated no apparent toxicity.[272] Later, the cycloaddition reaction was further modified,[273–276] and now commercialized copper-free click chemistry kits are available, such as the Click-iT Protein Analysis Detection Kit (Thermo Fisher Scientific).[277]

Recently, azido-based metabolic labeling and copper-free click chemistry was utilized to study cell surface sialoglycoprotein expressions,[278] a useful example of metabolic labeling methods. In this study, SILAC, azido-based metabolic labeling, and click chemistry were combined to achieve an efficient site-specific glycoprotein enrichment, identification, and quantitation. The use of novel reagent DBCO-sulfo-biotin ensured the click chemistry happened only on the cell surface due to the fact that DBCO cannot penetrate the cell membrane. A following biorthogonal enrichment and glycosylation site analysis was achieved by avidin beads through a biotin–avidin binding on beads digestion, ^{18}O labeling, and LC–MS/MS analysis, as shown in Figure 13a. In addition, the combination of SILAC and azido-based glycoprotein enrichment (Figure 13b) allowed an efficient and accurate comparison between two breast cancer cell lines (Figure 13c). However, since click chemistry forms covalent bonds, it is not reversible to release intact glycopeptides. The solution was to apply enzymes to cut off glycans (e.g., sialidase); however, it would induce a loss of glycan structural information. Another limitation of metabolic labeling is that it can only be performed on living systems, which limits its application for clinical samples such as patient tissue and biofluids.

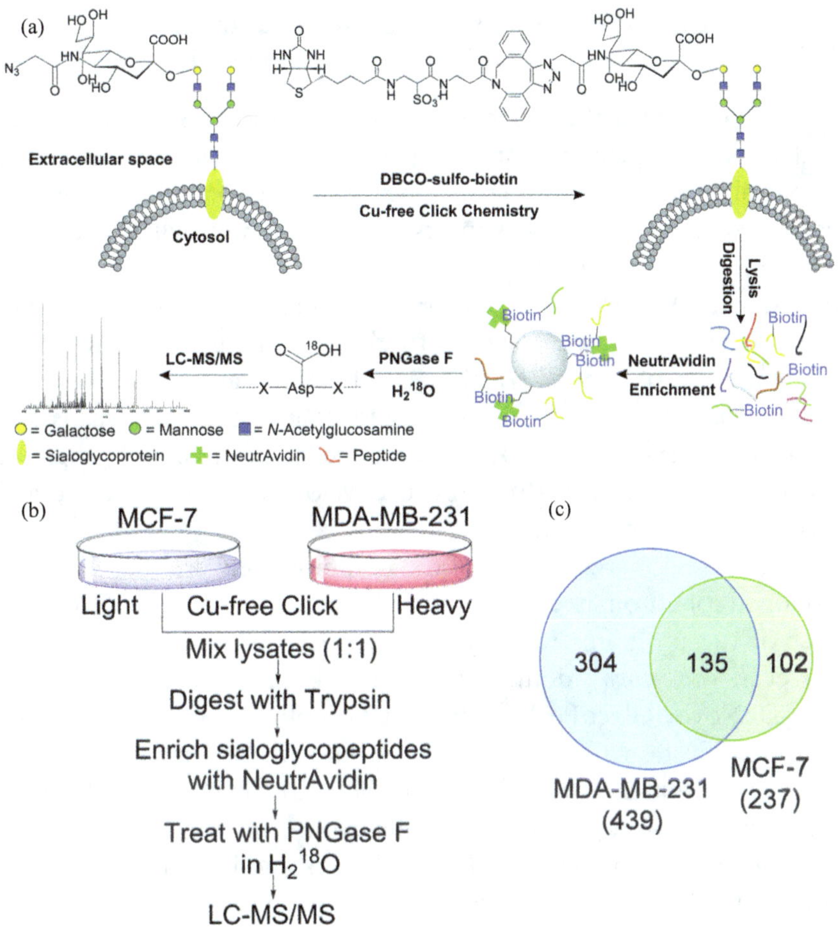

Figure 13. (a) The principle of the site-specific identification of the cell surface *N*-sialoglycoproteome by integrating metabolic labeling, copper-free click chemistry, and MS-based proteomics techniques. (b) Identification of the cell surface *N*-sialoglycoproteome site specifically between invasive MDA-MB-231 and non-invasive MCF-7 breast cancer cells. (c) Comparison of the *N*-sialoglycosylation sites identified on the cell surface of MDA-MB-231 and MCF-7 cells. Reproduced with permission from Ref. 278. Glycan symbols: fucose (▼); *N*-acetyl glucosamine (■); mannose (●); galactose (●), and sialic acid (◆).

13. Bioinformatics tools for the interpretation and quantification of glycans and glycopeptides

Because of the microheterogeneity of glycosylation, usually hundreds of glycans and thousands of glycopeptides can be identified from a biological system when performing glycomic or glycoproteomic analysis. Manual data processing is time-consuming when analyzing such large cohorts. This problem becomes much more serious when dealing with glycoproteomic data since the identification of glycopeptides needs to accurately assign both glycan structures and the amino acid sequences of peptide backbones. Therefore, automated software that can facilitate the identification and quantitation of glycans and glycopeptides is necessary. During past decades, a variety of programs with different algorithms have been developed to aid glycomic and glycoproteomic characterization.

Table 1 summarizes some of the common software that has been developed for automated glycomic and glycoproteomic identification. Most of the programs identify glycans based on a database search. However, *de novo* identification of glycans was achieved by GlycoDeNovo, GlycoPeakfinder, and Glycoforest, where the structural identification does not rely on a known database. Noticeably, MultiGlycan and MSfragger can achieve automated quantitation for glycomics and glycoproteomics, respectively. Most of this software is academic and open-source, but they have some limitations: namely, they usually focus on a specific study, are not user-friendly, and very few are regularly updated. In contrast, although commercialized software (e.g., SimGlycan, GlycoQuest, and Byonic) can be expensive, it usually provides a user-friendly interface, more functions, better compatibility, and is consistently updated to fix bugs and improve algorithms. Aside from these types of automated data processing software, there are several assistant software programs that can be helpful to facilitate manual checks and function as complementary tools to the aforementioned software. These tools include GlycoWorkbench [279–281] and Skyline,[282–284] among others.

The development of glycomic and glycoproteomic software has facilitated in-depth data analysis and interpretation for large biological

Table 1. Common software for glycomic and glycoproteomic data processing.

Name	Function	Algorithm	Type	Reference
GlycoMod	Identification	Database search	Glycomics	285
GlycoReSoft	Identification	Database search	Glycomics	286
Glyquest	Identification	Database search	Glycomics	287
SysBioWare	Identification	Database search	Glycomics	288
SimGlycan	Identification	Database search	Glycomics and glycoproteomics	289
MultiGlycan	Identification and quantitation	Database search	Glycomics	290
GlycanGUI	Identification and quantitation	Database search	Glycomics	291
GlycoDeNovo	Identification	*De novo*	Glycomics	292
GlycoPeakfinder	Identification	*De novo*	Glycomics	293
Glycoforest	Identification	*De novo*	Glycomics	294
GlycoPep ID	Identification	Database search	Glycoproteomics	295
GlycoMiner	Identification	Database search	Glycoproteomics	296
GlycoPeptideSearch	Identification	Database search	Glycoproteomics	297
GlypID 2	Identification	Database search	Glycoproteomics	298
GlycoPep Grader	Identification	Database search	Glycoproteomics	299

(Continued)

Table 1. (*Continued*)

Name	Function	Algorithm	Type	Reference
GPQuest	Identification	Database search	Glycoproteomics	300
GlycoSeq	Identification	Database search	Glycoproteomics	301
MSFragger	Identification and quantitation	Database search	Glycoproteomics	302
pGlyco	Identification	Database search	Glycoproteomics	303
Byonic	Identification	Database search	Glycoproteomics	304
GlycoProteome Analyzer	Identification	Database search	Glycoproteomics	305
SweetNET	Identification	Database search	Glycoproteomics	306
Glyco-DIA	Identification	Database search	Glycoproteomics	307
GlycoMaster DB	Identification	Database search	Glycoproteomics	308
GlycoPep Detector	Identification	Database search	Glycoproteomics	309

cohorts. However, although false discovery rate has been applied to most of these tools, the false positive rate cannot be neglected; it demands subsequent manual checks to ensure identification accuracy. In addition, automated isomeric assignment and quantitation is still a huge challenge. However, the fact that some software have the ability to assign different retention times to the same structural composition indicates the potential for identification and quantitation of glycan and glycopeptide isomers in future studies.

14. Conclusion and future directions

We have discussed the advantages of different MS-based sample preparation and analysis approaches. Together, these advanced enrichment, derivatization, labeling, separation, and fragmentation methods, along with corresponding data processing software, provide the ability to decipher glycome and glycoproteome alteration in different biological processes. Among all techniques, LC–MS is still the most common method for glycomic and glycoproteomic analysis. HILIC–LC–MS is the most widely used technique for native and reducing end-labeled glycan profiling, while C18–LC–MS is standard for permethylated glycan and glycopeptide profiling. However, for glycan and glycopeptide isomers, PGC and MGC provide extra efficiency in isomeric analysis of both glycans and glycopeptides. CE–MS suffers from a relatively low ID number due to its low flow rate and ultrahigh resolution, while IM–MS is limited to a low resolution. Nevertheless, CE and IM can be promising complement to LC–MS as they provide different mechanisms for additional dimensions of analysis.

Despite the abundance of methods used in glycomics and glycoproteomics, isomeric characterization from complex biological samples remains a great challenge. In future studies, novel derivatization and separation methods need to be developed to enhance isomeric separation, especially for glycopeptide isomers. New dissociation methods and the combination of different fragmentation methods should also be investigated to increase glycan cross-ring fragmentation, providing more isomeric information in tandem MS. In addition, more efficient enrichment methods are needed to enhance glycopeptide analysis. Finally, these novel methods should be utilized to analyze more biological and clinical samples from larger cohorts in the future to better understand the mechanisms and attributes of glycosylation in diseases.

Acknowledgment

The work is also supported by grants from the Robert A. Welch Foundation (No. D-0005), and The CH Foundation. This study is supported by grants from the National Institutes of Health (grant numbers R01GM112490 and R01GM130091).

Conflict of Interest

The authors declare that there is no conflict of interest.

References

1. Hart GW, Copeland RJ. Glycomics Hits the Big Time. *Cell.* 2010; 143(5):672–676.
2. Apweiler R, Hermjakob H, Sharon N. On the Frequency of Protein Glycosylation, as Deduced from Analysis of the SWISS-PROT Database. Dedicated to Prof. Akira Kobata and Prof. Harry Schachter on the Occasion of Their 65th Birthdays. *Biochim. Biophys. Acta. Gen. Subj.* 1999;1473(1):4–8.
3. Tzeng SF, Tsai CH, Chao TK, *et al.* O-Glycosylation-Mediated Signaling Circuit Drives Metastatic Castration-Resistant Prostate Cancer. *FASEB J.* 2018;32(12):6965–6759.
4. de Vreede G, Morrison HA, Houser AM, *et al.* A Drosophila Tumor Suppressor Gene Prevents Tonic TNF Signaling Through Receptor N-Glycosylation. *Dev. Cell.* 2018;45(5):595–605 e4.
5. Oyama M, Kariya Y, Kariya Y, *et al.* Biological Role of Site-Specific O-Glycosylation in Cell Adhesion Activity and Phosphorylation of Osteopontin. *Biochem. J.* 2018;475(9):1583–1595.
6. Singh C, Shyanti RK, Singh V, *et al.* Integrin Expression and Glycosylation Patterns Regulate Cell-Matrix Adhesion and Alter

with Breast Cancer Progression. *Biochem. Biophys. Res. Commun.* 2018;499(2):374–380.
7. Sperandio M, Gleissner CA, Ley K. Glycosylation in Immune Cell Trafficking. *Immunol. Rev.* 2009;230(1):97–113.
8. Veillon L, Huang Y, Peng W, *et al.* Characterization of Isomeric Glycan Structures by LC-MS/MS. *Electrophoresis.* 2017;38(17):2100–2114.
9. Sola RJ, Griebenow K. Effects of Glycosylation on the Stability of Protein Pharmaceuticals. *J. Pharm. Sci.* 2009;98(4):1223–1245.
10. Kizuka Y, Kitazume S, Taniguchi N. N-Glycan and Alzheimer's Disease. *Biochim. Biophys. Acta.* 2017;1861(10):2447–2454.
11. Schedin-Weiss S, Winblad B, Tjernberg LO. The Role of Protein Glycosylation in Alzheimer Disease. *FEBS J.* 2014;281(1):46–62.
12. Russell AC, Simurina M, Garcia MT, *et al.* The N-Glycosylation of Immunoglobulin G as a Novel Biomarker of Parkinson's Disease. *Glycobiology.* 2017;27(5):501–510.
13. Everest-Dass AV, Moh ESX, Ashwood C, *et al.* Human Disease Glycomics: Technology Advances Enabling Protein Glycosylation Analysis — Part 2. *Expert Rev. Proteomics.* 2018;15(4):341–352.
14. Bermingham ML, Colombo M, McGurnaghan SJ, *et al.* N-Glycan Profile and Kidney Disease in Type 1 Diabetes. *Diabetes Care.* 2018;41(1):79–87.
15. Chandrasekaran R, Lacy DB. The Role of Toxins in Clostridium Difficile Infection. *FEMS Microbiol. Rev.* 2017;41(6):723–750.
16. Orlova OV, Drutsa VL, Spirin PV, *et al.* The Role of HCV e2 Protein Glycosylation in Functioning of Virus Envelope Proteins in Insect and Mammalian Cells. *Acta. Naturae.* 2015;7(1):87–97.
17. Bagdonaite I, Wandall HH. Global Aspects of Viral Glycosylation. *Glycobiology.* 2018;28(7):443–467.
18. Mechref Y, Hu Y, Garcia A, *et al.* Identifying Cancer Biomarkers by Mass Spectrometry-Based Glycomics. *Electrophoresis.* 2012;33(12): 1755–1767.
19. Kailemia MJ, Park D, Lebrilla CB. Glycans and Glycoproteins as Specific Biomarkers for Cancer. *Anal. Bioanal. Chem.* 2017; 409(2):395–410.

20. Guo H, Abbott KL. Functional Impact of Tumor-Specific N-Linked Glycan Changes in Breast and Ovarian Cancers. *Adv. Cancer Res.* 2015;126:281–303.
21. Pan S, Brentnall TA, Chen R. Glycoproteins and Glycoproteomics in Pancreatic Cancer. *World J. Gastroenterol.* 2016;22(42):9288–9299.
22. Mehta A, Herrera H, Block T. Glycosylation and Liver Cancer. *Adv. Cancer Res.* 2015;126 257–279.
23. Wooding KM, Peng W, Mechref Y. Characterization of Pharmaceutical IgG and Biosimilars Using Miniaturized Platforms and LC-MS-MS. *Curr. Pharm. Biotechnol.* 2016;17(9):788–801.
24. Peng W, Goli M, Mirzaei P, *et al.* Revealing the Biological Attributes of N-Glycan Isomers in Breast Cancer Brain Metastasis Using Porous Graphitic Carbon (PGC) Liquid Chromatography-Tandem Mass Spectrometry (LC-MS/MS). *J. Proteome Res.* 2019;18(10):3731–3740.
25. Banazadeh A, Nieman R, Goli M, *et al.* Characterization of Glycan Isomers Using Magnetic Carbon Nanoparticles as a MALDI Co-Matrix. *RSC Adv.* 2019;9(35):20137–20148.
26. Pett C, Nasir W, Sihlbom C, *et al.* Effective Assignment of Alpha2,3/alpha2,6-Sialic Acid Isomers by LC-MS/MS-Based Glycoproteomics. *Angew. Chem. Int. Ed. Engl.* 2018;57(30):9320–9324.
27. Toegel S, Pabst M, Wu SQ, *et al.* Phenotype-Related Differential Alpha-2,6- or Alpha-2,3-sialylation of Glycoprotein N-Glycans in Human Chondrocytes. *Osteoarthr. Cartil.* 2010;18(2):240–248.
28. *Essentials of Glycobiology*, 3rd ed; Cold Spring Harbor, NY: Cold Spring Harbor Laboratory Press, 2015–2017.
29. Schnaar RL. Glycobiology Simplified: Diverse Roles of Glycan Recognition in Inflammation. *J. Leukoc. Biol.* 2016;99(6):825–838.
30. Dong X, Huang Y, Cho BG, *et al.* Advances in Mass Spectrometry-Based Glycomics. *Electrophoresis.* 2018;39(24):3063–3081.
31. Dell A, Morris HR. Glycoprotein Structure Determination by Mass Spectrometry. *Science.* 2001;291(5512):2351–2356.
32. Mechref Y, Novotny MV. Structural Investigations of Glycoconjugates at High Sensitivity. *Chem. Rev.* 2002;102(2):321–369.

33. Novotny MV, Mechref Y. New Hyphenated Methodologies in High-Sensitivity Glycoprotein Analysis. *J. Sep. Sci.* 2005;28(15):1956–1968.
34. Banazadeh A, Peng W, Veillon L, *et al.* Carbon Nanoparticles and Graphene Nanosheets as MALDI Matrices in Glycomics: A New Approach to Improve Glycan Profiling in Biological Samples. *J. Am. Soc. Mass Spectrom.* 2018;29(9):1892–1900.
35. Cho BG, Gutierrez Reyes CD, Mechref Y. N-Glycomics of Cerebrospinal Fluid: Method Comparison. *Molecules.* 2021;26(6):1712. doi: 10.3390/molecules26061712.
36. Wang T, Cai ZP, Gu XQ, *et al.* Discovery and Characterization of a Novel Extremely Acidic Bacterial N-Glycanase with Combined Advantages of PNGase F and A. *Biosci. Rep.* 2014;34(6):e00149.
37. Guo RR, Comamala G, Yang HH, *et al.* Discovery of Highly Active Recombinant PNGase H(+) Variants Through the Rational Exploration of Unstudied Acidobacterial Genomes. *Front. Bioeng. Biotechnol.* 2020;8:741.
38. Wang C, Qiang S, Jin W, *et al.* Reductive Chemical Release of N-Glycans as 1-Amino-Alditols and Subsequent 9-Fluorenylmethyloxycarbonyl Labeling for MS and LC/MS Analysis. *J. Proteom.* 2018;187:47–58.
39. Wilkinson H, Saldova R. Current Methods for the Characterization of O-Glycans. *J. Proteome Res.* 2020;19(10):3890–3905.
40. Zhang S, Go EP, Ding H, *et al.* Analysis of Glycosylation and Disulfide Bonding of Wild-Type SARS-CoV-2 Spike Glycoprotein. *BioRxiv.* 2021;96(3):e0162621.
41. Schiel JE, Smith NJ, Phinney KW. Universal Proteolysis and MS(n) for N- and O-Glycan Branching Analysis. *J. Mass Spectrom.* 2013;48(4): 533–538.
42. Jensen PH, Karlsson NG, Kolarich D, *et al.* Structural Analysis of N- and O-Glycans Released from Glycoproteins. *Nat. Protoc.* 2012;7(7): 1299–1310.
43. Yuan J, Wang C, Sun Y, *et al.* Nonreductive Chemical Release of Intact N-Glycans for Subsequent Labeling and Analysis by Mass Spectrometry. *Anal. Biochem.* 2014;462:1–9.

44. Kozak RP, Royle L, Gardner RA, *et al.* Improved Nonreductive O-Glycan Release by Hydrazinolysis with Ethylenediaminetetraacetic Acid Addition. *Anal. Biochem.* 2014;453:29–37.
45. Lauber MA, Yu YQ, Brousmiche DW, *et al.* Rapid Preparation of Released N-Glycans for HILIC Analysis Using a Labeling Reagent that Facilitates Sensitive Fluorescence and ESI-MS Detection. *Anal. Chem.* 2015;87(10):5401–5409.
46. Maley F, Trimble RB, Tarentino AL, *et al.* Characterization of Glycoproteins and Their Associated Oligosaccharides Through the Use of Endoglycosidases. *Anal. Biochem.* 1989;180(2):195–204.
47. Zhu R, Zhou S, Peng W, *et al.* Enhanced Quantitative LC-MS/MS Analysis of N-Linked Glycans Derived from Glycoproteins Using Sodium Deoxycholate Detergent. *J. Proteome Res.* 2018;17(8):2668–2678.
48. Zhang L, Wang C, Wu Y, *et al.* Microwave Irradiation-Assisted High-Efficiency N-Glycan Release Using Oriented Immobilization of PNGase F on Magnetic Particles. *J. Chromatogr. A.* 2020;1619:460934. doi: 10.1016/j.chroma.2020.460934.
49. Nordgren M, Nägeli A, Nyhlén H, *et al.* Mapping O-Glycosylation Sites Using OpeRATOR and LC-MS. *Methods Mol. Biol.* 2021;2271:155–167.
50. Huang Y, Konse T, Mechref Y, *et al.* Matrix-Assisted Laser Desorption/Ionization Mass Spectrometry Compatible Beta-Elimination of O-Linked Oligosaccharides. *Rapid Commun. Mass Spectrom.* 2002;16(12):1199–1204.
51. Fukuda M. Beta-Elimination for Release of O-GalNAc-Linked Oligosaccharides from Glycoproteins and Glycopeptides. *Curr. Protoc. Mol. Biol.* 2001, Chapter 17, Unit17.15B.
52. Turyan I, Hronowski X, Sosic Z, *et al.* Comparison of Two Approaches for Quantitative O-Linked Glycan Analysis Used in Characterization of Recombinant Proteins. *Anal. Biochem.* 2014;446:28–36.
53. Song X, Ju H, Lasanajak Y, *et al.* Oxidative Release of Natural Glycans for Functional Glycomics. *Nat. Methods.* 2016;13(6):528–534.
54. Fischler DA, Orlando R. N-Linked Glycan Release Efficiency: A Quantitative Comparison Between NaOCl and PNGase F Release Protocols. *J. Biomol. Tech.* 2019;30(4):58–63.

55. Zhang X, Reed CE, Birdsall RE, *et al.* High-Throughput Analysis of Fluorescently Labeled N-Glycans Derived from Biotherapeutics Using an Automated LC-MS-Based Solution. *SLAS Technol.* 2020;25(4): 380–387.
56. Kotsias M, Kozak RP, Gardner RA, *et al.* Improved and Semi-Automated Reductive β-Elimination Workflow for Higher Throughput Protein O-Glycosylation Analysis. *PLoS One.* 2019;14(1):e0210759.
57. Adamczyk B, Stöckmann H, O'Flaherty R, *et al.* High-Throughput Analysis of the Plasma N-Glycome by UHPLC. *Methods Mol. Biol.* 2017;1503:97–108.
58. Selman MH, Hemayatkar M, Deelder AM, *et al.* Cotton HILIC SPE Microtips for Microscale Purification and Enrichment of Glycans and Glycopeptides. *Anal. Chem.* 2011;83(7):2492–2499.
59. Kim Y-G, Jeong H-J, Jang K-S, *et al.* Rapid and High-Throughput Analysis of N-Glycans from Ovarian Cancer Serum Using a 96-Well Plate Platform. *Anal. Biochem.* 2009;391(2):151–153.
60. Hirabayashi J. Lectin-Based Structural Glycomics: Glycoproteomics and Glycan Profiling. *Glycoconj. J.* 2004;21(1–2):35–40.
61. Sun N, Yao J, Deng C. Designed Synthesis of Carbon-Functional Magnetic Graphene Mesoporous Silica Materials Using Polydopamine as Carbon Precursor for the Selective Enrichment of N-Linked Glycan. *Talanta.* 2016;148:439–443.
62. Qin H, Hu Z, Wang F, *et al.* Facile Preparation of Ordered Mesoporous Silica-Carbon Composite Nanoparticles for Glycan Enrichment. *Chem. Commun. (Camb).* 2013;49(45):5162–5164.
63. Sun NR, Yao JZ, Wang JW, *et al.* Magnetic Nanoporous Hybrid Carbon from Core–Shell Metal–Organic Frameworks for Glycan Extraction. *RSC Adv.* 2016;6(41):34434–34438.
64. Kayili HM, Ertürk AS, Elmacı G, *et al.* Poly(amidoamine) Dendrimer-Coated Magnetic Nanoparticles for the Fast Purification and Selective Enrichment of Glycopeptides and Glycans. *J. Sep. Sci.* 2019:3209–3216.
65. Kayili HM, Barlas N, Atakay M, *et al.* Fast Purification of Glycans and Glycopeptides Using Silk-Packed Micropipette Tip for Matrix-Assisted Laser Desorption/Ionization-Mass Spectrometry and High-Performance Liquid Chromatography-Fluorescence Detection Analysis. *Microchem. J.* 2018;139:492–499.

66. Hacı Mehmet K, İzzet A, Bekir S. A New Titania Glyco-Purification Tip for the Fast Enrichment and Efficient Analysis of Glycopeptides and Glycans by MALDI-TOF-MS. *J. Pharm. Biomed. Anal.* 2019;174: 191–197.
67. Sha Q, Wu Y, Wang C, *et al.* Cellulose Microspheres-Filled Pipet Tips for Purification and Enrichment of Glycans and Glycopeptides. *J. Chromatogr. A.* 2018;1569:8–16.
68. Dosekova E, Filip J, Bertok T, *et al.* Nanotechnology in Glycomics: Applications in Diagnostics, Therapy, Imaging, and Separation Processes. *Med. Res. Rev.* 2017;37(3):514–626.
69. Hennion MC. Solid-Phase Extraction: Method Development, Sorbents, and Coupling with Liquid Chromatography. *J. Chromatogr. A.* 1999;856(1–2):3–54.
70. Cho BG, Veillon L, Mechref Y. N-Glycan Profile of Cerebrospinal Fluids from Alzheimer's Disease Patients Using Liquid Chromatography with Mass Spectrometry. *J. Proteome Res.* 2019;18(10):3770–3779.
71. Abdul Rahman S, Bergström E, Watson CJ, *et al.* Filter-Aided N-Glycan Separation (FANGS): A Convenient Sample Preparation Method for Mass Spectrometric N-Glycan Profiling. *J. Proteome Res.* 2014;13(3):1167–1176.
72. Rehulka P, Zahradnikova M, Rehulkova H, *et al.* Microgradient Separation Technique for Purification and Fractionation of Permethylated N-Glycans Before Mass Spectrometric Analyses. *J. Sep. Sci.* 2018;41(9):1973–1982.
73. Wang Y, Zhao H, Tao J, *et al.* A New Method for Purifying N-Glycans Released from Milk Glycoprotein. *J. Proteomics.* 2021;245:104283. doi: 10.1016/j.jprot.2021.104283.
74. Naven T, Harvey D. Cationic Derivatization of Oligosaccharides with Girard's T Reagent for Improved Performance in Matrix-Assisted Laser Desorption/Ionization and Electrospray Mass Spectrometry. *Rapid Commun. Mass Spectrom.* 1996;10(7):829–834.
75. Lattová E, Perreault H. The Usefulness of Hydrazine Derivatives for Mass Spectrometric Analysis of Carbohydrates. *Mass Spectrom. Rev.* 2013;32(5):366–385.
76. Wang C, Wu Z, Yuan J, *et al.* Simplified Quantitative Glycomics Using the Stable Isotope Label Girard's Reagent p by Electrospray Ionization Mass Spectrometry. *J. Proteome Res.* 2014;13(2):372–384.

77. Bereman MS, Comins DL, Muddiman DC. Increasing the Hydrophobicity and Electrospray Response of Glycans Through Derivatization with Novel Cationic Hydrazides. *Chem. Commun. (Camb).* 2010;46(2): 237–239.
78. Walker SH, Budhathoki-Uprety J, Novak BM, *et al.* Stable-Isotope Labeled Hydrophobic Hydrazide Reagents for the Relative Quantification of N-Linked Glycans by Electrospray Ionization Mass Spectrometry. *Anal. Chem.* 2011;83(17):6738–6745.
79. Zhang Y, Wang B, Jin W, *et al.* Sensitive and Robust MALDI-TOF-MS Glycomics Analysis Enabled by Girard's Reagent T On-Target Derivatization (GTOD) of Reducing Glycans. *Anal. Chim. Acta.* 2019;1048:105–114.
80. Gao X, Lu Y, Wei M, *et al.* Matrix-Assisted Laser Desorption/ Ionization Time-of-Flight Mass Spectrometry Analysis of Human Milk Neutral and Sialylated Free Oligosaccharides Using Girard's Reagent P On-Target Derivatization. *J. Agric. Food Chem.* 2019; 67(32):8958–8966.
81. Zhang H, Shi X, Vu NQ, *et al.* On-Tissue Derivatization with Girard's Reagent P Enhances N-Glycan Signals for Formalin-Fixed Paraffin-Embedded Tissue Sections in MALDI Mass Spectrometry Imaging. *Anal. Chem.* 2020;92(19):13361–13368.
82. Zhang Y, Hu Z, Zhang C, *et al.* A Robust Glycan Labeling Strategy Using a New Cationic Hydrazide Tag for MALDI-MS-Based Rapid and Sensitive Glycomics Analysis. *Talanta.* 2020;219:121356. doi: 10.1016/j.talanta.2020.121356.
83. Gong B, Hoyt E, Lynaugh H, *et al.* N-Glycosylamine-Mediated Isotope Labeling for Mass Spectrometry-Based Quantitative Analysis of N-Linked Glycans. *Anal. Bioanal. Chem.* 2013;405(17):5825–5831.
84. Liu X, Zhang G, Chan K, *et al.* Microwave-Assisted Kochetkov Amination Followed by Permanent Charge Derivatization: A Facile Strategy for Glycomics. *Chem. Commun. (Camb).* 2010;46(39):7424–7426.
85. Cohen SA, Michaud DP. Synthesis of a Fluorescent Derivatizing Reagent, 6-Aminoquinolyl-N-hydroxysuccinimidyl Carbamate, and Its Application for the Analysis of Hydrolysate Amino Acids via High-Performance Liquid Chromatography. *Anal. Biochem.* 1993;211(2): 279–287.

86. Stöckmann H, Duke RM, Millán Martín S, *et al.* Ultrahigh Throughput, Ultrafiltration-Based n-Glycomics Platform for Ultraperformance Liquid Chromatography (ULTRA(3)). *Anal. Chem.* 2015;87(16): 8316–8322.
87. Wu Y, Sha Q, Du J, *et al.* Determination of N-Glycans by High Performance Liquid Chromatography Using 6-Aminoquinolyl-N-hydroxysuccinimidyl Carbamate as the Glycosylamine Labeling Reagent. *J. Chromatogr. A.* 2018;1535:114–122.
88. Segu Z, Stone T, Berdugo C, *et al.* A Rapid Method for Relative Quantification of N-Glycans from a Therapeutic Monoclonal Antibody During Trastuzumab Biosimilar Development. *MAbs.* 2020;12(1): 1750794. doi: 10.1080/19420862.2020.1750794.
89. Dong X, Peng W, Yu CY, *et al.* 8-plex LC-MS/MS Analysis of Permethylated N-Glycans Achieved by Using Stable Isotopic Iodomethane. *Anal. Chem.* 2019;91(18):11794–11802.
90. Lageveen-Kammeijer G. SM, de Haan N, Mohaupt P, *et al.* Highly Sensitive CE-ESI-MS Analysis of N-Glycans from Complex Biological Samples. *Nat. Commun.* 2019;10(1):2137. doi: 10.1038/s41467-019-09910-7.
91. Cheng M, Shu H, Peng Y, *et al.* Specific Analysis of α-2,3-Sialylated N-Glycan Linkage Isomers by Microchip Capillary Electrophoresis-Mass Spectrometry. *Anal. Chem.* 2021;93(13):5537–5546.
92. Hanamatsu H, Nishikaze T, Miura N, *et al.* Sialic Acid Linkage Specific Derivatization of Glycosphingolipid Glycans by Ring-Opening Aminolysis of Lactones. *Anal. Chem.* 2018;90(22):13193–13199.
93. Pongracz T, Wuhrer M, de Haan N. Expanding the Reaction Space of Linkage-Specific Sialic Acid Derivatization. *Molecules.* 2019;24(19): 3617. doi: 10.3390/molecules24193617.
94. Yang S, Jankowska E, Kosikova M, *et al.* Solid-Phase Chemical Modification for Sialic Acid Linkage Analysis: Application to Glycoproteins of Host Cells Used in Influenza Virus Propagation. *Anal. Chem.* 2017;89(17):9508–9517.
95. Peng Y, Wang L, Zhang Y, *et al.* Stable Isotope Sequential Derivatization for Linkage-Specific Analysis of Sialylated N-Glycan Isomers by MS. *Anal. Chem.* 2019;91(24):15993–16001.

96. Alpert AJ. Hydrophilic-Interaction Chromatography for the Separation of Peptides, Nucleic Acids and Other Polar Compounds. *J. Chromatogr.* 1990;499:177–196.
97. Takegawa Y, Deguchi K, Keira T, *et al.* Separation of Isomeric 2-Aminopyridine Derivatized N-Glycans and N-Glycopeptides of Human Serum Immunoglobulin G by Using a Zwitterionic Type of Hydrophilic-Interaction Chromatography. *J. Chromatogr. A.* 2006; 1113(1–2):177–181.
98. Ahn J, Bones J, Yu YQ, *et al.* Separation of 2-Aminobenzamide Labeled Glycans Using Hydrophilic Interaction Chromatography Columns Packed with 1.7 Microm Sorbent. *J. Chromatogr. B Analyt. Technol. Biomed. Life Sci.* 2010;878(3–4):403–408.
99. Yang Y, Lu Y, Liu Y, *et al.* Comparative Analysis of Yak Milk and Bovine Milk Glycoprotein N/O-Glycome by Online HILIC-UV-ESI-MS/MS. *Carbohydr. Polym.* 2022;278 118918.
100. O'Flaherty R, Simon Á, Alonso-Sampedro M, *et al.* Changes in Serum N-Glycome for Risk Drinkers: A Comparison with Standard Markers for Alcohol Abuse in Men and Women. *Biomolecules.* 2022;12(2):241.
101. Hou H, Yang H, Liu P, *et al.* Profile of Immunoglobulin G N-Glycome in COVID-19 Patients: A Case-Control Study. *Front. Immunol.* 2021;12:748566.
102. Zauner G, Deelder AM, Wuhrer M. Recent Advances in Hydrophilic Interaction Liquid Chromatography (HILIC) for Structural Glycomics. *Electrophoresis.* 2011;32(24):3456–3466.
103. Tao S, Huang Y, Boyes BE, *et al.* Liquid Chromatography-Selected Reaction Monitoring (LC-SRM) Approach for the Separation and Quantitation of Sialylated N-Glycans Linkage Isomers. *Anal. Chem.* 2014;86(21):10584–10590.
104. Jiang K, Wang C, Sun Y, *et al.* Comparison of Chicken and Pheasant Ovotransferrin N-Glycoforms via Electrospray Ionization Mass Spectrometry and Liquid Chromatography Coupled with Mass Spectrometry. *J. Agric. Food Chem.* 2014;62(29):7245–7254.
105. Abd Hamid UM, Royle L, Saldova R, *et al.* A Strategy to Reveal Potential Glycan Markers from Serum Glycoproteins Associated with Breast Cancer Progression. *Glycobiology.* 2008;18(12):1105–1118.

106. Terkelsen T, Haakensen VD, Saldova R, *et al.* N-Glycan Signatures Identified in Tumor Interstitial Fluid and Serum of Breast Cancer Patients: Association with Tumor Biology and Clinical Outcome. *Mol. Oncol.* 2018;12(6):972–990.
107. Doherty M, Theodoratou E, Walsh I, *et al.* Plasma N-Glycans in Colorectal Cancer Risk. *Sci. Rep.* 2018;8(1):8655.
108. Arnold JN, Saldova R, Galligan MC, *et al.* Novel Glycan Biomarkers for the Detection of Lung Cancer. *J. Proteome Res.* 2011;10(4): 1755–1764.
109. Raghunathan R, Polinski NK, Klein JA, *et al.* Glycomic and Proteomic Changes in Aging Brain Nigrostriatal Pathway. *Mol. Cell Proteomics.* 2018;17(9):1778–1787.
110. Molnarova K, Kozlík P. Comparison of Different HILIC Stationary Phases in the Separation of Hemopexin and Immunoglobulin G Glycopeptides and Their Isomers. *Molecules.* 2020;25(20):4655.
111. Moravcová D, Čmelík R, Křenková J. Separation of Labeled Isomeric Oligosaccharides by Hydrophilic Interaction Liquid Chromatography — the Role of Organic Solvent in Manipulating Separation Selectivity of the Amide Stationary Phase. *J. Chromatogr. A.* 2021;1651:462303. doi: 10.1016/j.chroma.2021.462303.
112. Zhang W, Wang T, Chen X, *et al.* Absolute Quantification of Twelve Oligosaccharides in Human Milk Using a Targeted Mass Spectrometry-Based Approach. *Carbohydr. Polym.* 2019;219:328–333.
113. Zhao J, Li S, Li C, *et al.* Identification of Low Abundant Isomeric N-Glycan Structures in Biological Therapeutics by LC/MS. *Anal. Chem.* 2016;88(14):7049–7059.
114. Mancera-Arteu M, Giménez E, Barbosa J, *et al.* Zwitterionic-Hydrophilic Interaction Capillary Liquid Chromatography Coupled to Tandem Mass Spectrometry for the Characterization of Human Alpha-Acid-Glycoprotein N-Glycan Isomers. *Anal. Chim. Acta.* 2017;991: 76–88.
115. Messina A, Palmigiano A, Esposito F, *et al.* HILIC-UPLC-MS for High Throughput and Isomeric N-Glycan Separation and Characterization in Congenital Disorders Glycosylation and Human Diseases. *Glycoconj. J.* 2021;38(2):201–211.

116. Wilkinson H, Thomsson KA, Rebelo AL, *et al.* The O-Glycome of Human Nigrostriatal Tissue and Its Alteration in Parkinson's Disease. *J. Proteome Res.* 2021;20(8):3913–3924.
117. Smith J, Millán-Martín S, Mittermayr S, *et al.* 2-Dimensional Ultra-High Performance Liquid Chromatography and DMT-MM Derivatization Paired with Tandem Mass Spectrometry for Comprehensive Serum N-Glycome Characterization. *Anal. Chim. Acta.* 2021;1179:338840.
118. West C, Elfakir C, Lafosse M. Porous Graphitic Carbon: A Versatile Stationary Phase for Liquid Chromatography. *J. Chromatogr. A.* 2010;1217(19):3201–3216.
119. Kawasaki N, Ohta M, Hyuga S, *et al.* Analysis of Carbohydrate Heterogeneity in a Glycoprotein Using Liquid Chromatography/Mass Spectrometry and Liquid Chromatography with Tandem Mass Spectrometry. *Anal. Biochem.* 1999;269(2):297–303.
120. Kawasaki N, Ohta M, Hyuga S, *et al.* Application of Liquid Chromatography/Mass Spectrometry and Liquid Chromatography with Tandem Mass Spectrometry to the Analysis of the Site-Specific Carbohydrate Heterogeneity in Erythropoietin. *Anal. Biochem.* 2000;285(1):82–91.
121. Karlsson NG, Wilson NL, Wirth HJ, *et al.* Negative Ion Graphitised Carbon Nano-Liquid Chromatography/Mass Spectrometry Increases Sensitivity for Glycoprotein Oligosaccharide Analysis. *Rapid Commun. Mass Spectrom.* 2004;18(19):2282–2292.
122. Niñonuevo M, An H, Yin H, *et al.* Nanoliquid Chromatography-Mass Spectrometry of Oligosaccharides Employing Graphitized Carbon Chromatography on Microchip with a High-Accuracy Mass Analyzer. *Electrophoresis.* 2005;26(19):3641–3649.
123. Hua S, Williams CC, Dimapasoc LM, *et al.* Isomer-Specific Chromatographic Profiling Yields Highly Sensitive and Specific Potential N-Glycan Biomarkers for Epithelial Ovarian Cancer. *J. Chromatogr. A.* 2013;1279:58–67.
124. Park D, Arabyan N, Williams CC, *et al.* Salmonella Typhimurium Enzymatically Landscapes the Host Intestinal Epithelial Cell (IEC) Surface Glycome to Increase Invasion. *Mol. Cell Proteomics.* 2016; 15(12):3653–3664.

125. Xu G, Goonatilleke E, Wongkham S, *et al.* Deep Structural Analysis and Quantitation of O-Linked Glycans on Cell Membrane Reveal High Abundances and Distinct Glycomic Profiles Associated with Cell Type and Stages of Differentiation. *Anal. Chem.* 2020;92(5):3758–3768.

126. Zhang T, Madunić K, Holst S, *et al.* Development of a 96-Well Plate Sample Preparation Method for Integrated N- and O-Glycomics Using Porous Graphitized Carbon Liquid Chromatography-Mass Spectrometry. *Mol. Omics.* 2020;16(4):355–363.

127. Madunić K, Zhang T, Mayboroda OA, *et al.* Colorectal Cancer Cell Lines Show Striking Diversity of Their O-Glycome Reflecting the Cellular Differentiation Phenotype. *Cell Mol. Life Sci.* 2021; 78(1):337–350.

128. She YM, Tam RY, Li X, *et al.* Resolving Isomeric Structures of Native Glycans by Nanoflow Porous Graphitized Carbon Chromatography-Mass Spectrometry. *Anal. Chem.* 2020;92(20):14038–14046.

129. Costello CE, Contado-Miller JM, Cipollo JF. A Glycomics Platform for the Analysis of Permethylated Oligosaccharide Alditols. *J. Am. Soc. Mass Spectr.* 2007;18(10):1799–1812.

130. Zhou S, Huang Y, Dong X, *et al.* Isomeric Separation of Permethylated Glycans by Porous Graphitic Carbon (PGC)-LC-MS/MS at High Temperatures. *Anal. Chem.* 2017;89(12):6590–6597.

131. Zhou S, Dong X, Veillon L, *et al.* LC-MS/MS Analysis of Permethylated N-Glycans Facilitating Isomeric Characterization. *Anal. Bioanal. Chem.* 2017;409(2):453–466.

132. Huang Y, Zhou S, Zhu J, *et al.* LC-MS/MS Isomeric Profiling of Permethylated N-Glycans Derived from Serum Haptoglobin of Hepatocellular Carcinoma (HCC) and Cirrhotic Patients. *Electrophoresis.* 2017;38(17):2160–2167.

133. Dong X, Mondello S, Kobeissy F, *et al.* Serum Glycomics Profiling of Patients with Primary Restless Legs Syndrome Using LC-MS/MS. *J. Proteome Res.* 2020;19(8):2933–2941.

134. Dong X, Mondello S, Kobeissy F, *et al.* LC-MS/MS Glycomics of Idiopathic Rapid Eye Movement Sleep Behavior Disorder. *Electrophoresis.* 2018;39(24):3096–3103.

135. Cho BG, Peng W, Mechref Y. Separation of Permethylated O-Glycans, Free Oligosaccharides, and Glycosphingolipid-Glycans Using Porous

Graphitized Carbon (PGC) Column. *Metabolites.* 2020;10(11):433. doi: 10.3390/metabo10110433.

136. Dong X, Zhou S, Mechref Y. LC-MS/MS Analysis of Permethylated Free Oligosaccharides and N-Glycans Derived from Human, Bovine, and Goat Milk Samples. *Electrophoresis.* 2016;37(11):1532–1548.
137. Chen CH, Lin YP, Ren CT, *et al.* Enhancement of Fucosylated N-Glycan Isomer Separation with an Ultrahigh Column Temperature in Porous Graphitic Carbon Liquid Chromatography-Mass Spectrometry. *J. Chromatogr. A.* 2020;1632:461610.
138. Ashwood C, Pratt B, MacLean BX, *et al.* Standardization of PGC-LC-MS-Based Glycomics for Sample Specific Glycotyping. *Analyst.* 2019;144(11):3601–3612.
139. Gautam S, Peng W, Cho BG, *et al.* Glucose Unit Index (GUI) of Permethylated Glycans for Effective Identification of Glycans and Glycan Isomers. *Analyst.* 2020;145(20):6656–6667.
140. Bapiro TE, Richards FM, Jodrell DI. Understanding the Complexity of Porous Graphitic Carbon (PGC) Chromatography: Modulation of Mobile-Stationary Phase Interactions Overcomes Loss of Retention and Reduces Variability. *Anal. Chem.* 2016;88(12):6190–6194.
141. Gautam S, Banazadeh A, Cho BG, *et al.* Mesoporous Graphitized Carbon Column for Efficient Isomeric Separation of Permethylated Glycans. *Anal. Chem.* 2021;93(12):5061–5070.
142. Žuvela P, Skoczylas M, Jay Liu J, *et al.* Column Characterization and Selection Systems in Reversed-Phase High-Performance Liquid Chromatography. *Chem. Rev.* 2019;119(6):3674–3729.
143. Hu Y, Zhou S, Khalil SI, *et al.* Glycomic Profiling of Tissue Sections by LC-MS. *Anal. Chem.* 2013;85(8):4074–4079.
144. Alley WRJ, Vasseur JA, Goetz JA, *et al.* N-Linked Glycan Structures and Their Expressions Change in the Blood Sera of Ovarian Cancer Patients. *J. Proteome Res.* 2012;11:2282–2300.
145. Mechref Y, Hu Y, Garcia A, *et al.* Defining Putative Glycan Cancer Biomarkers by MS. *Bioanalysis.* 2012;4:2457–2469.
146. Hu Y, Mechref Y. Comparing MALDI-MS, RP-LC-MALDI-MS and RP-LC-ESI-MS Glycomic Profiles of Permethylated N-Glycans Derived from Model Glycoproteins and Human Blood Serum. *Electrophsresis.* 2012;33:1768–1777.

147. Gaunitz S, Tjernberg LO, Schedin-Weiss S. The N-Glycan Profile in Cortex and Hippocampus Is Altered in Alzheimer Disease. *J. Neurochem.* 2020;159(2):292–304.
148. Zhou S, Hu Y, Mechref Y. High-Temperature LC-MS/MS of Permethylated Glycans Derived from Glycoproteins. *Electrophoresis.* 2016;37(11):1506–1513.
149. Hanisch FG, Müller S. Analysis of Methylated O-Glycan Alditols by Reversed-Phase NanoLC Coupled CAD-ESI Mass Spectrometry. *Methods Mol. Biol.* 2009;534:107–115.
150. Rozing G. Micropillar Array Columns for Advancing Nanoflow HPLC. *Microchem. J.* 2021;170:106629. doi: 10.1016/j.microc.2021.106629.
151. Cho BG, Jiang P, Goli M, *et al.* Using Micro Pillar Array Columns (μPAC) for the Analysis of Permethylated Glycans. *Analyst.* 2021; 146(13):4374–4383.
152. Cho BG, Gautam S, Peng W, *et al.* Direct Comparison of N-Glycans and Their Isomers Derived from Spike Glycoprotein 1 of MERS-CoV, SARS-CoV-1, and SARS-CoV-2. *J. Proteome Res.* 2021;20(9): 4357–4365.
153. Jooß K, Meckelmann SW, Klein J, *et al.* Capillary Zone Electrophoresis Coupled to Drift Tube Ion Mobility-Mass Spectrometry for the Analysis of Native and APTS-Labeled N-Glycans. *Anal. Bioanal. Chem.* 2019;411(24):6255–6264.
154. Khatri K, Klein JA, Haserick JR, *et al.* Microfluidic Capillary Electrophoresis–Mass Spectrometry for Analysis of Monosaccharides, Oligosaccharides, and Glycopeptides. *Anal. Chem.* 2017;89(12): 6645–6655.
155. Nagy G, Attah IK, Garimella SVB, *et al.* Unraveling the Isomeric Heterogeneity of Glycans: Ion Mobility Separations in Structures for Lossless Ion Manipulations. *Chem. Commun.* 2018;54(83):11701–11704.
156. Ben Faleh A, Warnke S, Rizzo TR. Combining Ultrahigh-Resolution Ion-Mobility Spectrometry with Cryogenic Infrared Spectroscopy for the Analysis of Glycan Mixtures. *Anal. Chem.* 2019;91(7):4876–4882.

157. Bansal P, Yatsyna V, AbiKhodr AH, *et al.* Using SLIM-Based IMS-IMS Together with Cryogenic Infrared Spectroscopy for Glycan Analysis. *Anal. Chem.* 2020;92(13):9079–9085.
158. Warnke S, Ben Faleh A, Scutelnic V, *et al.* Separation and Identification of Glycan Anomers Using Ultrahigh-Resolution Ion-Mobility Spectrometry and Cryogenic Ion Spectroscopy. *J. Am. Soc. Mass Spectrom.* 2019; 30(11):2204–2211.
159. Yalovenko N, Yatsyna V, Bansal P, *et al.* Analyzing Glycans Cleaved from a Biotherapeutic Protein Using Ultrahigh-Resolution Ion Mobility Spectrometry Together with Cryogenic Ion Spectroscopy. *Analyst.* 2020;145(20):6493–6499.
160. Lane CS, McManus K, Widdowson P, *et al.* Separation of Sialylated Glycan Isomers by Differential Mobility Spectrometry. *Anal. Chem.* 2019;91(15):9916–9924.
161. Xie C, Wu Q, Zhang S, *et al.* Improving Glycan Isomeric Separation via Metal Ion Incorporation for Drift Tube Ion Mobility-Mass Spectrometry. *Talanta.* 2020;211:120719. doi: 10.1016/j.talanta.2020. 120719.
162. Pallister EG, Choo MSF, Walsh I, *et al.* Utility of Ion-Mobility Spectrometry for Deducing Branching of Multiply Charged Glycans and Glycopeptides in a High-Throughput Positive Ion LC-FLR-IMS-MS Workflow. *Anal. Chem.* 2020;92(23):15323–15335.
163. Wei J, Tang Y, Ridgeway ME, *et al.* Accurate Identification of Isomeric Glycans by Trapped Ion Mobility Spectrometry-Electronic Excitation Dissociation Tandem Mass Spectrometry. *Anal. Chem.* 2020;92(19): 13211–13220.
164. Sastre Toraño J, Aizpurua-Olaizola O, Wei N, *et al.* Identification of Isomeric N-Glycans by Conformer Distribution Fingerprinting Using Ion Mobility Mass Spectrometry. *Chemistry (Easton).* 2021;27(6): 2149–2154.
165. Guntner AS, Thalhamer B, Klampfl C, *et al.* Collision Cross Sections Obtained with Ion Mobility Mass Spectrometry as New Descriptor to Predict Blood-Brain Barrier Permeation by Drugs. *Sci. Rep.* 2019; 9(1):19182.

166. Marchand A, Livet S, Rosu F, *et al.* Drift Tube Ion Mobility: How to Reconstruct Collision Cross Section Distributions from Arrival Time Distributions? *Anal. Chem.* 2017;89(23):12674–12681.
167. Lanucara F, Holman SW, Gray CJ, *et al.* The Power of Ion Mobility-Mass Spectrometry for Structural Characterization and the Study of Conformational Dynamics. *Nat. Chem.* 2014;6(4):281–294.
168. Gelb AS, Lai R, Li H, *et al.* Composition and Charge State Influence on the Ion-Neutral Collision Cross Sections of Protonated N-Linked Glycopeptides: An Experimental and Theoretical Deconstruction of Coulombic Repulsion vs. Charge Solvation Effects. *Analyst.* 2019;144(19):5738–5747.
169. Garabedian A, Benigni P, Ramirez CE, *et al.* Towards Discovery and Targeted Peptide Biomarker Detection Using nanoESI-TIMS-TOF MS. *J. Am. Soc. Mass Spectrom.* 2018;29(5):817–826.
170. Meier F, Brunner A-D, Koch S, *et al.* Online Parallel Accumulation 2013; Serial Fragmentation (PASEF) with a Novel Trapped Ion Mobility Mass Spectrometer. *Mol. Cell Proteomics.* 2018;17(12):2534–2545.
171. Meier F, Brunner A-D, Frank M, *et al.* diaPASEF: Parallel Accumulation–Serial Fragmentation Combined with Data-Independent Acquisition. *Nat. Methods.* 2020;17(12):1229–1236.
172. Nys G, Nix C, Cobraiville G, *et al.* Enhancing Protein Discoverability by Data Independent Acquisition Assisted by Ion Mobility Mass Spectrometry. *Talanta.* 2020;213:120812.
173. Jeanne Dit Fouque K, Kaplan D, Voinov VG, *et al.* Proteoform Differentiation Using Tandem Trapped Ion Mobility, Electron Capture Dissociation, and ToF Mass Spectrometry. *Anal. Chem.* 2021; 93(27):9575–9582.
174. Giles K, Ujma J, Wildgoose J, *et al.* A Cyclic Ion Mobility-Mass Spectrometry System. *Anal. Chem.* 2019;91(13):8564–8573.
175. Sanda M, Morrison L, Goldman R. N- and O-Glycosylation of the SARS-CoV-2 Spike Protein. *Anal. Chem.* 2021;93(4):2003–2009.
176. Pathak P, Baird MA, Shvartsburg AA. High-Resolution Ion Mobility Separations of Isomeric Glycoforms with Variations on the Peptide and Glycan Levels. *J. Am. Soc. Mass Spectrom.* 2020;31(7):1603–1609.
177. Wu R, Chen X, Wu WJ, *et al.* Fine Adjustment of Gas Modifier Loadings for Separation of Epimeric Glycopeptides Using Differential

Ion Mobility Spectrometry Mass Spectrometry. *Rapid Commun. Mass Spectrom.* 2020;34(9):e8751.

178. Ashline DJ, Yu Y, Lasanajak Y, *et al.* Structural Characterization by Multistage Mass Spectrometry (MSn) of Human Milk Glycans Recognized by Human Rotaviruses. *Mol. Cell Proteomics.* 2014; 13(11):2961–2974.
179. Mitchell Wells J, McLuckey SA. Collision-Induced Dissociation (CID) of Peptides and Proteins. *Methods Enzymol.* 2005;402:148–185. Academic Press.
180. Sleno L, Volmer DA. Ion Activation Methods for Tandem Mass Spectrometry. *J. Mass Spectrom.* 2004;39(10):1091–1112.
181. Han L, Costello CE. Mass Spectrometry of Glycans. *Biochemistry (Mosc.).* 2013;78(7):710–720.
182. Gutierrez Reyes CD, Jiang P, Donohoo K, *et al.* Glycomics and Glycoproteomics: Approaches to Address Isomeric Separation of Glycans and Glycopeptides. *J. Sep. Sci.* 2021;44(1):403–425.
183. Kurz S, Sheikh MO, Lu S, *et al.* Separation and Identification of Permethylated Glycan Isomers by Reversed Phase NanoLC-NSI-MS(n). *Mol. Cell Proteomics.* 2021;20:100045.
184. Yu X, Jiang Y, Chen Y, *et al.* Detailed Glycan Structural Characterization by Electronic Excitation Dissociation. *Anal. Chem.* 2013;85(21): 10017–10021.
185. Tang Y, Wei J, Costello CE, *et al.* Characterization of Isomeric Glycans by Reversed Phase Liquid Chromatography-Electronic Excitation Dissociation Tandem Mass Spectrometry. *J. Am. Soc. Mass Spectrom.* 2018;29(6):1295–1307.
186. Tang Y, Pu Y, Gao J, *et al.* De Novo Glycan Sequencing by Electronic Excitation Dissociation and Fixed-Charge Derivatization. *Anal. Chem.* 2018;90(6):3793–3801.
187. Wei J, Tang Y, Bai Y, *et al.* Toward Automatic and Comprehensive Glycan Characterization by Online PGC-LC-EED MS/MS. *Anal. Chem.* 2020;92(1):782–791.
188. Brodbelt JS. Photodissociation Mass Spectrometry: New Tools for Characterization of Biological Molecules. *Chem. Soc. Rev.* 2014; 43(8):2757–2783.
189. Ropartz D, Lemoine J, Giuliani A, *et al.* Deciphering the Structure of Isomeric Oligosaccharides in a Complex Mixture by Tandem Mass

Spectrometry: Photon Activation with Vacuum Ultra-Violet Brings Unique Information and Enables Definitive Structure Assignment. *Anal. Chim. Acta.* 2014;807:84–95.

190. Ropartz D, Giuliani A, Fanuel M, *et al.* Online Coupling of High-Resolution Chromatography with Extreme UV Photon Activation Tandem Mass Spectrometry: Application to the Structural Investigation of Complex Glycans by Dissociative Photoionization. *Anal. Chim. Acta.* 2016;933:1–9.
191. Ropartz D, Giuliani A, Hervé C, *et al.* High-Energy Photon Activation Tandem Mass Spectrometry Provides Unprecedented Insights into the Structure of Highly Sulfated Oligosaccharides Extracted from Macroalgal Cell Walls. *Anal. Chem.* 2015;87(2):1042–1049.
192. Riggs DL, Hofmann J, Hahm HS, *et al.* Glycan Isomer Identification Using Ultraviolet Photodissociation Initiated Radical Chemistry. *Anal. Chem.* 2018;90(19):11581–11588.
193. Klein DR, Leach FE, 3rd, Amster IJ, *et al.* Structural Characterization of Glycosaminoglycan Carbohydrates Using Ultraviolet Photodissociation. *Anal. Chem.* 2019;91(9):6019–6026.
194. Crittenden CM, Escobar EE, Williams PE, *et al.* Characterization of Antigenic Oligosaccharides from Gram-Negative Bacteria via Activated Electron Photodetachment Mass Spectrometry. *Anal. Chem.* 2019; 91(7):4672–4679.
195. Klein DR, Powers MJ, Trent MS, *et al.* Top-Down Characterization of Lipooligosaccharides from Antibiotic-Resistant Bacteria. *Anal. Chem.* 2019;91(15):9608–9615.
196. An HJ, Lebrilla CB. Structure Elucidation of Native N- and O-Linked Glycans by Tandem Mass Spectrometry (Tutorial). *Mass Spectrom. Rev.* 2011;30(4):560–578.
197. Tan Y, Polfer NC. Linkage and Anomeric Differentiation in Trisaccharides by Sequential Fragmentation and Variable-Wavelength Infrared Photodissociation. *J. Am. Soc. Mass Spectrom.* 2015;26(2):359–368.
198. Depraz Depland A, Renois-Predelus G, Schindler B, *et al.* Identification of Sialic Acid Linkage Isomers in Glycans Using Coupled InfraRed Multiple Photon Dissociation (IRMPD) Spectroscopy and Mass Spectrometry. *Int. J. Mass Spectrom.* 2018;434:65–69.

199. Renois-Predelus G, Schindler B, Compagnon I. Analysis of Sulfate Patterns in Glycosaminoglycan Oligosaccharides by MSn Coupled to Infrared Ion Spectroscopy: The Case of GalNAc4S and GalNAc6S. *J. Am. Soc. Mass Spectrom.* 2018;29(6):1242–1249.
200. Schindler B, Laloy-Borgna G, Barnes L, *et al.* Online Separation and Identification of Isomers Using Infrared Multiple Photon Dissociation Ion Spectroscopy Coupled to Liquid Chromatography: Application to the Analysis of Disaccharides Regio-Isomers and Monosaccharide Anomers. *Anal. Chem.* 2018;90(20):11741–11745.
201. Jaravel L, Schindler B, Randon J, *et al.* Off-Line Coupling of Capillary Isotachophoresis Separation to IRMPD Spectroscopy for Glycosaminoglycans Analysis: Application to the Chondroitin Sulfate Disaccharides Model Solutes. *J. Chromatogr. A.* 2020;1617:460782.
202. Zhu F, Clemmer DE, Trinidad JC. Characterization of Lectin Binding Affinities via Direct LC-MS Profiling: Implications for Glycopeptide Enrichment and Separation Strategies. *Analyst.* 2017;142(1):65–74.
203. Alpert AJ. Electrostatic Repulsion Hydrophilic Interaction Chromatography for Isocratic Separation of Charged Solutes and Selective Isolation of Phosphopeptides. *Anal. Chem.* 2008;80(1):62–76.
204. Bermudez A, Pitteri SJ. Enrichment of Intact Glycopeptides Using Strong Anion Exchange and Electrostatic Repulsion Hydrophilic Interaction Chromatography. *Methods Mol. Biol. (Clifton, N.J.).* 2021;2271:107–120.
205. Cui Y, Tabang DN, Zhang Z, *et al.* Counterion Optimization Dramatically Improves Selectivity for Phosphopeptides and Glycopeptides in Electrostatic Repulsion-Hydrophilic Interaction Chromatography. *Anal. Chem.* 2021;93(22):7908–7916.
206. Pu C, Zhao H, Hong Y, *et al.* Facile Preparation of Hydrophilic Mesoporous Metal–Organic Framework via Synergistic Etching and Surface Functionalization for Glycopeptides Analysis. *Anal. Chem.* 2020;92(2):1940–1947.
207. Yang S-S, Wang C, Yu X, *et al.* A Hydrophilic Two-Dimensional Titanium-Based Metal-Organic Framework Nanosheets for Specific Enrichment of Glycopeptides. *Anal. Chim. Acta.* 2020;1119:60–67.

208. Ali MM, Zhu Z, Hussain D, *et al.* Flexible and Hierarchical Metal-Organic Framework Composite as Solid-Phase Media for Facile Affinity-Tip Fabrication to Selectively Enrich Glycopeptides and Phosphopeptides. *Talanta.* 2021;233:122576.
209. Lu Q, Chen C, Xiong Y, *et al.* High-Efficiency Phosphopeptide and Glycopeptide Simultaneous Enrichment by Hydrogen Bond–based Bifunctional Smart Polymer. *Anal. Chem.* 2020;92(9):6269–6277.
210. Bi C, Liang Y, Shen L, *et al.* Maltose-Functionalized Hydrophilic Magnetic Nanoparticles with Polymer Brushes for Highly Selective Enrichment of N-Linked Glycopeptides. *ACS Omega.* 2018;3(2): 1572–1580.
211. Zhang N, Hu X, Chen H, *et al.* Specific Enrichment and Glycosylation Discrepancy Profiling of Cellular Exosomes Using a Dual-Affinity Probe. *Chem. Commun.* 2021.
212. Liu L, Zhu B, Fang Z, *et al.* Automated Intact Glycopeptide Enrichment Method Facilitating Highly Reproducible Analysis of Serum Site-Specific N-Glycoproteome. *Anal. Chem.* 2021;93(20): 7473–7480.
213. Bai H, Pan Y, Qi L, *et al.* Development a Hydrazide-Functionalized Thermosensitive Polymer Based Homogeneous System for Highly Efficient N-Glycoprotein/Glycopeptide Enrichment from Human Plasma Exosome. *Talanta.* 2018;186:513–520.
214. Xiao H, Chen W, Smeekens JM, *et al.* An Enrichment Method Based on Synergistic and Reversible Covalent Interactions for Large-Scale Analysis of Glycoproteins. *Nat. Commun.* 2018;9(1):1692.
215. Peng W, Gutierrez Reyes CD, Gautam S, *et al.* MS-Based Glycomics and Glycoproteomics Methods Enabling Isomeric Characterization. *Mass Spectrom. Rev.* 2021;42(2):577–616.
216. Suttapitugsakul S, Tong M, Sun F, *et al.* Enhancing Comprehensive Analysis of Secreted Glycoproteins from Cultured Cells Without Serum Starvation. *Anal. Chem.* 2021;93(4):2694–2705.
217. Shah P, Wang X, Yang W, *et al.* Integrated Proteomic and Glycoproteomic Analyses of Prostate Cancer Cells Reveal Glycoprotein Alteration in Protein Abundance and Glycosylation. *Mol. Cell Proteomics.* 2015;14(10):2753–2763.

218. de Haan N, Reiding KR, Haberger M, *et al.* Linkage-specific Sialic Acid Derivatization for MALDI-TOF-MS Profiling of IgG Glycopeptides. *Anal. Chem.* 2015;87(16):8284–8291.
219. de Haan N, Yang S, Cipollo J, *et al.* Glycomics Studies Using Sialic Acid Derivatization and Mass Spectrometry. *Nat. Rev. Chem.* 2020;4(5):229–242.
220. Yang L, Sun Z, Zhang L, *et al.* Chemical Labeling for Fine Mapping of IgG N-Glycosylation by ETD-MS. *Chem. Sci.* 2019;10(40): 9302–9307.
221. Yang S, Wu WW, Shen R, *et al.* Optimization of O-GIG for O-Glycopeptide Characterization with Sialic Acid Linkage Determination. *Anal. Chem.* 2020;92(16):10946–10951.
222. Yang S, Wu WW, Shen RF, *et al.* Identification of Sialic Acid Linkages on Intact Glycopeptides via Differential Chemical Modification Using IntactGIG-HILIC. *J. Am. Soc. Mass Spectrom.* 2018;29(6):1273–1283.
223. Wang W, Kałuża A, Nouta J, *et al.* High-Throughput Glycopeptide Profiling of Prostate-Specific Antigen from Seminal Plasma by MALDI-MS. *Talanta.* 2021;222:121495.
224. Zhong J, Huang Y, Mechref Y. Derivatization of Sialylated Glycopeptides (DOSG) Enabling Site-Specific Isomeric Profiling Using LC-MS/MS. *Anal. Chem.* 2021;93(14):5763–5772.
225. Bantscheff M, Lemeer S, Savitski MM, *et al.* Quantitative Mass Spectrometry in Proteomics: Critical Review Update from 2007 to the Present. *Anal. Bioanal. Chem.* 2012;404(4):939–965.
226. Ye H, Boyne MT, 2nd, Buhse LF, *et al.* Direct Approach for Qualitative and Quantitative Characterization of Glycoproteins Using Tandem Mass Tags and an LTQ Orbitrap XL Electron Transfer Dissociation Hybrid Mass Spectrometer. *Anal. Chem.* 2013;85(3):1531–1539.
227. Fang J, Sheng X, Bao H, *et al.* Comparative Analysis of Intact Glycopeptides from Mannose Receptor Among Different Breast Cancer Subtypes Using Mass Spectrometry. *Talanta.* 2021;223(Pt 1): 121676.
228. Chen Z, Yu Q, Hao L, *et al.* Site-Specific Characterization and Quantitation of N-Glycopeptides in PKM2 Knockout Breast Cancer

Cells Using DiLeu Isobaric Tags Enabled by Electron-Transfer/Higher-Energy Collision Dissociation (EThcD). *Analyst.* 2018; 143(11):2508–2519.

229. Sun Z, Ji G, Wang G, *et al.* One Step Carboxyl Group Isotopic Labeling for Quantitative Analysis of Intact N-Glycopeptides by Mass Spectrometry. *Chem. Commun. (Camb.).* 2021;57(34):4154–4157.
230. Sun F, Wu R. Chapter Ten — Systematic and Site-Specific Analysis of N-Glycoproteins on the Cell Surface by Integrating Bioorthogonal Chemistry and MS-Based Proteomics. In *Methods Enzymol*, Garcia BA, Ed. 2019;626:223–247. Academic Press.
231. Cho KC, Chen L, Hu Y, *et al.* Developing Workflow for Simultaneous Analyses of Phosphopeptides and Glycopeptides. *ACS Chem. Biol.* 2019;14(1):58–66.
232. Fang P, Ji Y, Silbern I, *et al.* A Streamlined Pipeline for Multiplexed Quantitative Site-Specific N-Glycoproteomics. *Nat. Commun.* 2020; 11(1):5268.
233. Gutierrez-Reyes CD, Jiang P, Atashi M, *et al.* Advances in Mass Spectrometry-Based Glycoproteomics: An Update Covering the Period 2017–2021. *Electrophoresis.* 2022;43(1–2):370–387.
234. Myers SA, Daou S, Affarel B, *et al.* Electron Transfer Dissociation (ETD): The Mass Spectrometric Breakthrough Essential for O-GlcNAc Protein Site Assignments — a Study of the O-GlcNAcylated Protein Host Cell Factor C1. *Proteomics.* 2013;13(6):982–991.
235. Zhurov KO, Fornelli L, Wodrich MD, *et al.* Principles of Electron Capture and Transfer Dissociation Mass Spectrometry Applied to Peptide and Protein Structure Analysis. *Chem. Soc. Rev.* 2013; 42(12):5014–5030.
236. Riley NM, Coon JJ. The Role of Electron Transfer Dissociation in Modern Proteomics. *Anal. Chem.* 2018;90(1):40–64.
237. Riley NM, Malaker SA, Driessen MD, *et al.* Optimal Dissociation Methods Differ for N- and O-Glycopeptides. *J. Proteome Res.* 2020; 19(8):3286–3301.
238. Escobar EE, King DT, Serrano-Negrón JE, *et al.* Precision Mapping of O-Linked N-Acetylglucosamine Sites in Proteins Using Ultraviolet Photodissociation Mass Spectrometry. *J. Am. Chem. Soc.* 2020;142(26): 11569–11577.

239. Lee JY, Lee HK, Park GW, *et al.* Characterization of Site-Specific N-Glycopeptide Isoforms of α-1-Acid Glycoprotein from an Interlaboratory Study Using LC–MS/MS. *J. Proteome Res.* 2016; 15(12):4146–4164.
240. Ye H, Boyne MT, Buhse LF, *et al.* Direct Approach for Qualitative and Quantitative Characterization of Glycoproteins Using Tandem Mass Tags and an LTQ Orbitrap XL Electron Transfer Dissociation Hybrid Mass Spectrometer. *Anal. Chem.* 2013;85(3):1531–1539.
241. Yin H, An M, So PK, *et al.* The Analysis of Alpha-1-Antitrypsin Glycosylation with Direct LC-MS/MS. *Electrophoresis.* 2018;39(18): 2351–2361.
242. Gutierrez Reyes CD, Huang Y, Atashi M, *et al.* PRM-MS Quantitative Analysis of Isomeric N-Glycopeptides Derived from Human Serum Haptoglobin of Patients with Cirrhosis and Hepatocellular Carcinoma. *Metabolites.* 2021;11(8):563.
243. Mechref Y. Use of CID/ETD Mass Spectrometry to Analyze Glycopeptides. *Curr. Protoc. Protein Sci.* 2012, Chapter 12, Unit-12.11.11.
244. Singh C, Zampronio CG, Creese AJ, *et al.* Higher Energy Collision Dissociation (HCD) Product Ion-Triggered Electron Transfer Dissociation (ETD) Mass Spectrometry for the Analysis of N-Linked Glycoproteins. *J. Proteome Res.* 2012;11(9):4517–4525.
245. Zhu J, Chen Z, Zhang J, *et al.* Differential Quantitative Determination of Site-Specific Intact N-Glycopeptides in Serum Haptoglobin Between Hepatocellular Carcinoma and Cirrhosis Using LC-EThcD-MS/MS. *J. Proteome Res.* 2019;18(1):359–371.
246. Yu Q, Wang B, Chen Z, *et al.* Electron-Transfer/Higher-Energy Collision Dissociation (EThcD)-Enabled Intact Glycopeptide/Glycoproteome Characterization. *J. Am. Soc. Mass Spectrom.* 2017;28(9):1751–1764.
247. Glover MS, Yu Q, Chen Z, *et al.* Characterization of Intact Sialylated Glycopeptides and Phosphorylated Glycopeptides from IMAC Enriched Samples by EThcD Fragmentation: Toward Combining Phosphoproteomics and Glycoproteomics. *Int. J. Mass Spectrom.* 2018; 427:35–42.
248. Walker SH, Taylor AD, Muddiman DC. Individuality Normalization when Labeling with Isotopic Glycan Hydrazide Tags (INLIGHT):

A Novel Glycan-Relative Quantification Strategy. *J. Am. Soc. Mass Spectrom.* 2013;24(9):1376–1384.

249. Ong SE, Blagoev B, Kratchmarova I, *et al.* Stable Isotope Labeling by Amino Acids in Cell Culture, SILAC, as a Simple and Accurate Approach to Expression Proteomics. *Mol. Cell Proteom.* 2002; 1(5):376–386.
250. Clark DJ, Mei Y, Sun S, *et al.* Glycoproteomic Approach Identifies KRAS as a Positive Regulator of CREG1 in Non-Small Cell Lung Cancer Cells. *Theranostics.* 2016;6(1):65–77.
251. Deeb SJ, Cox J, Schmidt-Supprian M, *et al.* N-Linked Glycosylation Enrichment for In-Depth Cell Surface Proteomics of Diffuse Large B-Cell Lymphoma Subtypes. *Mol. Cell Proteom.* 2014;13(1): 240–251.
252. Taga Y, Kusubata M, Ogawa-Goto K, *et al.* Site-Specific Quantitative Analysis of Overglycosylation of Collagen in Osteogenesis Imperfecta Using Hydrazide Chemistry and SILAC. *J. Proteome Res.* 2013; 12(5):2225–2232.
253. Ong SE, Mann M. A Practical Recipe for Stable Isotope Labeling by Amino Acids in Cell Culture (SILAC). *Nat. Protoc.* 2006;1(6): 2650–2660.
254. Ji Y, Wei S, Hou J, *et al.* Integrated Proteomic and N-Glycoproteomic Analyses of Doxorubicin Sensitive and Resistant Ovarian Cancer Cells Reveal Glycoprotein Alteration in Protein Abundance and Glycosylation. *Oncotarget.* 2017;8(8):13413–13427.
255. Mahal LK, Yarema KJ, Bertozzi CR. Engineering Chemical Reactivity on Cell Surfaces Through Oligosaccharide Biosynthesis. *Science.* 1997;276(5315):1125–1128.
256. Sletten EM, Bertozzi CR. Bioorthogonal Chemistry: Fishing for Selectivity in a Sea of Functionality. *Angew. Chem. Int. Ed. Engl.* 2009; 48(38):6974–6998.
257. Hang HC, Yu C, Kato DL, *et al.* A Metabolic Labeling Approach Toward Proteomic Analysis of Mucin-Type O-Linked Glycosylation. *Proc. Natl. Acad. Sci. USA.* 2003;100(25):14846–14851.
258. Boyce M, Carrico IS, Ganguli AS, *et al.* Metabolic Cross-Talk Allows Labeling of O-Linked Beta-N-Acetylglucosamine-Modified Proteins

via the N-Acetylgalactosamine Salvage Pathway. *Proc. Natl. Acad. Sci. USA*. 2011;108(8):3141–3146.

259. Rabuka D, Hubbard SC, Laughlin ST, *et al.* A Chemical Reporter Strategy to Probe Glycoprotein Fucosylation. *J. Am. Chem. Soc.* 2006;128(37):12078–12079.
260. Sawa M, Hsu TL, Itoh T, *et al.* Glycoproteomic Probes for Fluorescent Imaging of Fucosylated Glycans in Vivo. *Proc. Natl. Acad. Sci. USA*. 2006;103(33):12371–12376.
261. Vocadlo DJ, Hang HC, Kim EJ, *et al.* A Chemical Approach for Identifying O-GlcNAc-Modified Proteins in Cells. *Proc. Natl. Acad. Sci. USA*. 2003;100(16):9116–9121.
262. Gurcel C, Vercoutter-Edouart AS, Fonbonne C, *et al.* Identification of New O-GlcNAc Modified Proteins Using a Click-Chemistry-Based Tagging. *Anal. Bioanal. Chem.* 2008;390(8):2089–2097.
263. Sarkar AK, Fritz TA, Taylor WH, *et al.* Disaccharide Uptake and Priming in Animal Cells: Inhibition of Sialyl Lewis X by Acetylated Gal Beta 1-->4GlcNAc Beta-O-naphthalenemethanol. *Proc. Natl. Acad. Sci. USA*. 1995;92(8):3323–3327.
264. Saxon E, Bertozzi CR. Cell Surface Engineering by a Modified Staudinger Reaction. *Science*. 2000;287(5460):2007–2010.
265. Saxon E, Luchansky SJ, Hang HC, *et al.* Investigating Cellular Metabolism of Synthetic Azidosugars with the Staudinger Ligation. *J. Am. Chem. Soc.* 2002;124(50):14893–14902.
266. Palaniappan KK, Bertozzi CR. Chemical Glycoproteomics. *Chem. Rev.* 2016;116(23):14277–14306.
267. Wang Q, Chan TR, Hilgraf R, *et al.* Bioconjugation by Copper(I)-Catalyzed Azide-Alkyne [3 + 2] Cycloaddition. *J. Am. Chem. Soc.* 2003; 125(11):3192–3193.
268. Pan PW, Zhang Q, Hou J, *et al.* Cell Surface Glycoprotein Profiling of Cancer Cells Based on Bioorthogonal Chemistry. *Anal. Bioanal. Chem.* 2012;403(6):1661–1670.
269. Woo CM, Felix A, Zhang L, *et al.* Isotope-Targeted Glycoproteomics (IsoTaG) Analysis of Sialylated N- and O-Glycopeptides on an Orbitrap Fusion Tribrid Using Azido and Alkynyl Sugars. *Anal. Bioanal. Chem.* 2017;409(2):579–588.

270. Link AJ, Tirrell DA. Cell Surface Labeling of *Escherichia coli* via Copper(I)- Catalyzed [3+2] Cycloaddition. *J. Am. Chem. Soc.* 2003; 125(37):11164–11165.
271. Agard NJ, Prescher JA, Bertozzi CR. A Strain-Promoted [3 + 2] Azide-Alkyne Cycloaddition for Covalent Modification of Biomolecules in Living Systems. *J. Am. Chem. Soc.* 2004;126(46):15046–15047.
272. Chang PV, Prescher JA, Sletten EM, *et al.* Copper-Free Click Chemistry in Living Animals. *Proc. Natl. Acad. Sci. USA.* 2010; 107(5):1821–1826.
273. Agard NJ, Baskin JM, Prescher JA, *et al.* A Comparative Study of Bioorthogonal Reactions with Azides. *ACS Chem. Biol.* 2006; 1(10):644–648.
274. Baskin JM, Prescher JA, Laughlin ST, *et al.* Copper-Free Click Chemistry for Dynamic in Vivo Imaging. *Proc. Natl. Acad. Sci. USA.* 2007;104(43):16793–16797.
275. Sletten EM, Bertozzi CR. A Hydrophilic Azacyclooctyne for Cu-Free Click Chemistry. *Org. Lett.* 2008;10(14):3097–3099.
276. Ning X, Guo J, Wolfert MA, *et al.* Visualizing Metabolically Labeled Glycoconjugates of Living Cells by Copper-Free and Fast Huisgen Cycloadditions. *Angew. Chem. Int. Ed. Engl.* 2008;47(12):2253–2255.
277. Roper SM, Zemskova M, Neely BA, *et al.* Targeted Glycoprotein Enrichment and Identification in Stromal Cell Secretomes Using Azido Sugar Metabolic Labeling. *Proteomics Clin. Appl.* 2013; 7(5–6):367–371.
278. Chen W, Smeekens JM, Wu R. Systematic and Site-Specific Analysis of N-Sialoglycosylated Proteins on the Cell Surface by Integrating Click Chemistry and MS-Based Proteomics. *Chem. Sci.* 2015;6(8): 4681–4689.
279. Ceroni A, Maass K, Geyer H, *et al.* GlycoWorkbench: A Tool for the Computer-Assisted Annotation of Mass Spectra of Glycans. *J. Proteome Res.* 2008;7(4):1650–1659.
280. Damerell D, Ceroni A, Maass K, *et al.* The GlycanBuilder and GlycoWorkbench Glycoinformatics Tools: Updates and New Developments. *Biol. Chem.* 2012;393(11):1357–1362.

281. Damerell D, Ceroni A, Maass K, *et al.* Annotation of Glycomics MS and MS/MS Spectra Using the GlycoWorkbench Software Tool. *Methods Mol. Biol.* 2015;1273:3–15.
282. MacLean B, Tomazela DM, Shulman N, *et al.* Skyline: An Open Source Document Editor for Creating and Analyzing Targeted Proteomics Experiments. *Bioinformatics.* 2010;26(7):966–968.
283. Loziuk PL, Hecht ES, Muddiman DC. N-Linked Glycosite Profiling and Use of Skyline as a Platform for Characterization and Relative Quantification of Glycans in Differentiating Xylem of *Populus trichocarpa. Anal. Bioanal. Chem.* 2017;409(2):487–497.
284. Ashwood C, Lin CH, Thaysen-Andersen M, *et al.* Discrimination of Isomers of Released N- and O-Glycans Using Diagnostic Product Ions in Negative Ion PGC-LC-ESI-MS/MS. *J. Am. Soc. Mass Spectr.* 2018; 29(6):1194–1209.
285. Cooper CA, Gasteiger E, Packer NH. GlycoMod — a Software Tool for Determining Glycosylation Compositions from Mass Spectrometric Data. *Proteomics.* 2001;1(2):340–349.
286. Maxwell E, Tan Y, Tan Y, *et al.* GlycReSoft: A Software Package for Automated Recognition of Glycans from LC/MS Data. *PloS One.* 2012;7(9):e45474. doi: 10.1371/journal.pone.0045474.
287. Gao HY. Generation of Asparagine-Linked Glycan Structure Databases and Their Use. *J. Am. Soc. Mass Spectr.* 2009;20(9):1739–1742.
288. Vakhrushev SY, Dadimov D, Peter-Katalinić J. Software Platform for High-Throughput Glycomics. *Anal. Chem.* 2009;81(9):3252–3260.
289. Apte A, Meitei NS. Bioinformatics in Glycomics: Glycan Characterization with Mass Spectrometric Data Using SimGlycan. *Methods Mol. Biol.* 2010;600:269–281.
290. Hu Y, Zhou S, Yu CY, *et al.* Automated Annotation and Quantitation of Glycans by Liquid Chromatography/Electrospray Ionization Mass Spectrometric Analysis Using the MultiGlycan-ESI Computational Tool. *Rapid Commun. Mass Sp.* 2015;29(1):135–142.
291. Zhang R, Peng W, Gautam S, *et al.* GlycanGUI: Automated Glycan Annotation and Quantification Using Glucose Unit Index. *Front. Chem.* 2021;9:707382.

292. Hong P, Sun H, Sha L, *et al.* GlycoDeNovo — an Efficient Algorithm for Accurate de novo Glycan Topology Reconstruction from Tandem Mass Spectra. *J. Am. Soc. Mass Spectr.* 2017;28(11):2288–2301.
293. Maass K, Ranzinger R, Geyer H, *et al.* "Glyco-peakfinder" — de novo Composition Analysis of Glycoconjugates. *Proteomics.* 2007;7(24): 4435–4444.
294. Horlacher O, Jin C, Alocci D, *et al.* Glycoforest 1.0. *Anal. Chem.* 2017;89(20):10932–10940.
295. Irungu J, Go EP, Dalpathado DS, *et al.* Simplification of Mass Spectral Analysis of Acidic Glycopeptides Using GlycoPep ID. *Anal. Chem.* 2007;79(8):3065–3074.
296. Ozohanics O, Krenyacz J, Ludányi K, *et al.* GlycoMiner: A New Software Tool to Elucidate Glycopeptide Composition. *Rapid Commun. Mass Spectrom.* 2008;22(20):3245–3254.
297. Pompach P, Chandler KB, Lan R, *et al.* Semi-Automated Identification of N-Glycopeptides by Hydrophilic Interaction Chromatography, Nano-Reverse-Phase LC-MS/MS, and Glycan Database Search. *J. Proteome Res.* 2012;11(3):1728–1740.
298. Mayampurath AM, Wu Y, Segu ZM, *et al.* Improving Confidence in Detection and Characterization of Protein N-Glycosylation Sites and Microheterogeneity. *Rapid Commun. Mass Spectrom.* 2011;25(14): 2007–2019.
299. Woodin CL, Hua D, Maxon M, *et al.* GlycoPep Grader: A Web-Based Utility for Assigning the Composition of N-Linked Glycopeptides. *Anal. Chem.* 2012;84(11):4821–4829.
300. Toghi Eshghi S, Shah P, Yang W, *et al.* GPQuest: A Spectral Library Matching Algorithm for Site-Specific Assignment of Tandem Mass Spectra to Intact N-Glycopeptides. *Anal. Chem.* 2015;87(10): 5181–5188.
301. Yu CY, Mayampurath A, Zhu R, *et al.* Automated Glycan Sequencing from Tandem Mass Spectra of N-Linked Glycopeptides. *Anal. Chem.* 2016;88(11):5725–5732.
302. Polasky DA, Yu F, Teo GC, *et al.* Fast and Comprehensive N- and O-Glycoproteomics Analysis with MSFragger-Glyco. *Nat. Methods.* 2020;17(11):1125–1132.

303. Zeng WF, Liu MQ, Zhang Y, *et al.* pGlyco: A Pipeline for the Identification of Intact N-Glycopeptides by Using HCD- and CID-MS/MS and MS3. *Sci. Rep.* 2016;6:25102. doi: 10.1038/srep25102.
304. Bern M, Kil YJ, Becker C. Byonic: Advanced Peptide and Protein Identification Software. *Curr. Protoc. Bioinformatics.* 2012, Chapter 13, Unit13.20. doi: 10.1002/0471250953.bi1320s40.
305. Park GW, Kim JY, Hwang H, *et al.* Integrated GlycoProteome Analyzer (I-GPA) for Automated Identification and Quantitation of Site-Specific N-Glycosylation. *Sci. Rep.* 2016;6:21175.
306. Nasir W, Toledo AG, Noborn F, *et al.* SweetNET: A Bioinformatics Workflow for Glycopeptide MS/MS Spectral Analysis. *J. Proteome Res.* 2016;15(8):2826–2840.
307. Ye Z, Mao Y, Clausen H, *et al.* Glyco-DIA: A Method for Quantitative O-Glycoproteomics with in Silico-Boosted Glycopeptide Libraries. *Nat. Methods.* 2019;16(9):902–910.
308. He L, Xin L, Shan B, *et al.* GlycoMaster DB: Software to Assist the Automated Identification of N-Linked Glycopeptides by Tandem Mass Spectrometry. *J. Proteome Res.* 2014;13(9):3881–3895.
309. Zhu Z, Hua D, Clark DF, *et al.* GlycoPep Detector: A Tool for Assigning Mass Spectrometry Data of N-Linked Glycopeptides on the Basis of Their Electron Transfer Dissociation Spectra. *Anal. Chem.* 2013;85(10):5023–5032.

https://doi.org/10.1142/9789811283376_0008

Chapter 8

Mapping Subcellular Proteomic Landscapes with Mass Spectrometry

Trisha Tucholski, Caleb Bridgwater, and Aleksandra Nita-Lazar

Functional Cellular Networks Section, Laboratory of Immune System Biology, National Institute of Allergy and Infectious Diseases, National Institutes of Health, Bethesda, MD 20892, USA

Contents

Abstract

Protein functions in the cell depend heavily on their localization and their subcellular compartment, where the proteins reside and can change on protein translocation. Quantitative mass spectrometry (MS)–based approaches enable systems-level investigations of proteomes. This chapter focuses on MS-based spatial proteomic approaches available for characterizing defined subsets of whole-cell and tissue proteomes, including the latest methods for multiplexed identification and quantification of proteins from subcellular compartments and the software that has enabled interpretation of the information-rich data generated from such analyses. Spatial proteomic studies pertaining to specific subcellular compartments, such as membrane- and non-membrane-bound organelles and protein complexes of varying sizes are also described.

1. Introduction

Cells carry out a wide array of biomolecular and biochemical processes simultaneously. This sophisticated coordination is made possible by sequestering these activities to discrete locations within the cell containing the appropriate substrate concentrations, pH, or redox conditions.[1] Eukaryotic cells are organized into compartments, including organelles that are bound by membranes, such as the nucleus and mitochondria; non-membrane-bound organelles, such as native condensates; and macromolecular complexes, such as ribosomes, proteasomes, and other large protein complexes. Each subcellular structure is composed of proteins that execute distinct functions. Some proteins reside in specific organelles, while some proteins transition between compartments.[2] Disruption to the organization of proteins within distinct subcellular compartments or dysregulation of protein translocation has been linked to disease.[3–12] Therefore, identifying and accurately quantifying proteins and their proteoforms in subcellular compartments is an important first step in our quest to understand biology and disease at the systems level.

Quantitative mass spectrometry (MS)–based proteomics has emerged as an important tool enabling systems biology investigations. We can now accurately and precisely identify and quantify thousands of proteins and their post-translational modifications (PTMs) from large sample cohorts in a multiplexed and high-throughput manner, which has enabled the large-scale analysis of whole-cell and tissue proteomes.[13–16] MS-based spatial proteomics, which identifies and quantifies proteins in discrete locations within the cell using a variety of biochemical fractionation, affinity purification, and labeling techniques, has been developed substantially over the past 15 years to aid the goal of defining subcellular proteomic landscapes.[2,17–19] Herein, we will describe the MS-based spatial proteomic approaches available for characterizing defined subsets of whole-cell and tissue proteomes, including the latest methods for multiplexed identification and quantification of proteins from subcellular compartments and the software that has enabled the interpretation of the information-rich data generated from such analyses. We will also describe spatial proteomic studies pertaining to specific subcellular compartments, such as membrane- and non-membrane-bound organelles and protein complexes of varying sizes. While there are useful spatial proteomic imaging techniques, this chapter will be limited to MS-based approaches. However, imaging-based strategies, such as high-content microscopy and antibody-based imaging, are complementary approaches for studying protein localization and have been demonstrated and discussed elsewhere.[18,20–22]

2. Mass spectrometry–based workflows for spatial proteomics

A critical first step in any MS-based spatial proteomic workflow is cell lysis. Importantly, the integrity of organelles and suborganellar structures of interest must be preserved to ensure they can be fractionated. To achieve this, most methods opt to homogenize cells or tissue in a detergent-free buffer containing enzyme inhibitors (typically, protease and phosphatase inhibitors) to halt cellular processes, using either ball-bearing or Dounce homogenization. Next, subcellular structures are enriched or isolated from the resulting homogenate. Methods for separating compartments from whole-cell homogenate include (but are not limited to) conventional centrifugation-based biochemical fractionation, affinity purification, and fluorescent-assisted organelle sorting and have been reviewed elsewhere in detail.[23] Centrifuge-based biochemical fractionation enriches subcellular structures based on the characteristics, such as size, shape, and density. Popular options in this category include density-gradient centrifugation and differential centrifugation.[23] Affinity-based purification of organelles, which employs antibodies against native or tagged organellar surface proteins to enrich subcellular structures, has also been used for the analysis of subcellular proteomes with high specificity.[2,23–26] Once subcellular structures have been isolated or enriched, proteins are digested, separated using offline and/or online liquid chromatography (LC), and subsequently analyzed using high-resolution tandem MS (MS/MS or MS^n). Since the subcellular location of proteins is often modulated through the addition or removal of chemical modifications, it is advantageous to perform PTM-level analyses. Peptides that are post-translationally modified can be enriched prior to LC–MS/MS analysis using selective PTM enrichment strategies. Information about protein identity and quantity in each fraction is then inferred from the LC–MS/MS data using a variety of sophisticated data analysis tools.[27] Protein contents in the fractions can be analyzed in series (label-free) or in parallel (multiplexed using *in vitro* isobaric mass labels). *In vivo* labeling options, such as stable isotope labeling using amino acids in cell culture (SILAC),[28] are also an option for multiplexed measurements of different conditions but have been less utilized for global organellar maps. Over the years, advances in multiplexing capabilities and MS data acquisition have improved the overall throughput, quantitative accuracy, and resolution of spatial proteomics.[29–31] Implementation of machine-learning-based data analysis pipelines has also significantly improved interpretation of the high-dimensional spatial proteomic datasets.[32–35] MS-based proteomic strategies for protein identification and quantification in different subcellular compartments are discussed below and summarized in Table 1.

2.1. *Protein correlation profiling*

Protein correlation profiling (PCP) operates based on the premise that proteins residing together in a subcellular compartment will exhibit a similar distribution profile across a set of collected fractions that have been enriched based on biophysical characteristics (size, density, or shape) of the compartments.[36] This concept was first demonstrated for centrosomes enriched from cell lysate using sucrose density gradient centrifugation.[36] Peptides of known centrosomal proteins showed similar abundance profiles across five collected fractions, whereas contaminant proteins exhibited different profiles and therefore could be distinguished from centrosomal proteins. The proteomes of several subcellular structures, including proteasomes, ribosomes, and transport vesicles have been profiled using PCP to answer targeted questions.[36–43] However, PCP has also been applied on the whole cell level by applying density gradient centrifugation to cell lysate collected from mouse liver tissue. In this study, protein abundance profiles were generated across 32 fractions, and 1400 proteins were localized to ten different subcellular compartments.[44] The subcellular contents of six different mouse organs were surveyed using the PCP approach.[45] This approach has mostly employed label-free MS applied to analyze organelle-enriched fractions; however, PCP has also been applied to SILAC-labeled samples to study stoichiometry and temporal changes to protein–protein interactions.[46,47] More recently, Mann and coworkers have incorporated phosphopeptide correlation profiling to study protein relocation together with

Table 1. Available MS-based approaches for delineating subcellular proteomes and local interactome networks. The lists of cell lines, organisms/living systems, and subcellular locales studied using each approach are not exhaustive, but rather representative examples. Cell lines are human unless otherwise specified.

Approach	Generations	Cell lines	Organisms/living systems	Subcellular locales
Protein correlation profiling (PCP)[36–48]	N/A	Lymphoblastic KE37 cells,[36] HeLa,[39,41,45,46] *Drosophila* S2,[40,41] HEK-293T[43]	Parasites (*Plasmodium falciparum*),[38] yeast (*Saccharomyces cerevisiae*),[37,42] mice (brain, heart, kidney, liver, lung, placenta)[44,45,48]	Multiple subcellular compartments,[44,45] centrosome,[36] mitochondrial contact site complex,[37] proteasome,[38] clathrin-coated vesicles,[39,41] lipid droplets,[40,48] mitochondria,[42] lysosomes,[43] protein complexes[46,47]
Localization of proteins by isotope tagging (LOPIT)[49–62]	LOPIT,[49–55] hyperLOPIT,[56–58,60,62] LOPIT-DC[59,61]	Murine pluripotent embryonic stem cells (E14TG2a),[56,57] U2OS,[59] MRC5 lung fibroblasts (infected with human cytomegalovirus),[58] chicken DT40 lymphocytes,[52] HeLa,[61] Flp-In T-REx 293,[61] human THP-1 monocytes[62]	Plants (*Arabidopsis thaliana*),[49–51,54,55] fruit flies (*Drosophila melanogaster*) embryos,[53] parasites (*Toxoplasma gondii*)[60]	Multiple subcellular compartments,[49–53,56,60,62,63] Golgi apparatus,[55,61] trans-Golgi network[54]

Dynamic organellar maps (DOM)[64,65]	N/A	HeLa,[64] murine primary neurons[65]		Multiple subcellular compartments[64,65]
Affinity purification (AP-MS)[66–71]	N/A	HEK-293T,[66,68–71] HeLa,[67–70] Huh-7,[68–70] A549,[69,70] Caco-2,[70] primary cells (monocyte derived-macrophages, umbilical vein endothelial cells),[68] Vero (*Cercopithecus aethiops*),[69,70] BHK (*Mesocricetus auratus*),[69] Aag2 (*Aedes aegypti*),[69] C6/36 (*Aedes albopictus*)[69]	Fruit flies (*D. melanogaster*)[69]	Interactomes[66,67] Host–pathogen protein–protein interactomes[68–71]
BioPlex[72–74]	BioPlex,[72] BioPlex 2.0,[73] BioPlex 3.0[74]	HEK-293T,[72–74] HCT116,[73,74] HeLa,[73] MCF10A[73]		Interactomes[72–74]
Proximity labeling				

(*Continued*)

Table 1. (*Continued*)

Approach	Generations	Cell lines	Organisms/living systems	Subcellular locales
Ascorbate peroxidase (APEX)[75–80]	N/A	HEK-293T,[75–80] U2OS,[80] Flp-In T-REx 293[80]		Mitochondria[76,77]
Biotin identification (BioID)[81–89]	BioID[81,82] BioID2[84] TurboID and miniTurbo[87]	HEK-293T,[82,85–88] HeLa,[85,86,88] T-REx 293,[83] RPE-1[83]	Bacteria (*Escherichia coli*),[81] worms (*Caenorhabditis elegans*),[87] fruit flies,[87] mice[89]	Cilium–centrosome interface,[83] plasma membrane,[88] Golgi apparatus,[86] mRNA granules[85]
RaPID[90]	N/A	HEK-293T,[90] Huh-7[90]		RNA–protein interactions[90]
Hybrid approaches[91,92]	MAC-tag,[91] GFP (green fluorescent protein)-tag/imaging[92]	HeLa,[91] U2OS,[91] DU-195,[91] Flp-In T-REx 293[91]	Algae (*Chlamydomonas reinhardtii*)[92]	Pyrenoids[92]

organelle-specific phosphorylation changes in a mouse model for non-alcoholic fatty acid liver disease.[48]

2.2. *Localization of proteins by isotope tagging*

One of the most extensively employed spatial proteomic workflows to date is the localization of proteins by isotope tagging (LOPIT), which operates similarly to PCP but utilizes *in vitro* isobaric labeling for multiplexed analysis of organelle-enriched compartments.[49,50] There have been three iterations of LOPIT: original LOPIT,[49,50] hyperplexed LOPIT (hyperLOPIT),[56,57] and most recently, LOPIT with differential centrifugation (LOPIT-DC).[59] First-generation LOPIT applied biochemical fractionation using density gradient ultracentrifugation to separate components of whole-cell lysate by their characteristic densities. The approach was combined with multiplexed high-resolution MS to simultaneously quantify and determine the steady-state distribution of hundreds of proteins in subcellular compartments per experiment. Proteins in each fraction were digested, and the peptides were subsequently labeled *in vitro* with isobaric mass tags, such as iTRAQ or tandem mass tags (TMTs), which allow for multiplexed analysis of the fractions and simultaneous relative quantification of peptides across the fractions.[93] Importantly, organelles are not purified in any given fraction but instead are differentially enriched and have distinct enrichment profiles across fractions. Therefore, proteins residing in the same organelle are expected to distribute similarly across fractions, allowing their assignment to organelles using classification algorithms based on similarities of their enrichment profiles with known organellar markers. Ten years later, enhanced multiplexing capability, advanced MS data acquisition, and multivariate analysis led to the introduction of hyperLOPIT, an expansion of the original LOPIT workflow, which allowed for thousands of proteins to be simultaneously quantified and localized per experiment.[56,57] Specifically, the implementation of TMT 10-plex labeling expanded the multiplexing capacity of a given experiment to ten samples.[29] This allowed for a larger number of subcellular fractions to be collected, increasing spatial resolution of the analysis. The accuracy of TMT-based quantitation was improved with the implementation of the synchronous precursor selection MS^3 (SPS-MS^3), an MS data acquisition strategy that combines the gains in quantitative accuracy and precision of MS^3 without loss of sensitivity.[30] The increased number of fractions collected allowed for the resolution of suborganellar structures and large protein complexes. This is a vast improvement on the original LOPIT, which completely excluded certain subcellular niches, like the nucleus and cytosol. With the ability to include more fractions in the multiplexed analysis, hyperLOPIT allows for the resolution of suborganellar structures, such as *cis*- and *trans*-Golgi networks. Proteins from large complexes, like the mitochondrial adenosine triphosphate (ATP)–synthase complex and the COP9 signalosome, were also found to cluster together when visualized with principal component analysis (PCA). HyperLOPIT combined with time-course analysis has allowed for the generation of spatiotemporal proteomic profiles.[58,62] Cristea and coworkers implemented the hyperLOPIT workflow across an infection time course to study protein translocation during human cytomegalovirus infection.[58] Mulvey *et al.* also applied this strategy to study proinflammatory response to lipopolysaccharide in human monocytes.[62] The most recent iteration of the LOPIT workflow swaps density gradient ultracentrifugation for differential centrifugation as a means of biochemical fractionation, inspired by the dynamic organellar map (DOM) workflow (discussed in Section 2.3). While density gradient ultracentrifugation depends on the sedimentation of proteins and organellar structures that are denser than the suspension, differential centrifugation has no such physically imposed lower limit and operates to separate all structures by their sedimentation rates. This version of the workflow, called LOPIT-DC, capitalizes on the resolution and quantitative accuracy gains of hyperLOPIT with sample multiplexing and SPS-MS^3 but requires less sample and presents an overall simpler protocol using differential ultracentrifugation. Most recently, LOPIT-DC was combined with a proximity labeling–based assay to characterize the contents of trafficking vesicles redirected to mitochondria.[61] LOPIT and all its iterations have enabled an abundance of different experiments on different cell lines and tissues and have provided important biological insights.[51–55,58,60]

The SubCellBarCode technique also used subcellular fractionation and TMT labeling to investigate protein subcellular location across five different human cancer cell lines.[94] This allowed for a comparison of subcellular location for proteins from over 12,000 genes. They used these data to understand how alternative splicing and protein domains affected protein localization and complex formation.

2.3. *Dynamic organellar maps*

The DOM workflow uses differential ultracentrifugation, a simpler alternative to density gradient fractionation, to obtain organelle-enriched fractions. The DOM strategy has been adapted for SILAC, TMT, or label-free quantitation.[64,65] The first demonstration of DOM utilized SILAC labeling in HeLa cells, which resulted in a database of 8700 proteins with localization and absolute copy number.[64] Postnuclear cell homogenate from SILAC heavy cells was used as a total organellar reference, while postnuclear homogenate from SILAC light cells was subjected to five subsequent differential centrifugation steps to collect organelle-enriched fractions. Each of the six light organelle-enriched fractions were spiked with the heavy total organellar reference prior to MS analysis. Label-free quantitative data were also generated for nuclear, organellar, and cytosolic fractions. PCA analysis revealed the resolution of all major organelles, including sub-organellar structures such as endoplasmic reticulum (ER) lumen and ER membrane. Using the "proteomic ruler approach,"[95] which estimates absolute copy number without the use of spiked isotope-labeled standards, they followed up with epidermal growth factor (EGF) stimulation to demonstrate the capability to capture physiological protein movement with the DOM approach. For untreated cells, epidermal growth factor receptor (EGFR) was localized to the plasma membrane (PM) cluster and localized to the endosomal cluster in the EGF-stimulated cells, recapitulating the well-documented endosomal uptake of EGFR. DOM was extended to tissue using both label-free and TMT labeling to map the spatial proteome of mouse primary neurons.[65]

2.4. *Affinity purification–based mass spectrometry and proximity labeling*

Proteins and their interacting partners make up a local spatial proteome that can be probed using affinity purification–based mass spectrometry (AP-MS). AP-MS allows the collection of information about protein interactions by using a target or "bait" protein to pull down interacting proteins and has proved a useful tool for deciphering cellular spatial organization.[96] To date, a few groups have undertaken large-scale interactome studies to characterize global protein–protein interaction networks.[66,67,72–74] Notably, Mann and coworkers generated a library of 1125 green fluorescent protein (GFP)-tagged proteins expressed in HeLa cells and used an antibody against GFP to probe bait–protein interactions.[67] Metrics such as abundance, stoichiometry, and specificity of interactions created a multidimensional interactome map with 28,500 interactions between 5400 human proteins.[67] Huttlin *et al.* similarly utilized high-throughput AP-MS to create protein–protein interaction resource that they called BioPlex.[72] They identified 23,744 interactions among 7668 proteins in human HEK cells by performing immunoprecipitation (IP) pull-downs for 2594 FLAG-HA-tagged baits derived from the lentiviral expression library of human open reading frames.[72,97] This study laid the grounds for BioPlex 2.0 and 3.0, which collectively represent the largest human interactome databases, to date.[73,74] The BioPlex databases represent an enormous amount of resources and experimental time but provide an important resource for researchers seeking to understand protein interactomes. Importantly, they validated thousands of protein interactions through comparison between two different human cell lines, demonstrating the utility of such datasets for future interaction-based biological inquiries.[74] AP-MS has also been implemented on a smaller scale to determine protein–protein interactions for a single target protein or set of proteins of interest. For instance, AP-MS has been used to identify interactions between human and viral proteins from the

Ebola virus,[68] flaviviruses,[69] and most recently, coronaviruses,[70,71] with the purpose of identifying druggable targets and antiviral treatments.

Spatial information can also be collected using AP-MS coupled with proximity-labeling strategies.[98,99] Popular methods include engineered soybean ascorbate peroxidase (APEX) and proximity-dependent biotin identification (BioID),[81,82] which utilizes various mutants of biotin ligases derived from *Escherichia coli*. For these methods, information about proteins proximal to complexes or structures of interest can be collected by tagging a bait protein with an enzyme capable of biotinylating proteins that come into its proximity. Any protein that was biotinylated can then be pulled down using affinity purification with streptavidin beads and subsequently analyzed by LC–MS/MS. Proximity labeling can be applied to understand specific protein–protein interactions, but these methods have also been applied to delineate protein interactomes within specific organelles such as mitochondria,[76,77] PM,[88] Golgi apparatus,[86] RNA granules,[85] and the centrosome–cilia interface.[83]

Because APEX proximity labeling can be accomplished on a sub-minute timescale, it has been used to delineate protein interaction cascades that occur quickly, such as G-protein-coupled receptor (GPCR) signaling.[79,80] Using APEX and TMT labeling, Paek *et al.* demonstrated time-resolved measurements of GPCR signaling and internalization in living cells.[80] Lobingier *et al.* similarly combined quantitative MS and APEX-mediated labeling to study GPCRs, using multiple protein–APEX constructs as spatial references to generate protein interaction networks with high spatial and temporal resolution.[79] In addition, APEX has been combined with RNA–protein chemical crosslinking to map RNA associated with subcellular compartments such as the nucleus, cytosol, mitochondrial matrix, ER, and suborganellar spaces such as mitochondrial–ER junctions and nuclear lamina.[78] The versatility of APEX-based labeling has also enabled its use as a reporter for electron microscopy.[75] One drawback of APEX-based proximity labeling is its use of H_2O_2, which exhibits high toxicity in living cells.

Unlike APEX-based techniques, BioID does not use toxic reagents and is therefore much more compatible with use in living systems, enabling *in vivo* studies. Compared to APEX, BioID is also a simpler technique since it only requires a biotin supply for labeling. Using directed evolution, two new mutants of biotin ligase, TurboID and miniTurbo, were developed by Ting and coworkers. These new enzymes catalyze proximity labeling reactions in under 10 min compared to the first- and second-generation BioID enzymes[81,82,84] and another comparable biotin ligase-based system, RaPID,[90] typically requires 18+ h for labeling.[87] Improved catalytic activity allowed the extension of TurboID proximity labeling to living organisms, flies, and worms.[87] Non-toxic, fast, and efficient proximity-labeling approaches like TurboID will surely enable many more *in vivo* experiments in the coming years. More recently, proximity labeling was applied *in vivo* to study the signaling dynamics in the basic unit of muscle contraction, the sarcomere, using CRISPR/Cas9 gene editing to generate titin-BioID knock-in mice.[89] Hybrid and orthogonal approaches to study protein–protein interactions and local spatial proteomes have also been developed. A versatile multiple approaches combined (MAC)-tag consists of an AP-MS compatible sequence, a biotin ligase, and a hemagglutinin (HA) epitope sequence allowing for AP-MS, BioID proximity labeling, and HA-based approaches (such as immunohistochemistry) in a single protein construct.[91] The MAC-tag approach was applied to characterize protein–protein interactions at 18 different subcellular locations, as well as three suborganellar locations. Using GFP-tagged proteins, researchers have also combined AP-MS with orthogonal imaging to survey algal pyrenoid proteins.[92]

3. Spatial proteomics of subcellular compartments

Levels of subcellular organization can be divided broadly into three groups: membrane-bound organelles, non-membrane-bound organelles, and macromolecular structures and protein complexes. This section will expand on the general MS-based spatial proteomic workflows described above by highlighting techniques used to profile proteomes of specific subcellular compartments. As discussed in the above sections, subcellular structures can be enriched through

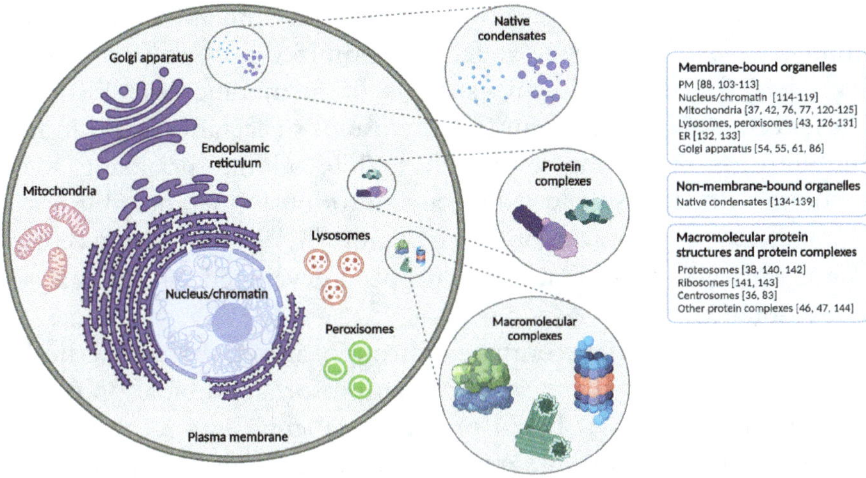

Figure 1. References to specific spatial proteomic studies organized by subcellular compartment or structure. Figure created with BioRender.com.

centrifuge-based biochemical fractionation methods, such as density gradient centrifugation or differential centrifugation. More specific and sensitive strategies for subcellular compartment isolation involve affinity-based purification targeting the epitopes of native or tagged compartment-specific proteins. Below, we will highlight some other clever strategies researchers have implemented to study subcellular proteomes.

3.1. *Membrane-bound organelles*

3.1.1. *Plasma membrane*

The PM is the outer bound of eukaryotic cells and sequesters intracellular activity from extracellular space. Structurally, the PM is a phospholipid bilayer composed of membrane-bound and membrane-associated proteins that function as ion channels, transporters, and cell-surface receptors. These proteins are important drug targets and can exist as integral membrane proteins or be anchored to the PM through lipid modifications.[100] PM proteins are challenging to study due to their hydrophobic properties and generally low-abundance relative to other proteins in the whole-cell proteome.[100,101] Nevertheless, a variety of optimized protocols have been developed for the proteomic analysis of membrane proteins.[101,102] These techniques typically involve the use of surfactants to aid in the solubilization of membrane proteins. While anionic surfactants such as sodium dodecyl sulfate (SDS) are notoriously incompatible with MS analysis, various acid-labile surfactants,[103–106] and most recently, a photocleavable surfactant[107,108] have been developed for MS analysis of membrane proteins.

PM proteins can be enriched through affinity-based methods, centrifuge-based biochemical fractionation, or a combination of the two approaches. BioID-based proximity labeling was recently used to isolate and identify integral PM proteins using a PM-localized biotin ligase.[88] Using this strategy, proteins in proximity to the lipid-modified biotin ligase were biotinylated. The PM was then enriched from whole-cell lysate using sucrose density-gradient fractionation and the biotinylated proteins were subsequently affinity-purified from the PM-enriched fraction with streptavidin beads. Weekes *et al.* also used biotinylation to isolate the PM by oxidizing and subsequently biotinylating sialic acid residues of glycosylated cell-surface proteins.[109] The biotinylated glycoproteins can then be pulled down using streptavidin. They used this strategy in combination with SILAC-based quantitative proteomics to identify proteins with cell surface expression that is regulated by the ER chaperone, gp96.[109] This technique for isolation of PM proteins was also used in a quantitative temporal viromics study to investigate host–pathogen interactions.[110] Specifically, human PM proteins were isolated at different time points postinfection with human cytomegalovirus and labeled using TMT for quantitative MS analysis.[110] Affinity-based methods of isolation are more specific than centrifuge-based methods, whereas traditional biochemical fractionation approaches often yield fractions with contamination from other subcellular compartments.[111,112] Regardless of the isolation technique, PM proteins remain an important target for analysis since many therapeutics target the cell-surface proteome.

3.1.2. *Nucleus and chromatin*

Each cell contains a nucleus housing tightly packaged nucleic acids, which encode genetic information. Proteins (e.g., transcription factors) frequently shuttle between the cytosol and nucleus to initiate and regulate gene expression. Key insights can be gleaned by studying the nuclear proteome with respect to disease since dysregulation of nuclear protein localization has been implicated in cellular dysfunction and cancer. Several centrifuge-based purification protocols and kits exist for the isolation of nuclei from whole-cell lysate.[113] Specific nuclear proteins can be further isolated by IP to identify protein — protein interactions.[114,115] One such approach, chromatin IP coupled to sequencing combined with MS-based proteomics, relies on the IP of chromatin following crosslinking with histones to generate a chromatin proteomic profile associated with specific gene sequences.[116] The evolution of techniques like this has yielded studies on the composition of native heterochromatin and euchromatin in recent years by using nucleosome-binding proteins as IP targets.[117] A comparison of the nuclear proteomes of acute myeloid leukemia (AML) blasts with normal human CD34+ cells helped to identify transcription factors that were misexpressed in AML.[118] Transcript levels measured by transcriptomics were not predictive of the transcription factor protein expression levels, highlighting the importance of integrating nuclear proteomics with transcriptomics.

3.1.3. *Mitochondria*

Mitochondria are key to energy metabolism and ATP production but are also important in apoptosis, ion homeostasis, and signaling.[119,120] Dysfunction of mitochondria has been linked to up to 50 diseases, so comparative analyses between health and disease states are important. APEX-proximity labeling has been used for mapping the mitochondrial proteome with suborganellar resolution, differentiating between proteins in the mitochondrial matrix and intermembrane space.[76,77] An important resource for the mitochondrial proteome is

MitoCarta.[121–123] Pagliarini *et al.* first created a mitochondrial protein inventory spanning 14 mouse tissues, cataloging 1098 mitochondrial genes. They used quantitative MS, machine learning, and GFP-based microscopy. Using improved methods, MitoCarta2.0 reconstructed the mouse mitochondrial catalog and included a human mitochondrial catalog, which laid an important framework for the mammalian mitochondrial proteome.[122] Most recently, they updated the resource, MitoCarta3.0, to 1136 mitochondrial genes and included subannotations for the mitochondrial matrix, inner membrane, intermembrane space, and outer membrane.[123] These resources combined will enable studies of mitochondrial biology and comparative proteomic analyses in the context of mitochondrial-related diseases. The Italian Mitochondrial Human Proteome Project Initiative set out to standardize mitochondrial proteomics. They evaluated three different enrichment protocols (differential centrifugation, sucrose gradient separation, and commercial isolation kit) for ten different cell lines with data from 17 different laboratories.[124] Based on their combined analyses, they provide useful guidelines for the enrichment and analysis of mitochondrial proteins from different cell lines, with no firm recommendation for the best enrichment protocol. PCP and proximity labeling have also been utilized to characterize both yeast and mammalian mitochondrial and submitochondrial proteomes.[37,42,76,77]

3.1.4. *Lysosomes, peroxisomes, endosomes, and exosomes*

The suffix "-some" in Latin denotes a body, in this case, separate identifiable vesicles within the cell that carry out functions that require biochemical conditions different from that of the cytosolic environment (e.g., concentration of reagents, pH). Throughout the cell, enclosed vesicles that comprise a variety of organelles carry out a myriad of functions. PM-derived vesicles include endosomes, lysosomes, peroxisomes, as well as exosomes that are secreted from the cell. Endosomes and exosomes are the vesicles involved in endocytosis and exocytosis, respectively. Endocytosis and exocytosis are important in the processes of pathogenesis and cell signaling. While many substances can pass through the PM via passive diffusion or with the aid of transmembrane transport membranes, the engulfment of a whole bacterium by a macrophage or the release of hormones and bulk transportation of signaling proteins from the cell involves either endocytosis or exocytosis. Lysosomes maintain an acidic internal pH, are involved in cell signaling, energy metabolism, and the degradation and recycling of cellular waste, and are a part of endocytosis, phagocytosis, and autophagy.[125] Peroxisomes play a role in sequestering oxidative reactions and are important for the detoxification of reactive oxygen species.[126] Analysis of this wide variety of membrane-bound organelles is possible by targeting biochemically distinct trafficking vesicles by their characteristic surface markers or contents. For instance, superparamagnetic iron oxide nanoparticles can be engineered to target specific subcellular compartments, such as late endosomes and lysosomes.[127] This approach has been established as a standardized magnetic isolation procedure for lysosomal proteomes and lipidomes.[128,129] Traditional subcellular fractionation approaches, such as differential centrifugation or sucrose density gradient fractionation, can also enrich lysosomes and other vesicles; however, with less specificity and more contamination compared to affinity-based approaches. Furthermore, researchers who seek to study lysosome metabolism have found biochemical fractionation methods to be too slow to preserve the lysosomal metabolome. For this reason, Abu-Remaileh *et al.* developed "LysoIP," which uses IP to pull down HA-tagged transmembrane protein 192, a protein that is retained on lysosomes.[130]

3.1.5. *Endoplasmic reticulum and Golgi apparatus*

The smooth and rough ER is responsible for lipid production as well as protein production, via ribosomes, within the rough ER. Their malleable nature can make discrete biochemical fractionation difficult since the properties of these organelles can be variable. For this reason, an approach that has emerged and gained favor is the organelle-specific labeling and isolation of said tags for proteomic analysis. One such approach was designed by creating ER-localizable reactive molecules (called ERMs), which spontaneously localize within the ER and allow for easy enrichment for studying the ER proteome.[131] Other popular approaches similarly entail the construct of fluorescent, GFP, RFP, and so on, tagged proteins for purification, though this requires targeted prior knowledge of protein localization.[132]

LOPIT is especially well-suited for studying the Golgi apparatus and its associated network and vesicles. Since the strategy does not require organelle purification and is well-suited for characterization of these transient and low-abundance structures which are otherwise difficult to isolate.[54,55,61] The Golgi apparatus and trans-Golgi network are the packaging centers of the cell, responsible for the secretory pathway and intracellular trafficking via vesicles. The Golgi apparatus has a well-conserved feature that lends itself to targeting in experiments: the Golgin coiled-coil proteins or Golgins. A structural protein of the Golgi stacks, with an integral role in vesicle shuttling, these structures can be identified in the mitochondrial fraction from LOPIT-DC fractionation to identify Golgi-origin vesicles in those fractions, as a way to overcome the transitory and short-lived nature of these vesicles that shuttle throughout the cell, as shown by Shin *et al.*[61] This provides an excellent example of the use of the LOPIT-DC procedure followed by protein correlation profiling to yield novel proteomic results.

3.2. *Non-membrane-bound organelles*

Without the defined barrier of a membrane, the isolation and subsequent proteomic analysis of native or biomolecular condensates can be challenging. Liquid–liquid phase separation (LLPS) is a physical phenomenon by which a solution self-separates into a dilute and condensed phase and is responsible for the suborganellar organization in many parts of the cell or within the cytosol.[133] These bodies are commonly seen in the nucleus for the distinct organization of chromatin but occur throughout the cell. RNA and RNA-binding proteins have also been reported as important components which contribute to the formation of native condensates.[134] These structures play important roles in disease states but can be difficult to target for proteomic analysis without the fractions available in membrane-bound organelles; therefore, AP-MS, proximity labeling, and microscopy have been essential in probing the composition, structure, and dynamics of native condensates.[135] Pathologic aggregates as native condensates have been a particular focus, like those involved in ALS or Tau aggregates. One study created fluorescent protein fusion constructs, that is, tau-GFP to pull down native condensates.[136] The list of known proteins and native condensate structures that involve themselves in LLPS grows every year, and databases have been made to catalog these interactions.[137,138]

3.3. *Macromolecular protein structures and protein complexes*

3.3.1. *Ribosomes and proteasomes*

Suborganellar structures like ribosomes and proteosomes encompass the beginning and end of protein lifecycles and are extremely important in various ribosomopathies, and diseases like cardiomyopathies or pathogenesis of infectious diseases.[139,140] Ribosomes are the mechanistic building blocks required for protein production in the ER as well as providing precursors to several PTM processes throughout the cell important in regulation, cell signaling, or protein degradation. Ribosomes are either free or fixed. The fixed ribosomes are membrane-bound and will be enriched in fractions containing the ER. Free ribosomes can be found throughout the cell. The ubiquitin–proteasome system is responsible for the post-translational tagging of proteins with ubiquitin units to mark them for digestion by proteasomes.[141] This process is incredibly important for the digestion of cell debris, recycling of amino acids, and clearance/resolution of infections or immune responses. Both ribosomes and proteasomes can be purified by centrifuge-based enrichment methods or affinity-based purification. One interesting application of ribosomal proteomics is the study of protein synthesis dynamics using an approach that combines ribosomal profiling, which sequences the mRNA fragments in actively translating ribosomes, and pulsed SILAC, which isotopically labels newly synthesized proteins for subsequent MS analysis.[142]

3.3.2. *Protein complexes*

Protein complexes are composed of various interacting proteoforms that have various structural and functional roles in the cells. A common approach for the study of protein complexes is AP-MS, which requires a known protein target.[137] Protein complexes have also been studied by MS using size-based fractionation, such as native size exclusion chromatography (SEC) or electrophoretic methods, such as clear native polyacrylamide electrophoresis.[96,143] These methods rely on the size-based fractionation of protein complexes from whole-cell lysate, followed by in-fraction or in-gel digestion of the complexes and LC–MS analysis. By nature, these strategies are excellent discovery-based tools because they are unbiased and do not require target-specific antibodies. Specifically, Kristensen *et al.* used SEC combined with PCP and SILAC labeling to quantitatively study protein–protein interactions and measure temporal changes in interactomes.[46,47]

4. Closing remarks

An inherent drawback of most quantitative proteomic experiments is the normalization of relative quantification between experimental replicates, fractions, or cell states. With the growing spectrum of isobaric tag technologies (iTRAQ, TMT, etc.) and other labeling techniques (SILAC), the field of proteomics is seeing a steady growth in multiplexing capabilities. Multiplexing allows for relative quantitation of the same protein across multiple fractions or conditions in a single MS run and reduces missing values, an issue inherent to label-free proteomics, with the added benefits of reduced MS analysis time and improved quantitative accuracy. Improved multiplexing capabilities have allowed for the simultaneous analysis of an increasing number of organelle-enriched fractions which have resulted in increased subcellular resolution. With the advent of TMT-pro (16-plex)[144] and potentially higher numbers of available channels in the future, more fractions and conditions can be compared at once, leading to higher-resolution analyses.

The sensitivity, resolution, and proteomic coverage achieved by methods involving TMT labeling (e.g., hyperLOPIT, SubCellBarcode) come at a cost of analysis throughput. Because subcellular fractions are pooled following labeling, extensive offline fractionation is required prior to MS analysis to achieve deep proteome coverage. Recognizing this throughput issue, Martinez-Val *et al.* recently developed a label-free method capable of profiling the proteome and phosphoproteome across six subcellular fractions in just five hours of MS time per sample without sacrificing proteomic coverage.[145] To do this, they used a sequential subcellular fractionation strategy combined with fast LC–MS/MS and direct data-independent acquisition for sensitive and reproducible subcellular maps of the proteome and phosphoproteome. With decreased analysis time, they demonstrate higher-complexity experiments with multiple biological replicates and time points for spatiotemporal analysis of EGFR signaling and stress conditions *in vitro* and *in vivo*. This method is significant because it generates data on phosphoproteomic dynamics, which is known to be important in driving protein translocation.

Alongside the growth in multiplexing capabilities, researchers are requiring more robust data analysis workflows to uncover biologically relevant findings in the LC–MS data. New advances with post-LC–MS data pipelines like the TRanslocation ANalysis of SPatIal pRoTEomics (TRANSPIRE)[146] can analyze data using machine learning to train probabilistic Gaussian process classifiers to predict protein distribution and translocation, which is more effective than traditional Bayesian models from fraction translocation profiles.

With the focus on MS-based methods, other spatiotemporal proteomic methods have been mentioned sparingly in this chapter; however, it is very important to stress that even though MS-based techniques are becoming more robust every year, many complementary methodologies are of great importance for verification and visualization of data in other ways. Electrophoretic gels are still a staple for the confirmation of protein–protein interactions and have used in the analysis of fractions as well and should never be discounted, though specific antibodies are required. Furthermore, the time-resolved dynamic resolution of immunofluorescence and imaging are valuable tools for the verification of MS-based discoveries and the illustration of these discoveries within the cell.

The arc of biomedical research is toward systems biology. In line with this, the mass spectrometer is perhaps one of the most powerful tools at our disposal. In a history that began with two-dimensional gel analysis and imprecise analysis of analytes and purified substances, the rapid growth over three decades of MS-based proteomics has been the story of growing complexity. The dynamic range and sensitivity of LC–MS platforms, coupled with growing computational power, topped by ever-evolving data analysis pipelines have given proteomics one of the most powerful toolkits for analyzing biological systems we could imagine. It is important to remember that proteomic researchers rely heavily on the advancements of genomics and transcriptomics. In this way, the strongest biological research is systems-based, incorporating all parts of the central dogma of biology. In the same spirit, spatiotemporal research is the best way to capture the full picture of disease pathogenesis or other pathologies. We hope this chapter has clearly illustrated the importance and the various applications of spatial proteomics at the subcellular level.

Acknowledgments

This research was supported by the Intramural Research Program of the National Institutes of Health and the National Institute of Allergy and Infectious Diseases.

References

1. Bauer NC, Doetsch PW, Corbett AH. Mechanisms Regulating Protein Localization. (1600-0854 (Electronic)).
2. Yates JR, 3rd, Gilchrist A, Howell KE, *et al.* Proteomics of Organelles and Large Cellular Structures. *Nat. Rev. Mol. Cell Biol.* Sep 2005; 6(9):702–714. doi: 10.1038/nrm1711.
3. Cody NA, Fau IC, Lécuyer E. The Many Functions of mRNA Localization During Normal Development and Disease: From Pillar to Post. (1759-7692 (Electronic)).

4. De Matteis MA, Luini A. Mendelian Disorders of Membrane Trafficking. *N. Engl. J. Med.* 2011;365(10):927–938 (accessed 2021-08-24).
5. Shin SJ, Smith JA, Rezniczek GA, *et al.* Unexpected Gain of Function for the Scaffolding Protein Plectin Due to Mislocalization in Pancreatic Cancer. *Proc. Natl. Acad. Sci.* 2013;110(48):19414.
6. Guardia CM, De Pace R, Mattera R, *et al.* Neuronal Functions of Adaptor Complexes Involved in Protein Sorting. (1873-6882 (Electronic)).
7. Meyer K, Kirchner M, Uyar B, *et al.* Mutations in Disordered Regions Can Cause Disease by Creating Dileucine Motifs. (1097-4172 (Electronic)).
8. Aridor M, Hannan LA. Traffic Jam: A Compendium of Human Diseases that Affect Intracellular Transport Processes. (1398-9219 (Print)).
9. Olkkonen VM, Ikonen E. When Intracellular Logistics Fails — Genetic Defects in Membrane Trafficking. (0021-9533 (Print)).
10. Valastyan JS, Lindquist S. Mechanisms of Protein-Folding Diseases at a Glance. (1754-8411 (Electronic)).
11. Aridor M, Hannan LA. Traffic Jams II: An Update of Diseases of Intracellular Transport. (1398-9219 (Print)).
12. Wang ET, Taliaferro JA-O, Lee JA-O, *et al.* Dysregulation of mRNA Localization and Translation in Genetic Disease. (1529-2401 (Electronic)).
13. Wilhelm M, Schlegl J, Hahne H, *et al.* Mass-Spectrometry-Based Draft of the Human Proteome. *Nature.* 2014;509(7502):582–587.
14. Sharma K, Schmitt S, Bergner CG, *et al.* Cell Type– and Brain Region–Resolved Mouse Brain Proteome. *Nat. Neurosci.* 2015;18(12): 1819–1831.
15. Aebersold R, Mann M. Mass-Spectrometric Exploration of Proteome Structure and Function. *Nature.* 2016;537(7620):347–355.
16. Doll S, Dreßen M, Geyer PE, *et al.* Region and Cell-Type Resolved Quantitative Proteomic Map of the Human Heart. *Nat. Commun.* 2017;8(1):1469.
17. Larance M, Lamond AI. Multidimensional Proteomics for Cell Biology. *Nat. Rev. Mol. Cell Biol.* 2015;16(5):269–280.

18. Lundberg E, Borner GHH. Spatial Proteomics: A Powerful Discovery Tool for Cell Biology. *Nat. Rev. Mol. Cell Biol.* 2019;5:285–302.
19. Andersen JS, Mann M. Organellar Proteomics: Turning Inventories into Insights. (1469-221X (Print)).
20. Uhlén M, Fagerberg L, Hallström BM, *et al.* Tissue-Based Map of the Human Proteome. *Science.* 2015;347(6220):1260419.
21. Mattiazzi Usaj M, Styles EB, Verster AJ, *et al.* High-Content Screening for Quantitative Cell Biology. *Trends Cell Biol.* 2016;26(8):598–611.
22. Thul PJ, Åkesson L, Wiking M, *et al.* A Subcellular Map of the Human Proteome. *Science.* 2017;356(6340):eaal3321.
23. Lee YH, Tan HT, Chung MCM. Subcellular Fractionation Methods and Strategies for Proteomics. *Proteomics.* 2010;10(22):3935–3956 (accessed 2021-08-24).
24. Nakai W, Yoshida T, Diez D, *et al.* A Novel Affinity-Based Method for the Isolation of Highly Purified Extracellular Vesicles. *Sci. Rep.* 2016; 6(1):33935.
25. Fritsch J, Tchikov V, Hennig L, *et al.* A Toolbox for the Immunomagnetic Purification of Signaling Organelles. *Traffic.* 2019;20(3):246–258. (accessed 2021-08-24).
26. Xiong J, He JA-O, Xie WA-O, *et al.* Rapid Affinity Purification of Intracellular Organelles Using a Twin Strep Tag. LID — 10.1242/ jcs.235390 [doi] LID - jcs235390. (1477-9137 (Electronic)).
27. Sinitcyn P, Rudolph JD, Cox J. Computational Methods for Understanding Mass Spectrometry–Based Shotgun Proteomics Data. *Annu. Rev. Biomed. Data Sci.* 2018;1(1):207–234 (accessed 2021-08-24).
28. Ong S-E, Mann M. A Practical Recipe for Stable Isotope Labeling by Amino Acids in Cell Culture (SILAC). *Nat. Protoc.* 2006;1(6): 2650–2660.
29. McAlister GC, Huttlin EL, Haas W, *et al.* Increasing the Multiplexing Capacity of TMTs Using Reporter Ion Isotopologues with Isobaric Masses. *Anal. Chem.* Sep 4, 2012;84(17):7469–7478. doi: 10.1021/ ac301572t.
30. McAlister GC, Nusinow DP, Jedrychowski MP, *et al.* MultiNotch MS3 Enables Accurate, Sensitive, and Multiplexed Detection of Differential

Expression Across Cancer Cell Line Proteomes. *Anal. Chem.* 2014;86(14):7150–7158.
31. Ting L, Rad R, Gygi SP, *et al.* MS3 Eliminates Ratio Distortion in Isobaric Multiplexed Quantitative Proteomics. (1548-7105 (Electronic)).
32. Gatto L, Vizcaíno JA, Hermjakob H, *et al.* Organelle Proteomics Experimental Designs and Analysis. *Proteomics.* Nov 2010;10(22): 3957–3969. doi: 10.1002/pmic.201000244.
33. Gatto L, Breckels LM, Burger T, *et al.* A Foundation for Reliable Spatial Proteomics Data Analysis. *Mol. Cell. Proteomics.* 2014;13(8):1937–1952.
34. Gatto L, Breckels LM, Wieczorek S, *et al.* Mass-Spectrometry-Based Spatial Proteomics Data Analysis Using pRoloc and pRolocdata. *Bioinformatics (Oxford, England).* 2014;30(9):1322–1324.
35. Breckels LM, Holden SB, Wojnar D, *et al.* Learning from Heterogeneous Data Sources: An Application in Spatial Proteomics. *PLoS Comput. Biol.* 2016;12(5):e1004920–e1004920.
36. Andersen JS, Wilkinson CJ, Mayor T, *et al.* Proteomic Characterization of the Human Centrosome by Protein Correlation Profiling. *Nature.* 2003;6966:570–574.
37. Harner M, Körner C, Walther D, *et al.* The Mitochondrial Contact Site Complex, a Determinant of Mitochondrial Architecture. *EMBO J.* 2011;21:4356–4370.
38. Sessler N, Krug K, Nordheim A, *et al.* Analysis of the Plasmodium Falciparum Proteasome Using Blue Native PAGE and Label-Free Quantitative Mass Spectrometry. *Amino Acids.* Sep 2012; 43(3):1119–1129. doi: 10.1007/s00726-012-1296-9.
39. Borner GH, Antrobus R, Hirst J, *et al.* Multivariate Proteomic Profiling Identifies Novel Accessory Proteins of Coated Vesicles. *J. Cell Biol.* 2012;1:141–160.
40. Krahmer N, Hilger M, Kory N, *et al.* Protein Correlation Profiles Identify Lipid Droplet Proteins with High Confidence. *Mol. Cell. Proteomics.* 2013;5:1115–1126.
41. Borner GH, Hein MY, Hirst J, *et al.* Fractionation Profiling: A Fast and Versatile Approach for Mapping Vesicle Proteomes and

Protein-Protein Interactions. *Mol. Biol. Cell.* © 2014 Borner *et al.* This article is distributed by The American Society for Cell Biology under license from the author(s). Two months after publication it is available to the public under an Attribution–Noncommercial–Share Alike 3.0 Unported Creative Commons License (http://creativecommons.org/licenses/by-nc-sa/3.0). 2014;20:3178–3194.

42. Morgenstern M, Stiller SB, Lübbert P, *et al.* Definition of a High-Confidence Mitochondrial Proteome at Quantitative Scale. *Cell Rep.* Jun 27, 2017;19(13):2836–2852. doi: 10.1016/j.celrep.2017.06.014.
43. Wyant GA, Abu-Remaileh M, Frenkel EM, *et al.* NUFIP1 Is a Ribosome Receptor for Starvation-Induced Ribophagy. *Science.* 2018;360(6390):751.
44. Foster LJ, de Hoog CL, Zhang Y, *et al.* A Mammalian Organelle Map by Protein Correlation Profiling. *Cell.* 2006;1:187–199.
45. Kislinger T, Cox B, Kannan A, *et al.* Global Survey of Organ and Organelle Protein Expression in Mouse: Combined Proteomic and Transcriptomic Profiling. *Cell.* 2006;1:173–186.
46. Kristensen AR, Gsponer J, Foster LJ. A High-Throughput Approach for Measuring Temporal Changes in the Interactome. *Nat. Methods.* 2012;9(9):907–909.
47. Kristensen AR, Foster LJ. Protein Correlation Profiling-SILAC to Study Protein-Protein Interactions. *Methods Mol. Biol.* 2014;1188:263–270. doi: 10.1007/978-1-4939-1142-4_18.
48. Krahmer N, Najafi B, Schueder F, *et al.* Organellar Proteomics and Phospho-Proteomics Reveal Subcellular Reorganization in Diet-Induced Hepatic Steatosis. *Developmental Cell.* Oct 22, 2018;47(2):205–221.e7. doi: 10.1016/j.devcel.2018.09.017.
49. Dunkley TP, Watson R, Griffin JL, *et al.* Localization of Organelle Proteins by Isotope Tagging (LOPIT). *Mol. Cell. Proteomics.* 2004;11:1128–1134.
50. Sadowski PG, Dunkley TP, Shadforth IP, *et al.* Quantitative Proteomic Approach to Study Subcellular Localization of Membrane Proteins. *Nat. Protoc.* 2006;4:1778–1789.
51. Dunkley TPJ, Hester S, Shadforth IP, *et al.* Mapping the Arabidopsis Organelle Proteome. *Proc. Natl. Acad. Sci.* 2006;103(17):6518.

52. Hall SL, Hester S, Griffin JL, *et al.* The Organelle Proteome of the DT40 Lymphocyte Cell Line. (1535-9484 (Electronic)).
53. Tan DJ, Dvinge H, Christoforou A, *et al.* Mapping Organelle Proteins and Protein Complexes in Drosophila Melanogaster. (1535-3893 (Print)).
54. Groen AJ, Sancho-Andrés G, Breckels LM, *et al.* Identification of Trans-Golgi Network Proteins in *Arabidopsis thaliana* Root Tissue. *J. Proteome Res.* 2014;13(2):763–776.
55. Nikolovski N, Rubtsov D, Segura MP, *et al.* Putative Glycosyltransferases and Other Plant *Golgi apparatus* Proteins are Revealed by LOPIT Proteomics. (1532-2548 (Electronic)).
56. Christoforou A, Mulvey CM, Breckels LM, *et al.* A Draft Map of the Mouse Pluripotent Stem Cell Spatial Proteome. *Nat Commun.* 2016;7:9992.
57. Mulvey CM, Breckels LM, Geladaki A, *et al.* Using hyperLOPIT to Perform High-Resolution Mapping of the Spatial Proteome. *Nat. Protoc.* 2017;6:1110–1135.
58. Jean Beltran PM, Mathias RA, Cristea IM. A Portrait of the Human Organelle Proteome in Space and Time During Cytomegalovirus Infection. *Cell Syst.* Oct 26, 2016;3(4):361–373.e6. doi: 10.1016/j.cels.2016.08.012.
59. Geladaki A, Kočevar Britovšek N, Breckels LM, *et al.* Combining LOPIT with Differential Ultracentrifugation for High-Resolution Spatial Proteomics. *Nat. Commun.* 2019;1:331.
60. Barylyuk K, Koreny L, Ke H, *et al.* A Comprehensive Subcellular Atlas of the Toxoplasma Proteome via hyperLOPIT Provides Spatial Context for Protein Functions. *Cell Host Microbe.* 2020;28(5):752–766.e9.
61. Shin JJH, Crook OM, Borgeaud AC, *et al.* Spatial Proteomics Defines the Content of Trafficking Vesicles Captured by Golgin Tethers. *Nat. Commun.* Nov 25, 2020;11(1):5987. doi: 10.1038/s41467-020-19840-4.
62. Mulvey CM, Breckels LM, Crook OM, *et al.* Spatiotemporal Proteomic Profiling of the Pro-Inflammatory Response to Lipopolysaccharide in the THP-1 Human Leukaemia Cell Line. *Nat. Commun.* Oct 1, 2021;12(1):5773. doi: 10.1038/s41467-021-26000-9.

63. Mulvey CM, Breckels LM, Geladaki A, *et al.* Using hyperLOPIT to Perform High-Resolution Mapping of the Spatial Proteome. *Nat. Protoc.* 2017;12(6):1110–1135.
64. Itzhak DN, Tyanova S, Cox J, *et al.* Global, Quantitative and Dynamic Mapping of Protein Subcellular Localization. *Elife.* Jun 9, 2016;5. doi: 10.7554/eLife.16950.
65. Itzhak DN, Davies C, Tyanova S, *et al.* A Mass Spectrometry-Based Approach for Mapping Protein Subcellular Localization Reveals the Spatial Proteome of Mouse Primary Neurons. *Cell Rep.* Sep 12, 2017;20(11):2706–2718. doi: 10.1016/j.celrep.2017.08.063.
66. Ewing RM, Chu P, Elisma F, *et al.* Large-Scale Mapping of Human Protein-Protein Interactions by Mass Spectrometry. *Mol. Syst. Biol.* 2007;3:89.
67. Hein MY, Hubner NC, Poser I, *et al.* A Human Interactome in Three Quantitative Dimensions Organized by Stoichiometries and Abundances. *Cell.* 2015;3:712–723. © 2015 Elsevier Inc.
68. Batra J, Hultquist JF, Liu D, *et al.* Protein Interaction Mapping Identifies RBBP6 as a Negative Regulator of Ebola Virus Replication. *Cell.* Dec 13, 2018;175(7):1917–1930.e13. doi: 10.1016/j.cell.2018.08.044.
69. Shah PS, Link N, Jang GM, *et al.* Comparative Flavivirus-Host Protein Interaction Mapping Reveals Mechanisms of Dengue and Zika Virus Pathogenesis. *Cell.* Dec 13, 2018;175(7):1931–1945.e18. doi: 10.1016/j.cell.2018.11.028.
70. Gordon David E, Hiatt J, Bouhaddou M, *et al.* Comparative Host-Coronavirus Protein Interaction Networks Reveal Pan-Viral Disease Mechanisms. *Science.* Dec 4, 2020;370(6521):eabe9403. doi: 10.1126/science.abe9403.
71. Gordon DE, Jang GM, Bouhaddou M, *et al.* A SARS-CoV-2 Protein Interaction Map Reveals Targets for Drug Repurposing. *Nature.* Jul 1, 2020;583(7816):459–468. doi: 10.1038/s41586-020-2286-9.
72. Huttlin EL, Ting L, Bruckner RJ, *et al.* The BioPlex Network: A Systematic Exploration of the Human Interactome. *Cell.* Jul 16, 2015;162(2):425–440.
73. Huttlin EL, Bruckner RJ, Paulo JA, *et al.* Architecture of the Human Interactome Defines Protein Communities and Disease Networks. *Nature.* 2017;7655:505–509.

74. Huttlin EL, Bruckner RJ, Navarrete-Perea J, *et al.* Dual Proteome-Scale Networks Reveal Cell-Specific Remodeling of the Human Interactome. *Cell.* 2021;184(11):3022–3040.e28.
75. Martell JD, Deerinck TJ, Sancak Y, *et al.* Engineered Ascorbate Peroxidase as a Genetically Encoded Reporter for Electron Microscopy. *Nat. Biotechnol.* Nov 2012;30(11):1143–1148. doi: 10.1038/nbt.2375.
76. Rhee H-W, Zou P, Udeshi Namrata D, *et al.* Proteomic Mapping of Mitochondria in Living Cells via Spatially Restricted Enzymatic Tagging. *Science.* Mar 15, 2013;339(6125):1328–1331. doi: 10.1126/science.1230593.
77. Hung V, Zou P, Rhee H-W, *et al.* Proteomic Mapping of the Human Mitochondrial Intermembrane Space in Live Cells via Ratiometric APEX Tagging. *Mol. Cell.* Jul 17, 2014;55(2):332–341. doi: 10.1016/j.molcel.2014.06.003.
78. Kaewsapsak P, Shechner DM, Mallard W, *et al.* Live-Cell Mapping of Organelle-Associated RNAs via Proximity Biotinylation Combined with Protein-RNA Crosslinking. *Elife.* Dec 14, 2017;6. doi: 10.7554/eLife.29224.
79. Lobingier BT, Hüttenhain R, Eichel K, *et al.* An Approach to Spatiotemporally Resolve Protein Interaction Networks in Living Cells. *Cell.* Apr 6, 2017;169(2):350–360.e12. doi: 10.1016/j.cell.2017.03.022.
80. Paek J, Kalocsay M, Staus DP, *et al.* Multidimensional Tracking of GPCR Signaling via Peroxidase-Catalyzed Proximity Labeling. *Cell.* Apr 6, 2017;169(2):338–349.e11. doi: 10.1016/j.cell.2017.03.028.
81. Choi-Rhee E, Schulman H, Cronan JE. Promiscuous Protein Biotinylation by *Escherichia coli* Biotin Protein Ligase. *Protein Sci.* Nov 2004;13(11):3043–3050. doi: 10.1110/ps.04911804.
82. Roux KJ, Kim DI, Raida M, *et al.* A Promiscuous Biotin Ligase Fusion Protein Identifies Proximal and Interacting Proteins in Mammalian Cells. *J. Cell. Biol.* Mar 19, 2012;196(6):801–810. doi: 10.1083/jcb.201112098.
83. Gupta Gagan D, Coyaud É, Gonçalves J, *et al.* A Dynamic Protein Interaction Landscape of the Human Centrosome-Cilium Interface. *Cell.* Dec 3, 2015;163(6):1484–1499. doi: 10.1016/j.cell.2015.10.065.

84. Kim DI, Jensen SC, Noble KA, *et al.* An Improved Smaller Biotin Ligase for BioID Proximity Labeling. *Mol. Biol. Cell.* Apr 15, 2016;27(8):1188–1196. doi: 10.1091/mbc.E15-12-0844.
85. Youn J-Y, Dunham WH, Hong SJ, *et al.* High-Density Proximity Mapping Reveals the Subcellular Organization of mRNA-Associated Granules and Bodies. *Mol. Cell.* Feb 1, 2018;69(3):517–532.e11. doi: 10.1016/j.molcel.2017.12.020.
86. Chan CJ, Le R, Burns K, *et al.* BioID Performed on Golgi Enriched Fractions Identify C10orf76 as a GBF1 Binding Protein Essential for Golgi Maintenance and Secretion. *Mol. Cell. Proteomics.* Nov 2019;18(11):2285–2297. doi: 10.1074/mcp.RA119.001645.
87. Branon TC, Bosch JA, Sanchez AD, *et al.* Efficient Proximity Labeling in Living Cells and Organisms with TurboID. *Nat. Biotechnol.* 2018;36(9):880–887.
88. Akdag M, Yunt ZS, Kamacioglu A, *et al.* Proximal Biotinylation-Based Combinatory Approach for Isolating Integral Plasma Membrane Proteins. *J. Proteome Res.* Aug 7, 2020;19(8):3583–3592. doi: 10.1021/acs.jproteome.0c00113.
89. Rudolph F, Fink C, Hüttemeister J, *et al.* Deconstructing Sarcomeric Structure–Function Relations in Titin-BioID Knock-in Mice. *Nat. Commun.* Jun 19, 2020;11(1):3133. doi: 10.1038/s41467-020-16929-8.
90. Ramanathan M, Majzoub K, Rao DS, *et al.* RNA-Protein Interaction Detection in Living Cells. *Nat. Methods.* Mar 2018;15(3):207–212. doi: 10.1038/nmeth.4601.
91. Liu X, Salokas K, Tamene F, *et al.* An AP-MS- and BioID-Compatible MAC-Tag Enables Comprehensive Mapping of Protein Interactions and Subcellular Localizations. *Nat. Commun.* Mar 22, 2018;9(1):1188. doi: 10.1038/s41467-018-03523-2.
92. Mackinder LCM, Chen C, Leib RD, *et al.* A Spatial Interactome Reveals the Protein Organization of the Algal CO2-Concentrating Mechanism. *Cell.* Sep 21, 2017;171(1):133–147.e14. doi: 10.1016/j.cell.2017.08.044.
93. Thompson A, Schäfer J, Kuhn K, *et al.* Tandem Mass Tags: A Novel Quantification Strategy for Comparative Analysis of Complex Protein Mixtures by MS/MS. *Anal. Chem.* 2003;75(8):1895–1904.

94. Orre LM, Vesterlund M, Pan Y, *et al.* SubCellBarCode: Proteome-Wide Mapping of Protein Localization and Relocalization. *Mol. Cell.* Jan 3, 2019;73(1):166–182.e7. doi: 10.1016/j.molcel.2018.11.035.
95. Wiśniewski JR, Hein MY, Cox J, *et al.* A "Proteomic Ruler" for Protein Copy Number and Concentration Estimation Without Spike-in Standards. *Mol. Cell. Proteomics.* Dec 2014;13(12):3497–506. doi: 10.1074/mcp.M113.037309.
96. Gingras A-C, Gstaiger M, Raught B, *et al.* Analysis of Protein Complexes Using Mass Spectrometry. *Nat. Rev. Mol. Cell. Biol.* 2007;8(8):645–654.
97. Yang X, Boehm JS, Yang X, *et al.* A Public Genome-Scale Lentiviral Expression Library of Human ORFs. *Nat. Methods.* 2011;8(8): 659–661.
98. Kim DI, Roux KJ. Filling the Void: Proximity-Based Labeling of Proteins in Living Cells. *Trends Cell Biol.* Nov 1, 2016;26(11):804–817. doi: 10.1016/j.tcb.2016.09.004
99. Lönn P, Landegren U. Close Encounters — Probing Proximal Proteins in Live or Fixed Cells. *Trends Biochem. Sci.* Jul 1, 2017;42(7):504–515. doi: 10.1016/j.tibs.2017.05.003.
100. Tan S, Tan HT, Chung MC. Membrane Proteins and Membrane Proteomics. *Proteomics.* Oct 2008;8(19):3924–3932. doi: 10.1002/pmic.200800597.
101. Wu CC, Yates JR, 3rd. The Application of Mass Spectrometry to Membrane Proteomics. *Nat. Biotechnol.* Mar 2003;21(3):262–267. doi: 10.1038/nbt0303-262.
102. Zhao Y, Zhang W, Kho Y, *et al.* Proteomic Analysis of Integral Plasma Membrane Proteins. *Anal. Chem.* Apr 1, 2004;76(7):1817–1823. doi: 10.1021/ac0354037.
103. Yu YQ, Gilar M, Lee PJ, *et al.* Enzyme-Friendly, Mass Spectrometry-Compatible Surfactant for In-Solution Enzymatic Digestion of Proteins. *Anal. Chem.* Nov 1, 2003;75(21):6023–6028. doi: 10.1021/ac0346196.
104. Yu YQ, Gilar M, Kaska J, *et al.* A Rapid Sample Preparation Method for Mass Spectrometric Characterization of N-Linked Glycans. *Rapid Commun. Mass Spectrom.* 2005;19(16):2331–2336. doi: 10.1002/rcm.2067.

105. Chang YH, Gregorich ZR, Chen AJ, *et al.* New Mass-Spectrometry-Compatible Degradable Surfactant for Tissue Proteomics. *J. Proteome Res.* Mar 6, 2015;14(3):1587–1599. doi: 10.1021/pr5012679.
106. Saveliev SV, Woodroofe CC, Sabat G, *et al.* Mass Spectrometry Compatible Surfactant for Optimized In-Gel Protein Digestion. *Anal. Chem.* Jan 15, 2013;85(2):907–914. doi: 10.1021/ac302423t.
107. Brown KA, Chen B, Guardado-Alvarez TM, *et al.* A Photocleavable Surfactant for Top-Down Proteomics. *Nat. Methods.* May 1, 2019;16(5):417–420. doi: 10.1038/s41592-019-0391-1.
108. Brown K, Tucholski T, Eken C, *et al.* High-Throughput Proteomics Enabled by a Photocleavable Surfactant. *Angew. Chem. Int. Ed.* Mar 13, 2020;59(22):8406–8410.
109. Weekes MP, Antrobus R, Talbot S, *et al.* Proteomic Plasma Membrane Profiling Reveals an Essential Role for gp96 in the Cell Surface Expression of LDLR Family Members, Including the LDL Receptor and LRP6. *J. Proteome Res.* 2012;11(3):1475–1484. doi: 10.1021/pr201135e.
110. Weekes MP, Tomasec P, Huttlin EL, *et al.* Quantitative Temporal Viromics: An Approach to Investigate Host-Pathogen Interaction. *Cell.* 2014;157(6):1460–1472. doi: 10.1016/j.cell.2014.04.028.
111. Elschenbroich S, Kim Y, Medin JA, *et al.* Isolation of Cell Surface Proteins for Mass Spectrometry-Based Proteomics. *Expert Rev. Proteomics.* Feb 1, 2010;7(1):141–154. doi: 10.1586/epr.09.97.
112. Smolders K, Lombaert N, Valkenborg D, *et al.* An Effective Plasma Membrane Proteomics Approach for Small Tissue Samples. *Sci. Rep.* Jun 5, 2015;5:10917. doi: 10.1038/srep10917.
113. Komatsu S. Isolation, Purity Assessment, and Proteomic Analysis of Nuclei. *Methods Mol. Biol.* 2018;1696:81–90. doi: 10.1007/978-1-4939-7411-5_5.
114. Liu G, Lu Y, Thulasi Raman SN, *et al.* Nuclear-Resident RIG-I Senses Viral Replication Inducing Antiviral Immunity. *Nat. Commun.* 2018;9(1):3199.
115. Meyerson NR, Zhou L, Guo YR, *et al.* Nuclear TRIM25 Specifically Targets Influenza Virus Ribonucleoproteins to Block the Onset of RNA Chain Elongation. *Cell Host Microbe.* Nov 8, 2017;22(5):627–638.e7. doi: 10.1016/j.chom.2017.10.003.

116. Ji X, Dadon DB, Abraham BJ, *et al.* Chromatin Proteomic Profiling Reveals Novel Proteins Associated with Histone-Marked Genomic Regions. *Proc. Natl. Acad. Sci.* 2015;112(12):3841.
117. Iglesias N, Paulo JA, Tatarakis A, *et al.* Native Chromatin Proteomics Reveals a Role for Specific Nucleoporins in Heterochromatin Organization and Maintenance. *Mol. Cell.* 2020;77(1):51–66.e8 (accessed 2021-08-19).
118. Alanazi B, Munje CR, Rastogi N, *et al.* Integrated Nuclear Proteomics and Transcriptomics Identifies S100A4 as a Therapeutic Target in Acute Myeloid Leukemia. *Leukemia.* Feb 1, 2020;34(2):427–440. doi: 10.1038/s41375-019-0596-4.
119. Song J, Herrmann JM, Becker T. Quality Control of the Mitochondrial Proteome. *Nat. Rev. Mol. Cell Biol.* Jan 1, 2021;22(1):54–70. doi: 10.1038/s41580-020-00300-2.
120. Kasahara A, Scorrano L. Mitochondria: From Cell Death Executioners to Regulators of Cell Differentiation. *Trends Cell Biol.* Dec 2014;24(12):761–770. doi: 10.1016/j.tcb.2014.08.005.
121. Pagliarini DJ, Calvo SE, Chang B, *et al.* A Mitochondrial Protein Compendium Elucidates Complex I Disease Biology. (1097-4172 (Electronic)).
122. Calvo SE, Clauser KR, Mootha VK. MitoCarta2.0: An Updated Inventory of Mammalian Mitochondrial Proteins. *Nucleic Acids Res.* 2016;44(D1):D1251–D1257. doi: 10.1093/nar/gkv1003.
123. Rath S, Sharma R, Gupta R, *et al.* MitoCarta3.0: An Updated Mitochondrial Proteome Now with Sub-Organelle Localization and Pathway Annotations. (1362-4962 (Electronic)).
124. Alberio T, Pieroni L, Ronci M, *et al.* Toward the Standardization of Mitochondrial Proteomics: The Italian Mitochondrial Human Proteome Project Initiative. *J. Proteome Res.* Dec 1, 2017;16(12):4319–4329. doi: 10.1021/acs.jproteome.7b00350.
125. Bonam SR, Wang F, Muller S. Lysosomes as a Therapeutic Target. *Nat. Rev. Drug Discov.* Dec 1, 2019;18(12):923–948. doi: 10.1038/s41573-019-0036-1.
126. Kao Y-T, Gonzalez KL, Bartel B. Peroxisome Function, Biogenesis, and Dynamics in Plants. *Plant Physiol.* 2018;176(1):162–177. doi: 10.1104/pp.17.01050.

127. Tharkeshwar AK, Trekker J, Vermeire W, *et al.* A Novel Approach to Analyze Lysosomal Dysfunctions Through Subcellular Proteomics and Lipidomics: The Case of NPC1 Deficiency. *Sci. Rep.* Jan 30, 2017;7(1):41408. doi: 10.1038/srep41408.
128. Ponnaiyan S, Akter F, Singh J, *et al.* Comprehensive Draft of the Mouse Embryonic Fibroblast Lysosomal Proteome by Mass Spectrometry Based Proteomics. *Sci. Data.* Feb 26, 2020;7(1):68. doi: 10.1038/s41597-020-0399-5.
129. Thelen M, Winter D, Braulke T, *et al.* SILAC-Based Comparative Proteomic Analysis of Lysosomes from Mammalian Cells Using LC-MS/MS. *Methods Mol. Biol.* 2017;1594:1–18. doi: 10.1007/978-1-4939-6934-0_1.
130. Abu-Remaileh M, Wyant GA, Kim C, *et al.* Lysosomal Metabolomics Reveals V-ATPase- and mTOR-Dependent Regulation of Amino Acid Efflux from Lysosomes. *Science (New York, NY).* 2017;358(6364): 807–813. doi: 10.1126/science.aan6298.
131. Fujisawa A, Tamura T, Yasueda Y, *et al.* Chemical Profiling of the Endoplasmic Reticulum Proteome Using Designer Labeling Reagents. *J. Am. Chem. Soc.* Dec 12, 2018;140(49):17060–17070. doi: 10.1021/jacs.8b08606.
132. Yamanaka T, Nishiyama R, Shimogori T, *et al.* Proteomics-Based Approach Identifies Altered ER Domain Properties by ALS-Linked VAPB Mutation. *Sci. Rep.* May 6, 2020;10(1):7610. doi: 10.1038/s41598-020-64517-z.
133. Alberti S, Gladfelter A, Mittag T. Considerations and Challenges in Studying Liquid-Liquid Phase Separation and Biomolecular Condensates. *Cell.* 2019;176(3):419–434.
134. Roden C, Gladfelter AS. RNA Contributions to the Form and Function of Biomolecular Condensates. *Nat. Rev. Mol. Cell Biol.* Mar 1, 2021;22(3):183–195. doi: 10.1038/s41580-020-0264-6.
135. Bracha D, Walls MT, Brangwynne CP. Probing and Engineering Liquid-Phase Organelles. *Nat. Biotechnol.* Dec 1, 2019;37(12):1435–1445. doi: 10.1038/s41587-019-0341-6.
136. Kanaan NM, Hamel C, Grabinski T, *et al.* Liquid-Liquid Phase Separation Induces Pathogenic Tau Conformations *In Vitro.* 2020;11(1):2809.

137. You K, Huang Q, Yu C, *et al.* PhaSepDB: A Database of Liquid–Liquid Phase Separation Related Proteins. *Nucleic Acids Res.* 2020;48(D1):D354–D359. doi: 10.1093/nar/gkz847.
138. Pancsa R, Vranken W, Mészáros B. Computational Resources for Identifying and Describing Proteins Driving Liquid–Liquid Phase Separation. *Brief. Bioinform.* 2021;22(5):bbaa408. doi: 10.1093/bib/bbaa408.
139. Gilda JE, Gomes AV. Proteasome Dysfunction in Cardiomyopathies. *J. Physiol.* 2017;595(12):4051–4071. doi: 10.1113/JP273607.
140. Kampen KR, Sulima SO, Vereecke S, *et al.* Hallmarks of Ribosomopathies. *Nucleic Acids Res.* 2020;48(3):1013–1028. doi: 10.1093/nar/gkz637.
141. Wang X, Guerrero C, Kaiser P, *et al.* Proteomics of Proteasome Complexes and Ubiquitinated Proteins. *Expert Rev. Proteomics.* Oct 2007;4(5):649–665. doi: 10.1586/14789450.4.5.649.
142. Liu TY, Huang HH, Wheeler D, *et al.* Time-Resolved Proteomics Extends Ribosome Profiling-Based Measurements of Protein Synthesis Dynamics. *Cell Syst.* Jun 28, 2017;4(6):636–644.e9. doi: 10.1016/j.cels.2017.05.001.
143. Gorka M, Swart C, Siemiatkowska B, *et al.* Protein Complex Identification and Quantitative Complexome by CN-PAGE. *Sci. Rep.* Aug 8, 2019;9(1):11523. doi: 10.1038/s41598-019-47829-7.
144. Thompson A, Wölmer N, Koncarevic S, *et al.* TMTpro: Design, Synthesis, and Initial Evaluation of a Proline-Based Isobaric 16-Plex Tandem Mass Tag Reagent Set. *Anal. Chem.* Dec 17, 2019;91(24):15941–15950. doi: 10.1021/acs.analchem.9b04474.
145. Martinez-Val A, Bekker-Jensen DB, Steigerwald S, *et al.* Spatial-Proteomics Reveals Phospho-Signaling Dynamics at Subcellular Resolution. *Nat. Commun.* Dec 7, 2021;12(1):7113. doi: 10.1038/s41467-021-27398-y.
146. Kennedy MA, Hofstadter WA, Cristea IM. TRANSPIRE: A Computational Pipeline to Elucidate Intracellular Protein Movements from Spatial Proteomics Data Sets. *J. Am. Soc. Mass Spectrom.* Jul 1, 2020;31(7):1422–1439. doi: 10.1021/jasms.0c00033.

https://doi.org/10.1142/9789811283376_0009

Chapter 9

Cardiovascular Proteomics Primer: Managing Expectations and Setting Yourself up for Success

Linda Berg Luecke,[1,2] Melinda Wojtkiewicz,[1]
Claudius Mahr[3], and Rebekah L. Gundry[1,3,*]

[1]*CardiOmics Program, Center for Heart and Vascular Research; and Department of Cellular and Integrative Physiology, University of Nebraska Medical Center, Omaha, NE*
[2]*Department of Biochemistry, Medical College of Wisconsin, Milwaukee, WI*
[3]*Medical City Healthcare, Dallas, TX*

Contents

*Corresponding author.
Rebekah L. Gundry, PhD, FAHA, Department of Cellular and Integrative Physiology, University of Nebraska Medical Center, 985850 Nebraska Medical Center, Omaha, NE, 68198-5850.

Abstract

To fully understand cardiovascular physiological and pathophysiological mechanisms, there is a need to know not only which proteins are present but also where they are localized, their proteoforms, and how these properties change as a result of pathogenesis. The scientific discipline of *proteomics* broadly includes applying research methods to determine the identification, abundance, localization, structure, function (physiological and pathophysiological), and interactions of proteins. Mass spectrometry–based proteomics can be applied to identify, characterize, and quantify thousands of proteins in a single experiment or used to perform targeted analyses of a single or few specific proteins to assess a wide variety of properties. Given the breadth of applications, one-size-fits-all approaches do not exist. Rather, the appropriate experimental design needs to be tailored to address specific biological questions. In this chapter, we introduce the reader to general concepts of proteomic analysis, emphasize keys to success associated with each aspect of the proteomic workflow, and describe reasonable expectations so readers are primed to discuss their research goals with mass spectrometry practitioners and develop a plan most suited to address their research question. This chapter is targeted to readers without prior proteomics experience and focuses on those applications we expect to be particularly interesting to clinician scientists who want to apply proteomic approaches to the analysis of human biological specimens. While many of the key points are generally applicable to any cell or tissue type, we include specific considerations that are relevant to unique challenges faced in cardiovascular research.

1. Introduction

Proteins are the workhorses of the cell and perform specific, critical functions in physiology and disease. Protein function is determined by multiple properties — including where a protein is localized within the cell, its three-dimensional structure, how it is post-translationally modified, and what it interacts with. These properties can change with experimental perturbations, during disease development, or because of disease. Consequently, to fully understand cardiovascular physiological and pathophysiological mechanisms, there is a need to know not only which proteins are present but also their specific location, their proteoforms [i.e., different forms of a protein produced from the genome with a variety of sequence variations, splice isoforms, and post-translational modifications (PTMs)],[1] and how these properties change — either as a result of disease development or to cause disease. The scientific discipline that is *proteomics* broadly includes applying research methods to determine the identification, abundance, localization, structure, function (physiological and pathophysiological), and interactions of proteins.

In this chapter, we provide an overview of considerations for using mass spectrometry (MS) for proteomic analysis of biological samples in the context of cardiovascular research. While aptamer and immunoassay-based technologies are emerging for cardiovascular blood biomarker surveillance applications (e.g., SomaScan, O-link[2,3]), these approaches are limited to the targets for which affinity reagents are available and specific.[4,5] Furthermore, these approaches are currently largely limited to assessing target abundance and are unable to inform subcellular localization, identify or localize PTMs, identify interaction partners, measure three-dimensional structure, or determine degradation products.[4] MS can address all these aspects, and more, and thus MS-based proteomics is the focus of this Cardiovascular Proteomics Primer.

This chapter is targeted to readers without prior proteomics experience and examples focus on those applications we expect will be particularly interesting to clinician scientists who want to apply proteomic approaches for the analysis of biological specimens. While many of the key points are generally applicable to any cell or tissue type, we include specific considerations relevant to unique challenges faced in cardiovascular research. Given the breadth of applications, a one-size-fits-all approach does not exist. This chapter does not attempt to present every available approach, as many excellent reviews are available.[6–14] Rather, we orient the reader to general concepts in proteomic analyses, emphasize keys to success associated with each aspect of the proteomic workflow, and describe reasonable expectations so readers are primed to discuss their research goals with MS practitioners and develop a plan most suited to address their research questions.

2. Terminology in MS

This section introduces terminology commonly used when discussing MS-based research and proteomics studies. MS is an analytical technique that can identify, characterize, and quantify biological molecules including metabolites, peptides, proteins, oligosaccharides, and lipids. All MS instruments analyze gas phase ions and have three major components: an ion source, mass analyzer(s), and a detector (Figure 1).[15–17] In the ion source, molecules are converted to gas phase ions. The way a sample is prepared and the presence of any contaminants in the sample can affect the efficiency of this ionization process, which is why attention to detail is required when preparing samples for MS analysis. In the analyzer, molecules are separated according to their mass-to-charge ratio (m/z). Mass analyzers are also used to perform gas phase dissociation of molecules (i.e., fragmentation). Tandem mass spectrometry (MS/MS) is the term used to describe two mass analysis events. A proteomics example of MS/MS is the analysis of the peptide (MS^1) followed by the analysis of the fragment ions generated from the gas phase dissociation of a peptide (MS/MS or MS^2). The information contained in the MS^1 and MS^2 spectra are used to identify, characterize, and quantify the molecule. MS instruments can have one or multiple analyzers capable of single or multiple different dissociation modes (Figure 1). Molecules can behave differently during the dissociation process based on the types of chemical bonds present and the type of dissociation method used. Thus, selecting the appropriate dissociation method is an important consideration when designing a proteomics experiment. In the final stage of MS, gas phase ions are recorded in the detector.

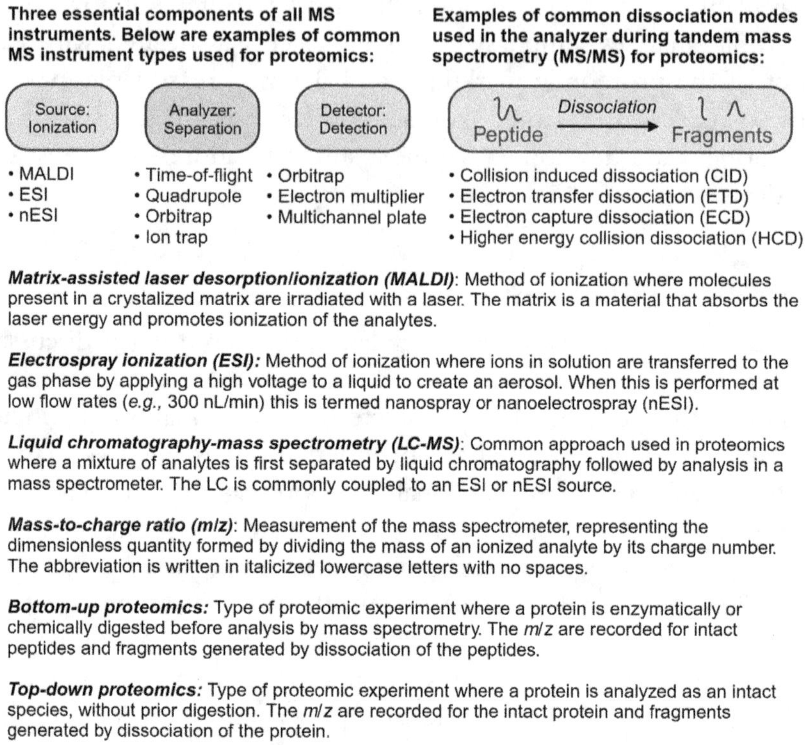

Figure 1. Overview of MS terminology. Top: overview of the three components of all MS instruments and examples of common MS types and dissociation modes used for proteomics. Some instruments will contain a single type of analyzer, while other instruments will contain multiple different analyzer types. Some instruments offer a single mode of dissociation, while other instruments offer multiple modes. Bottom: common terminology used in proteomics research.[18] Created with BioRender.com.

3. Overview of MS-based proteomics for cardiovascular research

MS can be applied to identify, characterize, and quantify thousands of proteins in a single experiment or can be used to perform targeted analyses of a single or few proteins to assess a wide variety of properties. The types of information that can be learned by applying MS

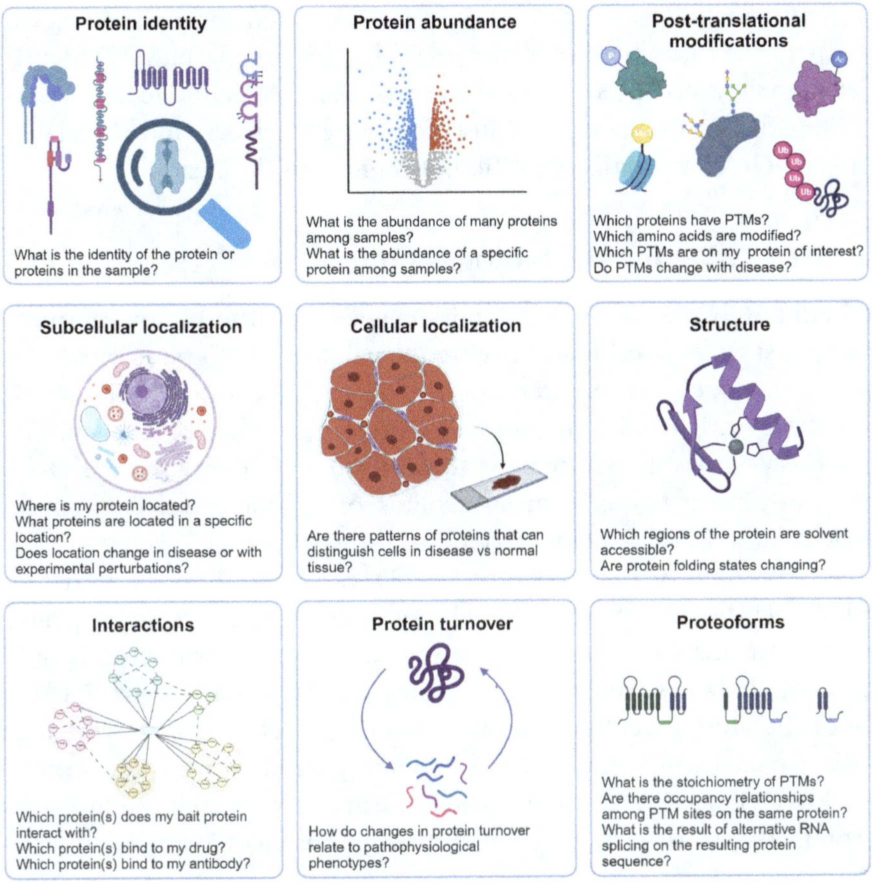

Figure 2. Examples of the types of information that can be learned by applying MS-based proteomic technologies for interrogating the proteome. Created with BioRender.com.

technologies for interrogating the proteome are vast, and a subset is highlighted in Figure 2. In cardiovascular research, MS-based proteomic approaches have led to the identification of biomarkers for cardiomyopathies,[19,20] sarcomeric proteoforms altered in hypertrophic cardiomyopathy and heart failure,[21,22] have revealed new cell-signaling roles of matrix metalloproteinase 9 during cardiac remodeling,[23] identified myofilament proteins that may play a compensatory role in heart failure,[24] and uncovered roles of neutrophil-derived protein S100A8/A9 in altering platelet reactivity in myocardial infarction.[25]

In-depth analysis of the cardiac proteome revealed pathophysiological mechanisms for ischemic[26,27] and dilated cardiomyopathies,[28,29] identified signaling proteins[30] from left ventricular cardiac tissue and metabolic pathways in healthy and disease mouse models,[31] and elucidated the protein composition of different regions of the heart.[32–34]

4. Step 1: Define the biological question

To effectively execute a proteomics experiment, the biological question must be defined first. Specifically, what needs to be learned? To define this question requires considering multiple aspects (Figure 3). Does the study need to measure one protein or many proteins simultaneously? Does the study need to perform a discovery screen? Does the study need to perform an analysis of only a preselected single protein or a set of proteins? Does the study focus on a specific subcellular proteome or secreted proteins? Does the study need to determine only the presence of a protein, or does it also need to determine relative or absolute abundance across multiple experimental conditions or disease states? Does the study need to characterize PTMs? Does the study need to determine which proteins or drugs interact with a protein of interest? Does the study need to measure protein turnover? Does the study need to measure how protein abundance, protein PTMs, or protein–protein interactions change among disease states?

Ultimately the answer to the question "What is your biological question?" will influence all aspects of experimental design. This is because the procedures used at each step of the proteomic workflow will introduce biases, so each aspect of the experimental design must be tailored to address the specific biological question. Consequently, defining the biological question in step #1 is key to selecting the appropriate experimental design in step #2. Proteomics can answer many types of biological questions. A non-exhaustive set of nine different biological questions especially relevant to cardiovascular research are summarized in Figure 4. Importantly, each of these nine questions requires a unique experimental design.

Figure 3. Examples of questions that are considered when defining a specific biological question that will be addressed using proteomics approaches.

Set yourself up for success: Start down the path to success by first defining the biological question in detail. Consult with the MS practitioner early on to learn about strategies that are available and practical. This process may reveal new possibilities as well as help you avoid problematic study designs. Expect this to be an interactive and, at times, iterative process to refine the biological question to one which is specific and addressable.

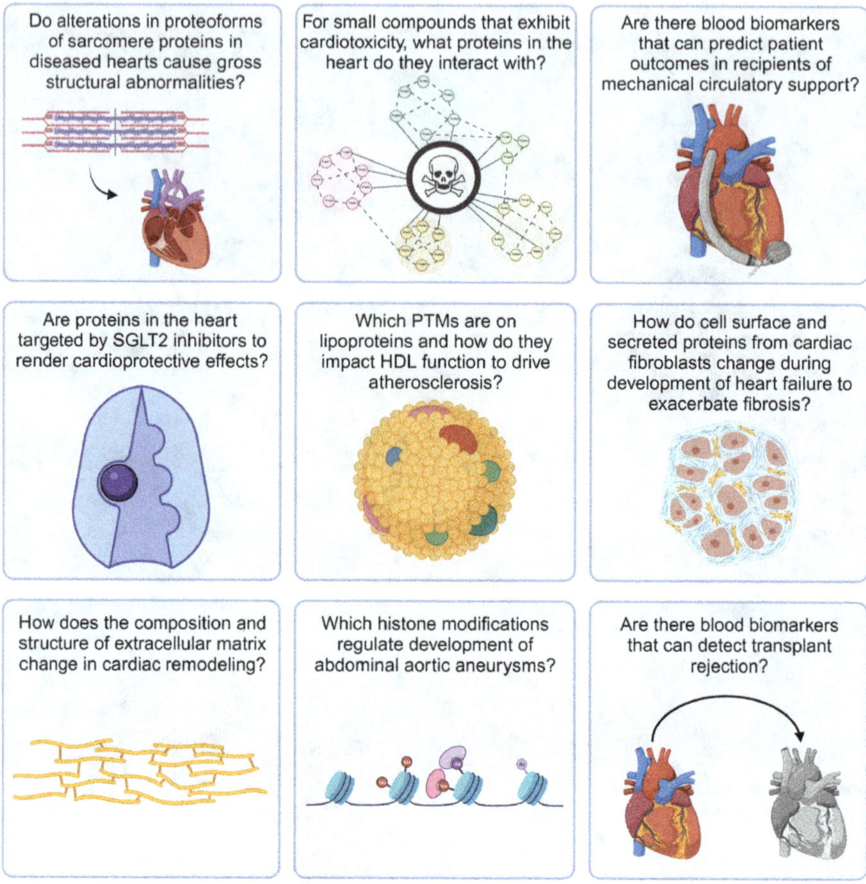

Figure 4. Examples of specific biological questions relevant to cardiovascular research that can be addressed using tailored proteomic workflows. Created with BioRender.com.

5. Step 2: Define the experimental design

5.1. *Five aspects of experimental design*

The biological question determines which experimental design is required to achieve the desired outcome. The experimental design includes five aspects: (1) sample source, (2) sample collection and storage, (3) sample preparation, (4) MS data acquisition, and (5) MS data analysis (Figure 5). These aspects are all interconnected. For example, capabilities of the MS instrument in step 4 will influence

Aspects of Experimental Design in Proteomics

① Sample Source ② Sample Collection & Storage ③ Sample Preparation ④ MS Data Acquisition ⑤ MS Data Analysis

Sample Source

- What is the relevant sample source to address the biological question?
- How many biological replicates are available and required?
- What are the appropriate biological controls?
- Will all specimens be collected simultaneously or over time?
- What sample source will provide sufficient amount of material for the specific type of experiment?
- What are the sampling biases associated with the human population or animal model being used?

Sample Collection & Storage

- Which type of sample contains the proteins of interest for the study? Tissue, blood, cell line?
- Do the samples already exist in a biobank, if so, what were the collection procedures used?
- What are the appropriate controls, including age, sex, and anatomic region (if tissue) match?
- What aliquot size can be obtained and how can freeze-thaw cycles be avoided?
- Does the sample need to be collected in a way that preserves cellular or tissue integrity?
- How will collection procedures avoid technical artifacts?

Sample Preparation

- Will this be a qualitative or quantitative experiment?
- Will a specific type of enrichment be required for the class of protein being studied?
- What types of experimental controls are available and necessary?
- Which types of PTMs or protein interactions need to be preserved?
- What is required to solubilize the protein or protein class of interest?
- Will the proteins be digested? If so, which enzyme or chemical will provide the necessary coverage?

MS Data Acquisition

- Is this a bottom-up or top-down experiment?
- What type of acquisition strategy is required - qualitative or quantitative?
- If a quantitative experiment, is this untargeted or targeted?
- Is the sample a simple or complex mixture?
- If PTMs are the focus, what type of fragmentation mode is appropriate?
- If MS imaging, what spatial resolution is required to observe the patterns of interest?

MS Data Analysis

- Do you have access to software that can analyze data generated by the MS method used?
- Is this a bottom-up or top-down experiment?
- What species is used and is there a good reference proteome database available?
- How was the sample prepared so that processing options are appropriately used?
- If studying PTMs, how will the confidence in identification and localization be determined?
- If performing quantification, what is the appropriate strategy for normalization?

Figure 5. Five aspects of experimental design for proteomics studies. For each step, examples of the details that need to be considered are listed.

how to collect and prepare the samples in steps 2 and 3. The software tools available for step 5 will influence how to collect the MS data in step 4. How the samples are processed in step 3 will define the MS data acquisition and analysis protocols in steps 4 and 5. Figure 5 provides examples of questions to contemplate when designing an experiment to answer a specific biological question. The remainder of this section describes examples of biases that are inherent to each of the five aspects of the experimental workflow. Being familiar with these biases is important for managing expectations of the results produced by applying different workflows. Also, this section briefly summarizes special considerations for quantitative workflows and highlights important questions to address when defining the experimental design.

5.2. *Inherent experimental biases*

5.2.1. *Sample source*

When using human specimens, sample source biases can include geographical selection for a population that is not broadly representative of the global population, survivorship bias, self-selection, and non-response bias.[35,36] Potential biases in the types of specimens available for research may limit the types of biological questions that can be addressed by proteomics. If animal models are used, biases can be introduced if animals are not randomly assigned to experimental or control groups. For some models, males are less expensive and easier to work with,[37] but sex-specific differences will be ignored in male-only studies.

5.2.2. *Sample collection*

Sample collection biases can be introduced through various model systems (e.g., transfection systems, mouse models, cell lines). For example, cell lines used as an alternative to primary isolated cells may not recapitulate human physiology or have PTMs that are not identical to those found in their primary cell counterpart and will affect which proteins and PTMs will be identified and prioritized. Collection of fresh tissue can enable myocardial cell type–specific studies, whereas formalin-fixed paraffin-embedded (FFPE) or flash frozen may not, which is described in more detail in Section 6.1.

5.2.3. *Sample preparation*

Sample preparation biases will affect the types and characteristics of proteins that will be measured. For example, approaches that use whole-cell lysate will be biased toward the identification of the most abundant, soluble proteins, making it very challenging to detect low-abundance PTMs that need to be specifically enriched to enhance detection.[38–41] Such an approach will also yield relatively low coverage of transmembrane proteins compared to approaches that enrich for that subproteome.[34] In turn, methodologies designed to enrich cell surface proteins will not provide coverage of nuclear or mitochondrial proteins. If using an antibody to enrich for a specific protein, it is important to know if the antibody will recognize all proteoforms of the protein or if it will be biased to selecting for a subset of proteoforms that either lack or contain specific PTMs (e.g., anti-CD133 antibodies that recognize glycosylation-dependent versus unmodified epitopes).[42]

5.2.4. *MS data acquisition*

Regarding data acquisition bias, the specific MS instrumentation and configuration available will define which types of acquisition are possible, which can in turn, limit the types of studies that can be performed. For example, specific ion dissociation methods can affect labile PTM bonds, which can present challenges to identification and site localization.[43] While some MS instruments are capable of both discovery and targeted analyses, others may be limited to targeted analyses only. If designing a quantitative study, strategies may differ in terms of sensitivity and quantitative accuracy.[44–46] To illustrate the importance of instrumentation selection when studying PTMs, consider histone PTMs. Histone PTMs play key roles in several cardiovascular disease processes including but not limited to abdominal aortic aneurysms, cardiovascular damage under high salt loading, and platelet aggregation.[47–49] Histone PTMs can be determined by MS, but an important consideration is that masses of several common modifications are similar (e.g., tri-methylation

(42.046950) and acetylation (42.010565) are within 0.036385 Da). In this case, MS instrumentation capable of resolving these masses with high accuracy is necessary for the accurate identification of these modifications.

5.2.5. *MS data analysis*

Biases during MS data analysis can result from the specific parameters used to process the data. For example, if a particular type of PTM is not expected or allowed during protein identification analyses, that PTM may go unidentified. If the reference proteome used for interrogation of the data is not representative of the proteome in the sample (e.g., incorrect species, or sample is contaminated with unknown additional species not contained in the reference proteome), proteins may go unidentified or there could be problems with false-positive assignments. Numerous algorithms exist for data analysis, and they differ with regard to how they assign identifications and how they assess confidence in protein identifications.[34,50–55] Similarly, there are many differences in how protein abundances are determined (e.g., how peaks are picked, and quantification is performed).[56,57] Moreover, some search engines are engineered to perform well for the analysis of specific PTMs or structural proteomics data while others may not be as well suited.[58–60] Ultimately, the features and performance metrics of the data analysis tool used to analyze the data will affect what is determined from the data.

5.3. *Important considerations for quantitative proteomics studies*

Protein quantification is a common goal for many proteomics studies and adds an additional layer of experimental design complexity on top of the considerations for protein identification, characterization, protein structure, and protein interactions. Importantly, MS is not inherently quantitative; thus, the experimental design must include this consideration from inception if the end goal is to compare the relative or absolute abundance of proteins or proteoforms among experimental groups. Protein enrichment, enzymatic digestion, sample clean-up, and sample handling can all introduce variations that will affect the relative or absolute abundance observed. Therefore, quantitative values cannot be asserted *post hoc* for experimental designs that do not include the proper procedures to enable direct quantitative comparisons from the beginning. Thus, consultation with the MS practitioner about the goals of the study must address the need for quantification, as it is a key consideration in study design.

5.4. *Discuss the details early and often*

The examples described above are only a small representation of the specific details that must be considered for any experiment and are provided to emphasize why defining the biological question at the beginning of a study is key to informing important details for all aspects of experimental design. For the highest chance of success, investigators should be prepared to define the biological question and discuss numerous technical details with the MS practitioner before sample collection begins (Table 1). Given the variety of applications and suitable methods, the specific requirements for each step will depend on answers to these questions. It is not possible to fully

Table 1. Questions that should be discussed with the MS practitioner to inform experimental design.

Question	Why it matters
What is the biological question?	Will inform the type of sample preparation, data acquisition, and data analysis methods required.
What species are the samples from?	The reference proteome for some model systems is less well characterized than others. This impacts data analysis. For example, rabbits and rats can be more challenging than humans and mice.
What types of samples are available? Blood, cells, tissue?	The type of sample available will determine what types of questions can be asked. While tissue or cells may be useful for studying physiological or disease processes, blood may be more apt for biomarker studies.

(*Continued*)

Table 1. (*Continued*)

Question	Why it matters
Were the samples collected and stored in the same manner?	Variations in sample collection and storage procedures can result in differences observed among samples that are not related to the underlying physiology being investigated.
How many samples are available?	The statistical power of the analyses will be related to the number of molecules measured across the number of biological replicates.
Will all samples be collected simultaneously, or will they be collected over time?	Samples for which data are acquired at different times may not be directly comparable to each other. For some study designs it may be necessary to collect and store samples until all are available for processing simultaneously.
How many samples can be obtained?	Different workflows will have different starting material requirements. If a very low abundance molecule is being targeted, it may require more starting material or some type of enrichment approach. Top-down approaches will require more protein than targeted bottom-up approaches.
How will starting material be normalized? By total protein? By total volume?	If the study goals include quantification, knowing how the starting material will be normalized will affect how the data will be analyzed and interpreted.
What positive and negative controls will be used?	Define this ahead of time to ensure the controls can be processed simultaneously with the experimental samples to avoid technical bias.
Is there a specific PTM that needs to be analyzed?	Most PTMs require specialized enrichment due to low abundance and preparation methods to preserve.
Is the protein of interest relatively high or low abundance in the relevant sample?	The relative abundance of the protein(s) of interest will affect the experimental design. If a low abundance protein or class of proteins is of interest, then an enrichment or fractionation method may be required to maximize experimental success.
What instrumentation is available?	Will inform what type of sample preparation, data acquisition, and data analysis methods are required.

Table 1. (*Continued*)

Question	Why it matters
Is equipment for automated sample preparation available?	Very large cohorts may benefit from automation for enhanced reproducibility.
What data analysis platforms are available?	Different platforms are available, and many are tailored to specific types of workflows.
Is this a qualitative or quantitative study?	Will inform what type of sample preparation, data acquisition, and data analysis methods are required.
What consumables and reagents will be used?	Many consumables contain contaminants from the manufacturing process (e.g., polyethylene glycol in molecular weight cutoff filters). Most MS practitioners are familiar with these types of common problems so having them review the specific tubes, tips, and filters you plan to use can avoid problems.

> ***Set yourself up for success***: Define the biological question and be prepared to discuss numerous technical details with the MS practitioner at the nascent stage of the research study. Every detail of the experimental design will impact the results.

delineate all details for all methods here, and readers are encouraged to refer to many excellent reviews.[15–17,61–63] The remaining sections of this chapter provide general guidelines and concepts to consider for each aspect of the experimental design.

6. Considerations for sample collection

6.1. *What to collect*

Modern proteomic technologies can be applied to nearly any biological material (Figure 6a shows a non-exclusive representation) to address a variety of question types (Figure 4). The choice of starting

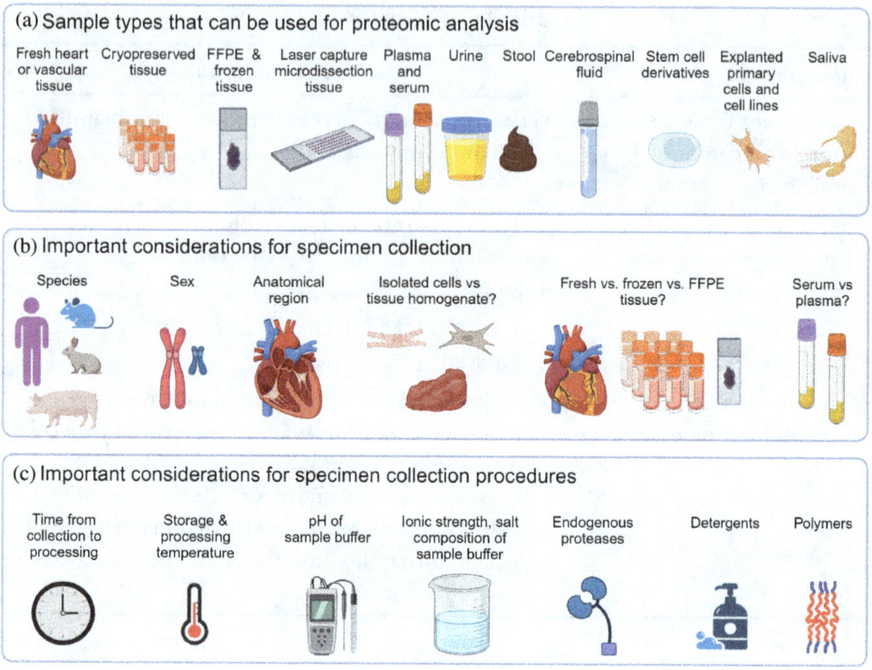

Figure 6. Overview of sample types and key considerations for specimen collection used for proteomic analysis. (a) Examples of sample types that can be used for proteomic analysis. (b) Important considerations when selecting which type of specimen to use. (c) Important considerations when planning specimen collection procedures. Created with BioRender.com.

material should be based on the research question. For example, blood is one of the easiest biological specimens to access for human research subjects. However, if the study aims to study a specific molecular process that occurs in the myocardium, then it might make more sense to directly study myocardial tissue, if available, as it may be more challenging to attribute observations from the blood directly to events in the tissue. But if the study aims to evaluate molecules in the circulation, either to understand heart/brain/kidney/lung signaling or to identify or measure biomarkers, blood can be an excellent sample source. Each cell type or sample type will have their own distinct proteome, so it is critical to select the source material most relevant to address the question. Importantly, sex-specific differences are increasingly being recognized in cardiovascular disease and biomarker research,[64–69] thus males and females should be included in the study unless a process that only occurs in one sex is being investigated.[70]

If using cardiac tissue, consider the anatomical region most relevant for the biological question, as this will inform what to collect. The appropriate region to use may appear obvious for some studies (e.g., left ventricular heart failure, atrial fibrillation), while other studies would benefit from the analysis of multiple regions (e.g., cardiomyopathies, congenital heart disease). Typically, it is not good practice to interchange left- and right-sided regions — for example, using right ventricle free wall as a control for left ventricle free wall. Rather, it is important to acknowledge that cells in different heart regions emerge from different developmental precursors (e.g., paraxial and extraembryonic mesoderm, neuroectoderm, and endoderm),[71] and each region has a different function (e.g., the right ventricle receives deoxygenated blood to be delivered to the lungs, whereas the left ventricle pumps oxygenated blood throughout the body). Emerging views confirm differences in the proteome of cells collected from different heart regions.[32,34,135] Consequently, matching anatomical regions in addition to age and sex are key to data interpretation.

6.2. *How to collect*

Before using any sample in a proteomics experiment, investigators must consider not only *what* to collect but also *how* to collect the samples (Figure 6b–6c). It is advisable to collect samples from animals or patients and process or preserve them as quickly as possible to maximize sample integrity and minimize artifacts that result from removing a biological material from its native environment (e.g., degradation, oxidation). How that is achieved will depend on practical aspects surrounding the type of specimens and their source — including whether they are animal or human and, if human, what procedures are in place or allowed during collection.

Considering the type of molecule to be studied (e.g., protein with a PTM), some may be more prone to molecular changes and/or autodigestion when they experience multiple freeze–thaw cycles.[72,73] Thus, studies involving large cohorts of clinical samples should pay careful attention to how specimens are acquired, aliquoted, stored frozen, thawed, and used. For example, if you obtain a 100-μL aliquot serum for each patient, what will you do with this aliquot when you thaw it? It is likely only a small portion of this would be used for a single proteomics experiment, so will the unused serum be refrozen? If so, it will be important to note the aliquots that undergo additional freeze–thaw cycles so that they can be accounted for during data analysis, as differences may occur when comparing data from samples that underwent a single freeze–thaw cycle to those from multiple cycles.[74]

6.3. *Special considerations for fresh tissues*

When using fresh tissues, it is typically advisable to process them as soon as they are retrieved to preserve specimen integrity. This is typically not a challenge when using animal models. However, when using human tissue specimens obtained from surgery, biopsy, or rapid autopsy, these processes will be impacted by the infrastructure and clinical protocols used in the hospital. It is common that human tissues are collected over time and stored until the requisite number of samples are available to facilitate simultaneous processing. In some hospitals, it may be possible to flash-freeze tissue immediately in the operating room. Although some scientific questions are answerable using flash-frozen tissue, this preservation method most likely excludes cell-type-specific proteomics. If the study design requires isolated cell types, we suggest collecting the tissue specimens directly in cardiac preservation solution [e.g., University of Wisconsin (UW) solution, RPS-96, dextran 40-based UW solution, histidine–tryptophan–ketoglutarate solution, Celsior solution[75]] on ice to preserve specimen integrity while they are transported to the research laboratory for further processing.

From this point forward, how samples are processed or archived will impact what they can be used for. For studies that analyze bulk tissue homogenate, samples can be flash-frozen using dry ice or liquid nitrogen methods.[72] Alternatively, for studies of individual cell types, cell-type-specific isolations can be performed immediately following the collection of fresh tissue.[76–82] If it is impractical to perform cell isolation immediately (e.g., nighttime surgery) or if you are receiving samples shipped from off-site collaborators, we recommend cryopreservation with slow freezing as this preserves the ability to isolate intact cardiomyocytes and cardiac fibroblasts in addition to performing analyses of bulk tissue homogenate.[34] In our experience, preservation in Allprotect Tissue Reagent (Qiagen) does not allow for cell-type isolation from human myocardium.[34] Importantly, some hospital systems require all tissues removed during the surgery to go to pathology prior to distribution to the research laboratory. Although not ideal due to the variability in time and temperature to which the specimen is exposed, if this is necessary, the length of time between surgical collection and processing for proteomics — especially the time spent at room temperature — should be recorded as this can adversely affect cellular and protein integrity.[83,84] Another consideration relevant to cardiovascular research is the time of day that surgeries are performed. If cardiac surgery is performed outside of normal working hours, will the collection procedure and timing be the same as those implemented during working hours?

6.4. *Special considerations for other sample types*

In addition to fresh tissue, archived surgical specimens preserved as flash-frozen tissue or FFPE tissues and serum, plasma, urine, stool, and cerebral spinal fluid can also be used in proteomics studies. For each of these specimen types, investigators should be familiar with specimen collection protocols. When using FFPE slides, differences in tissue quality caused by variations in fixation procedures used (e.g., type of fixative, temperature, stabilizers, penetration time) can lead to variability in the resulting proteomics data. For blood, it may not be feasible to expect that all samples in a large cohort are collected by the same phlebotomist, but ideally, they will follow the same standard operating procedure. Any differences in collection time can affect proteomic results especially if some blood samples are processed immediately and others are refrigerated for various lengths of time (e.g., overnight) prior to processing.[85–87] Overall, investigators should ensure that standardized procedures for collection and archiving are used for all samples within a study.[88] If using cultured cells such as cell lines, explanted primary cells, or human pluripotent stem cell derivatives, consider how extended culturing or cell density may affect your proteins of interest.[89,90] If studying extracellular vesicles from blood, it is important to implement collection procedures that allow for separation away from lipoprotein particles that are of the same size. If density gradient ultracentrifugation is the only separation method used, extracellular vesicles and lipoprotein particles will be co-isolated.[91,92]

6.5. *Work together with clinical teams to ensure specimen utility for proteomics*

Ultimately, if you plan to use human specimens in your studies, we encourage you to consult with the clinical teams to find ways to implement collection procedures that maximize specimen integrity and minimize variability. While some research institutions have established biobanks and protocols and are unable to accommodate changes to their sample collection procedures, others may be able to update their collection workflows to make them compatible with specialized approaches. We encourage investigators interested in using human biological specimens to find ways to work with appropriate surgery, pathology, and biobanking teams to explore what may be possible at their own institution. For example, it is advisable to minimize the number of freeze–thaw cycles when possible and ensure that specimens used for comparison are treated similarly. If you plan to use serum or plasma, the biobank may accommodate requests for smaller size aliquots (e.g., 10 μL instead of 100 μL) which would allow you to easily perform multiple analyses of the same sample without additional freeze–thaw cycles. This is particularly important to consider for multicenter studies, as any practice pattern variation in sample acquisition and storage will introduce confounding variables in the data obtained.

6.6. *Emerging technologies for sample collection*

The technologies and approaches available for sample collection for proteomics studies are continually evolving. Two particularly exciting developments are those related to microsampling and single-cell proteomics. Microsampling devices make it possible to collect sufficient biological material for a proteomics study from just a few drops of blood (e.g., Dried Blood Spot cards, Telimmune/Noviplex plasma cards) or other biological fluid (e.g., Mitra devices).[93,94] Due to their ease of use, these microsampling devices are expected to facilitate remote sampling, allowing patients to collect samples from the comfort of their own home and ship them to laboratories for analysis. Regarding single-cell proteomics, there have been tremendous advancements in this field over the past decade.[95,96] While it is exciting to see the possibilities enabled by this approach, there remain practical challenges to its implementation for cardiovascular research. Obtaining cultured cells or primary blood cells for single-cell analyses is relatively straightforward; however, obtaining single cardiac cells is considerably more challenging due to the unique nature of myocardial tissue. First, the myocardium is a dense tissue comprising multiple cell types that are difficult to dissociate depending on the disease state. Animal hearts can be perfused with solutions to aid in the disruption of cellular interactions,[97,98] but this is not feasible for human myocardium removed during surgery. Second, the large size of human cardiomyocytes (~100 × 20 μm) combined with their fragility present challenges to cell sorting using traditional flow cytometers. For this reason, specialized instrumentation capable of sorting large-sized cells (e.g., CellenONE from Cellenion, COPAS and BioSorter from Union Biometrica) is poised to play a role in advancing single-cell proteomic analyses of human heart tissue. Of course, the methods used to dissociate the tissue and collect single cells will impact the types of proteomic analyses that can be performed. For example, if the tissue is dissociated using enzymes with tryptic activity, this will cleave extracellular domains of cell surface proteins, preventing their detection by MS.

7. Considerations for sample preparation

7.1. *Overview of sample preparation*

Once the sample is collected, the next step in the experimental design is sample preparation. The purpose of sample preparation is to make the proteins amenable to MS analysis. The steps contained within the sample preparation include, but are not limited to, tissue homogenization or cell lysis, protein solubilization, chemical derivatization, organelle and/or protein enrichment, protein digestion, peptide enrichment, and sample clean-up. The specific procedures used for the sample preparation steps are defined, in part, by whether a peptide-centric (i.e., bottom-up) or protein-centric (i.e., top-down) experimental design is being used (Figure 7). Both workflows follow a common template. The appropriate specimen is collected, the sample is prepared, the peptides or proteins are analyzed by MS, and the MS data are evaluated. The most suitable approach will depend on the biological question. Bottom-up approaches are suitable to a wide array of biological questions, and MS instrumentation and numerous bioinformatic tools are available for both general and specific applications (reviewed in Refs 99–103). Disadvantages of

> ***Set yourself up for success***: Consult with the MS practitioner you will work with to conduct the studies *before* you begin sample collection. It might not be possible if using previously banked human samples, but it is important to familiarize yourself with collection procedures used in the hospital or the laboratory that will source the samples for your study. Ensure these methods are in line with best practices for proteomic analyses and that all samples used within a study are processed identically.

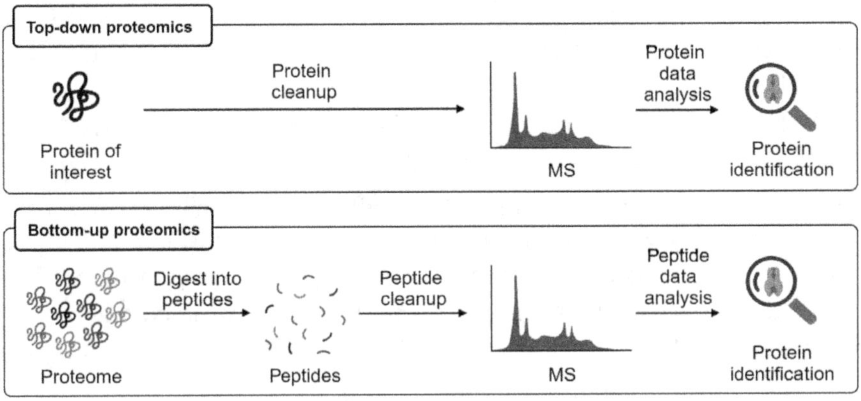

Figure 7. Overview of bottom-up and top-down proteomic workflows. Created with BioRender.com.

bottom-up approaches include that the relationship among peptides (e.g., modifications, truncations) is lost. In top-down approaches, the stoichiometry among PTMs, proteolytic cleavage products, and products of splicing events is preserved because the proteoforms are measured as intact proteins. However, compared to what is available for bottom-up proteomics, top-down proteomics is limited by specific MS instrumentation needs, fewer data analysis tools, and challenges to quantification.[63,104,105]

7.2. *Select the mode of sample disruption suited to the biological question*

Each step performed during sample preparation is tailored to the specific biological question, and suboptimal sample preparation strategies will adversely affect the quality of results. A typical first step during sample preparation involves a combination of homogenization and lysis depending on whether cells, tissue, or fluid are used. If this step is not optimized for the specific sample type and experimental question, this could lead to poor proteome coverage or additional technical variability. Cell lysis conditions will depend on whether subcellular organelles need to be enriched or whole-cell lysate is required.

Lysis of blood cells is relatively easy, while cardiac tissue and isolated cardiac cells require a more rigorous combination of homogenization and cell lysis.

For human myocardium and isolated cardiomyocytes, we use a combination of homogenization with bead beating followed by sonication,[34] though other approaches can be used successfully.[106,107] Blood vessels (e.g., aorta, pulmonary artery) may require pulverization in liquid nitrogen to adequately disrupt the tissue and provide coverage of extracellular matrix proteins like elastin, collagen, and proteoglycans.[108]

7.3. *Consider what needs to be preserved*

In addition to the mode of sample disruption, the composition of the lysis or homogenization buffer is critical. This is particularly important for studies that aim to measure protein–protein interactions and PTMs. Once cells or tissue are disrupted, proteins can contact other proteins that they may not have previously been in proximity to (e.g., nuclear proteins will now encounter cytoplasmic proteins). Regarding the study of PTMs, if phosphatase inhibitors are not used during sample homogenization and cell lysis, phosphoproteomic analyses may not be successful. Similarly, if deubiquitinase inhibitors are not used, studies of ubiquitin modifications may not be successful. If trypsin is used to detach adherent cultured cells, cell surface proteins may not be detected by MS as they get cleaved off during the detachment process. Finally, if the study aims to include the analysis of metabolites in addition to proteins, failure to quench metabolism immediately may lead to the loss of some metabolites. Thus, it is critical to think about what must be preserved during the sample preparation process. Is it a type of PTM that can be affected by specific buffer compositions or temperatures? Is it important to preserve protein–protein interactions or subcellular localization? If studying a transmembrane protein, how will the protein be solubilized? If non-MS compatible detergents are required to solubilize the protein, then how will contaminants be removed prior to MS? All the examples above would affect the choice of lysis buffer composition, the method of sample lysis, protein solubilization, and sample clean-up.

7.4. *Special considerations for enrichment or fractionation workflows*

If the study is designed to detect a protein that is present at relatively low abundance in the sample, is within a specific subcellular organelle, or has a specific PTM, the appropriate sample preparation workflow will likely require some sort of enrichment either for a subcellular organelle or a protein class of interest or fractionation to generate multiple lower complexity samples. Complex mixtures may pose challenges to coverage for lower abundance peptides, so it may be necessary to perform peptide fractionation to enhance the depth of proteome coverage. This requires more work and time to prepare the sample and to acquire the data and may not be feasible for very-low-abundance samples. Beads functionalized with an antibody to enrich a specific protein or streptavidin to enrich biotinylated peptides or proteins are key aspects of many types of enrichment workflows. However, one challenge in using these approaches is that the beads are prone to non-specific binding and can exhibit variability in binding properties among reagents.[109,110] Hence, evaluating different enrichment beads and enrichment conditions is an important step in developing a suitable workflow.

7.5. *Global internal standards and digestion considerations*

After sample solubilization, it may be appropriate to add an internal standard that is used to evaluate the success or repeatability of the entire experimental workflow. This is not always required but can be particularly advantageous when performing method development or quantitative analyses.[111–113] The choice of the appropriate internal standard will depend on the entire workflow and is typically representative of the type of protein the study aims to measure. Another step that may be included after solubilization and prior to digestion is total protein quantification using a fluorometric or absorbance assay. This will inform the amount of enzyme or chemical required to digest the sample for bottom-up workflows and can be used to ensure equal amounts of protein used in quantitative workflows.[114] If using a bottom-up workflow, the choice of enzyme or chemical for protein digestion will depend on the proteins being studied. Trypsin, which cleaves on the C-terminal side of lysine and arginine, is the most common enzyme used for bottom-up proteomic strategies. However, if the study focuses on a PTM expected to reside on a residue within a sequence of the protein that would not yield a peptide of suitable size for MS analysis after trypsin digestion, an alternative digestion approach should be considered.

7.6. *Modern options to facilitate sample processing*

Relevant to all workflows, recent advances in sample preparation can aid in throughput, maximize quality, and minimize variability. Many sample workflows can now be fully automated using standard laboratory liquid handling workstations[115] or those specifically designed for proteomics preparation [e.g., Evosep One (Evosep)].[116] In addition to automated workstations, recently developed products are designed to make the sample preparation process easier and more reproducible without requiring expertise in MS or proteomics. These include, but are not limited to, S-trap (ProtiFi), PreOmics iST kit (Preomics),

> ***Set yourself up for success***: Share all protocols — including all experimental details — you plan to use so the MS practitioner can review every reagent, consumable, and procedure you plan to include. It is easier and more efficient for the MS practitioner to identify potential problems to avoid prior to experimentation than it is to spend time and effort troubleshooting a failed experiment that will ultimately need to be repeated. The temperatures, buffer compositions, and procedures used during sample preparation should be chosen based on the scientific question.

High Throughput Sample Preparation Kit (Biognosys), and EasyPep kit (Thermo Fisher Scientific).[114,117,118] Using these products can help ensure the prepared samples will be suitable for MS analysis (e.g., good quality digestion and the final peptide sample is devoid of detergents, polymers, and other contaminants).

8. Considerations for data acquisition

8.1. *Overview of MS instrumentation*

There are many different types and configurations of MS systems. The most commonly used instruments for proteomics are matrix-assisted laser desorption ionization time-of-flight (MALDI-TOF) and liquid chromatography–mass spectrometry (LC–MS), though ion mobility, capillary zone electrophoresis, and other alternative ionization and separation techniques are gaining in popularity.[119,120] Capabilities vary among systems. Higher resolution instruments are well-suited to top-down analyses of intact proteins, while middle- to low-resolution instruments are better suited for analyzing peptides for bottom-up strategies. Fast scanning quadrupole instruments are workhorses in targeted workflows while hybrid instruments (i.e., combine multiple types of analyzers) are apt for both discovery and targeted methods.[121] MS instruments that perform dissociation or MS/MS may employ a single type of dissociation mode or multiple (Figure 1). Knowing what instrument is available for your study is important, as some applications may benefit from specific modes (e.g., electron-transfer/higher-energy collision dissociation for intact glycopeptides[122]). Some applications may be so specialized that it will be necessary to seek out collaborators at external institutions or commercial companies with the appropriate expertise, instrumentation, and software suited for that specific analysis (e.g., top-down analysis of very large proteins, >100 kDa). MS instrumentation is continually and rapidly advancing, bringing with it new capabilities. Thus, it is highly recommended to consult with the MS practitioner regarding available instrumentation.

8.2. *Data acquisition strategies*

Data acquisition strategies used for proteomics can be categorized as qualitative discovery (goal is to identify as many proteins as possible), quantitative discovery (goal is to quantify as many proteins as possible), or targeted (goal is to quantify a set of predefined proteins) (Figure 8a). In qualitative and quantitative discovery studies using a bottom-up strategy, proteins are digested into peptides and analyzed by MS. Depending on the complexity of the sample, thousands of proteins can be identified and quantified in a single sample. In targeted methods, the proteins of interest are chosen. Then, proteolytic peptides can be selected based on theoretical predictions or based on experimental observations from discovery workflows. Targeted workflows can be designed to target one or hundreds of proteins in a single analysis and offer more sensitivity and quantitative accuracy than untargeted methods.

8.3. *Blocking and randomization for LC–MS*

Quality controls are necessary for any system. However, if using LC–MS, the order in which samples are queued for data acquisition is especially important. Retention time standards are generally helpful, and required for some data analysis tools. When performing LC–MS analysis of bottom-up proteomics samples, it is not uncommon for the analysis time of a single sample to require 1–3 hours of run time. Consequently, studies with many samples can require weeks to months of continuous run time. The longer the acquisition queue, the more likely technical variation will appear (e.g., column performance, MS instrument drift from calibration). Thus, to avoid biases in data that are due to instrument drift and unrelated to changes in physiology, it is important to include acquisition designs that control for this.

Sample blocking and randomization is one way to achieve this.[44,123] A sample block is a defined set of samples that will be run consecutively (e.g., batch of 10 injections). In large cohorts, a quality control sample is typically acquired between blocks, and it may be

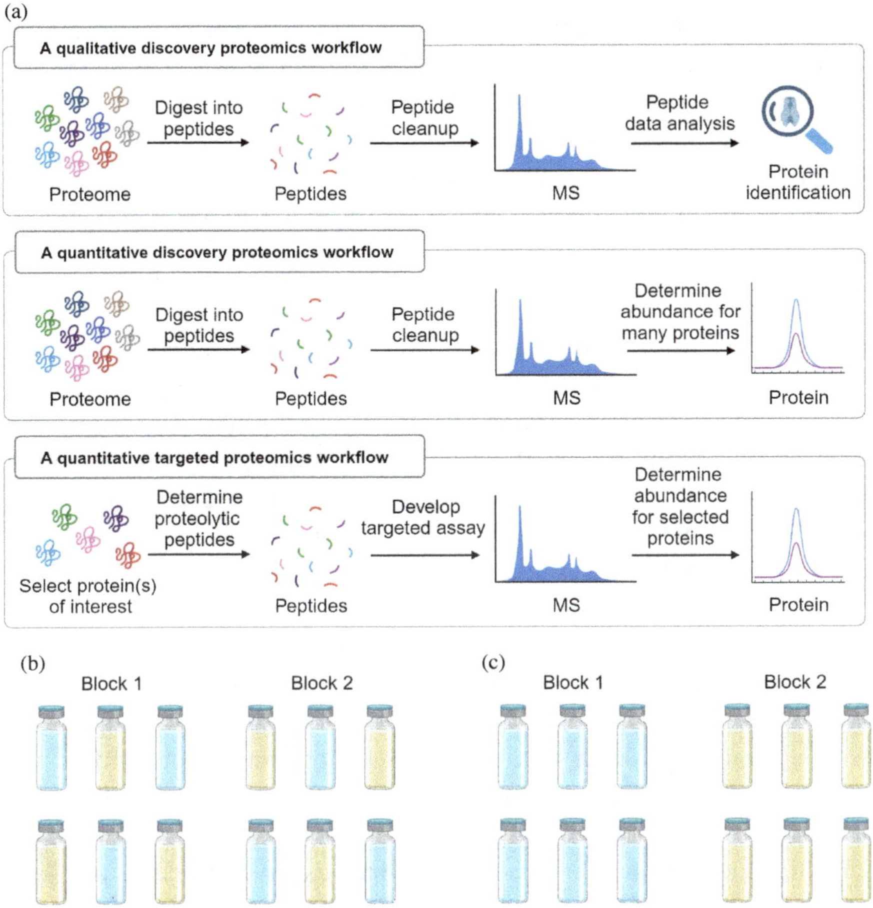

Figure 8. Examples of MS workflows and acquisition queues for LC–MS experiments. (a) Examples of three different types of bottom-up proteomics workflows. The top shows a qualitative workflow for discovering the identity of proteins in a sample. The middle shows a quantitative discovery workflow. The bottom shows a targeted workflow for the quantification of selected proteins of interest. (b) Example of an ideal block and randomization queue for LC–MS experiments. (c) Example of a poor block and randomization queue for LC–MS experiments. For (b) and (c), blue vials represent samples from the control group and yellow vials represent samples from the experimental group. Created with BioRender.com.

***Set yourself up for success*:** Ensure the data acquisition strategy is tailored to maximize accurate identification and/or quantification of your peptide, protein, or PTM-modified protein or peptides of interest. Results can vary depending on dissociation method used and ionization strategy. When using LC–MS, implement blocking, randomization, and quality control standards to ensure peak system performance and interpretability of the data.

necessary to run many blocks to acquire the data for an entire cohort. Within each block, all treatments or conditions should be represented, and the samples should be acquired in randomized order. For example, if a study included a total of 12 samples, 6 control and 6 experimental, it would be appropriate to acquire the data as two blocks of six injections. And the blocks would each include control and experimental samples acquired in randomized order (Figure 8b). An inappropriate acquisition queue would acquire data for all the controls in the first block followed by all the experimental samples in the second block (Figure 8c). We recommend routinely blocking and randomizing in this manner even if an experiment is not designed to be quantitative, as even method development and pilot studies can benefit from avoiding bias due to technical variability. In addition to blocking and randomization, the addition of retention time standards to the samples ensures that variations during acquisition are accounted for.

9. Considerations for data analysis

Data analysis is the final step of the experimental design, but like all other aspects, should be considered during the initial stages of study planning. Numerous analysis tools are available for interpreting raw MS spectra.[124–126] These tools typically include an algorithm to match theoretical data to observed data (i.e., to assign a peptide sequence to a mass spectrum), followed by an estimation of error in the assignment. These tools vary in the algorithm used, sensitivity, accuracy, processing time, types of data that can be analyzed, and specific

> ***Set yourself up for success***: Consult with the MS practitioner to determine the tools available for data analysis. Discuss whether data analysis to be performed requires advanced expertise or if a new user would be able to analyze data on their own. Consult with a statistician during the study design process. Be sure to explain anything special about your sample (e.g., does the recombinant protein have a sequence that differs from the canonical sequence? Is it possible that proteins from multiple species are present in the sample?)

parameters that can be customized. Some software tools are apt for protein identification, while others are solely applied to quantitative data. Some tools can be used to identify and localize any PTM, while others may have limitations in which PTMs can be analyzed. Commercial and free tools are available. Some offer easy–to-use intuitive interfaces while others have a steep learning curve for new users. Some tools include excellent data visualization features, while others do not. No matter which tool is available for the study, it is important to be aware of any chemical or genetic modifications or species that are possible in the sample, as this will define the search space and ultimately affect error estimation rates.[127] Finally, all software will deliver results with an associated confidence metric. It is up to the MS practitioner to determine the appropriate level of confidence required for different applications. Once the raw data are analyzed and a report of protein identifications or quantification is obtained, subsequent analyses typically include a range of ontology, functional, and statistical analyses and integration with other omics data types. Prior to publication, many journals now require that the raw and/or analyzed data are deposited in a public repository, and several large repositories are available for this purpose.[128]

10. What to avoid

MS practitioners are not magicians. Data quality is directly dependent on experimental design, sample quality, and data acquisition quality.

Every surface and chemical that interacts with the proteins in the sample and every environmental factor that the sample encounters has the potential to adversely affect results. Common laboratory reagents can render a sample unusable for MS analysis (e.g., polyethylene glycol, glycerol, detergents). In our experience, it is often more efficient to discuss all details of sample collection and preparation prior to beginning the study rather than trying to troubleshoot a failed experiment. Investigators should be prepared to provide details of all consumables, reagents, chemicals, and methods they propose to use so that the MS practitioner can review them. Often, they will identify potential problems (e.g., contaminants commonly found in molecular weight cutoff filters, non-specific binders from streptavidin bead–based enrichment, proteins solubilized in glycerol) in advance. By providing advice on how to avoid these common problems, the MS practitioner will be able to help investigators quickly generate MS-quality samples for analysis. Although not an exhaustive list, Table 2 outlines easily avoidable problems when performing proteomics experiments.

Table 2. What to be aware of when performing a proteomics experiment

Beware	Why it matters	Suggestions
Inserting pipettes into stock bottle of reagents, especially organic solvents.	Can introduce contamination.	Always pour from the stock bottle into a smaller container that is used for pipetting.
Glassware from dishwasher and glassware used for molecular biology experiments.	Can introduce contamination, detergents.	Use disposable low-binding plastic ware. Keep glassware used for preparing samples that go to MS separate from those used for other experiments.
Not wearing gloves. Touching your face, hair, and skin while preparing a sample even if wearing gloves.	Keratin can contaminate the sample and make it more difficult to detect some proteins of interest.	Always wear gloves. Do not touch your skin or hair or other surfaces while preparing the sample.

Table 2. (*Continued*)

Beware	Why it matters	Suggestions
Detergents and polymers (e.g., glycerol)	These contaminants can overwhelm peptide signals and are difficult or impossible to remove from the sample.	Avoid introducing these into the sample. Be aware that some reagents and consumables will be contaminated with these substances. Use MS-grade reagents and consumables.
Pipette tips from boxes where the lid is left open on the bench for long periods of time.	Dust and skin cells can clog nano-flow columns/tips.	Always close pipette tip boxes and other storage containers to prevent dust. Prepare samples in a laminar flow hood.
Trifluoroacetic acid	Suppresses electrospray ionization.	Use alternative ion-pairing reagents such as formic acid.
Molecular biology grade reagents	Insufficient purity.	Use MS-grade reagents during sample preparation.
Assuming all streptavidin beads perform equally or have equivalent specificity for biotinylated peptides or proteins	Performance characteristics can vary among bead types, manufacturers, and lots of the same bead product.	Use AVIDITY assay to evaluate the binding capacity of your lot of beads.[110] Use negative controls to aid in identifying which proteins captured are non-specific binders.
Relying solely on gels to determine protein purity	MS is more sensitive.	Use MS to check if the protein is truly pure before proceeding with more detailed studies of the purified protein.
Applying old protocols for modern studies	Missed cleavages can complicate data analysis and prevent quantification.	When peptide samples generated by old digestion protocols are analyzed using modern, sensitive MS instruments, it is not uncommon to find high missed cleavage rates compared to that observed when using older, less sensitive MS instrumentation.

> ***Set yourself up for success***: Be meticulous in handling samples for proteomics, starting from how you handle stock bottles of reagents, pipette tips, reagent containers, and specimens. Remember that everything the protein sample encounters can affect the result.

11. Expectation management

New methods, innovative approaches, and high-profile publications are exciting! These advancements enable new discoveries and inspire us to imagine what will be possible in the future. However, some of the studies using MS-based proteomics required multiple millions of dollars in instrumentation (e.g., sample preparation and MS), several highly trained MS experts, and months of high-end data acquisition and analysis to generate and analyze the data in a single study. Unfortunately, these details are not typically explicitly stated in the publication. Thus, it is possible that a published approach will not be realistic for your proposed study. It is important to have realistic expectations regarding what is possible considering the resources (e.g., personnel time, expertise, sample preparation equipment, MS instrumentation, and data analysis tools) that are available. Also, it is possible that a study published more than a few years ago will describe techniques that are no longer favored by present standards. For example, studies implementing single antibody-based immunoprecipitation for determining protein–protein interactions may face a more challenging course for publication than modern approaches.[129] Finally, it is not reasonable to expect that the results obtained from applying a specific method to your biological specimens will result in the exact same list of proteins as that published by another laboratory applying the same method. Unless all aspects of data acquisition have been harmonized among laboratories (e.g., LC column, MS instrument, identical sample), it is expected that there will be some variation in the number and identity of the proteins detected even for similar sample types.

Given the near-limitless variety and combinatorial nature of individual aspects that can be assembled into a proteomics workflow, there is commonly more than one approach that could be suitable to address a specific question. Consequently, consult with the MS practitioner you are working with to learn about the approaches that are most practical based on the expertise, instrumentation, and software available in the MS laboratory. It is not reasonable to expect that all MS practitioners will have standard operating procedures in place for all applications, although they likely will have them in place for the applications that they are most experienced with. Consequently, if you are proposing a type of study that is not standardized in that MS laboratory, you should expect to perform pilot studies to validate that the sample preparation and data acquisition methods are optimized before you perform a large-scale experiment.

Finally, no method is truly unbiased or comprehensive. Expecting this outcome will lead to disappointment. Instead, many biases are introduced throughout the workflow, and many of these are designed to enhance experimental success. For example, enriching for a specific subset of proteins helps to ensure low-abundance proteins can be identified. Digesting a protein with trypsin may prevent the identification of some peptides, but it adds the bias that most peptides will be generated with a lysine or arginine on the C-terminal end of the peptide, aiding the likelihood that the peptides can become positively charged during ionization and analyzed by MS.

> ***Set yourself up for success***: No two MS laboratories are the same. Each will have its own unique combination of expertise, standard operating procedures, instrumentation, and data analysis platforms. Consult with the MS practitioners you are working with at the nascent stage of a project to learn what types of analyses are possible and the requirements to perform them. Put in the effort to understand the limitations, biases, and strengths for each approach. Start the conversation with the "what" you need to learn and let that guide you to select the most suitable approach. Consult with an MS practitioner before you collect samples to avoid disappointment.

Accordingly, numerous types of proteomic workflows have been developed to address specific biological questions. The selection of the right strategies or combination of strategies to answer the biological question requires a blend of creativity, experience, access to key resources, and meticulous attention to detail. Importantly, new technologies and approaches are continually emerging, which in turn enable new questions to be addressed. Consequently, more possibilities exist today than just 5 years ago. Likewise, 10 years from now we will be able to routinely do things that we are only dreaming of today.

12. Example of how the biological question impacts every aspect of experimental design

To illustrate how the entire experimental design requires *a priori* determination of the procedures to be used at each step, we provide the example of protein palmitoylation. Palmitoylation (cysteine modified with lipid palmitate) is a PTM that regulates cardiac ion channel, adrenergic receptor, and phospholamban function.[130–132] Palmitoylation sites on modified proteins can be determined by MS, but the success rate for this type of experiment is affected by key details of the experimental design. Regarding sample source and sample collection, if using an overexpression system to purify the ion channel of interest, it is possible that the recombinant protein will not carry the same modifications as endogenous protein obtained from primary cells, so this will need to be verified. Regarding sample preparation, especially in standard bottom-up workflows, proteins are commonly reduced using dithiothreitol and digested with trypsin. This may result in palmitoyl losses due to the instability of palmitoylation with this reducing agent, which may pose challenges for the quantification of this PTM.[133] However, if sample preparation is performed in neutral Tris buffer with tris(2-carboxyethyl)phosphine as the reducing agent, the palmitoylation moieties are largely preserved.[133] Regarding data acquisition, if data are acquired using collision-induced dissociation (a mode used commonly in proteomic analyses), this may lead to loss of the palmitoyl group, making it difficult to determine palmitoylation sites. Alternatively, if electron transfer dissociation is used, the palmitoyl group can be largely preserved, enabling palmitoyl peptide analysis.[133] However, not all MS instruments can perform electron transfer dissociation. If a user does not have access to this dissociation mode, then an alternative strategy that incorporates bioorthogonal click chemistry can be used for the identification of palmitoylation using collision-induced dissociation.[134] However, click chemistry is not applicable for the analysis of all clinical samples (e.g., tissue) as it requires metabolic labeling in live cells. Regarding data analysis, the choice of software and analysis parameters will depend on how the data were acquired (e.g., direct detection of native palmitoylation versus bioorthogonal click chemistry, fragmentation mode).

Acknowledgments

This work was supported by the National Institutes of Health (R35-HL155460) to R.L.G. and American Heart Association (20PRE35200049) to L.B.L.. L.B.L. is a member of the MCW-MSTP, which is partially supported by a T32 grant from NIGMS (GM080202). Funding sources were not involved in study design, data collection, interpretation, analysis, or publication.

References

1. Smith LM, Kelleher NL. Proteoform: A Single Term Describing Protein Complexity. *Nat. Methods.* 2013;10:186–187.
2. Liu RX, Thiessen-Philbrook HR, Vasan RS, *et al.* Comparison of Proteomic Methods in Evaluating Biomarker-AKI Associations in Cardiac Surgery Patients. *Transl. Res.* 2021;238:49–62.
3. Raafs A, Verdonschot J, Ferreira JP, *et al.* Identification of Sex-Specific Biomarkers Predicting New-Onset Heart Failure. *ESC Heart Fail.* 2021;8:3512–3520.
4. Joshi A, Mayr M. In Aptamers They Trust: The Caveats of the SOMAscan Biomarker Discovery Platform from SomaLogic. *Circulation.* 2018;138:2482–2485.

5. Walker RG, Poggioli T, Katsimpardi L, *et al.* Biochemistry and Biology of GDF11 and Myostatin: Similarities, Differences and Questions for Future Investigation. *Circ. Res.* 2016;118:1125–1142.
6. Lam MPY, Ping P, Murphy E. Proteomics Research in Cardiovascular Medicine and Biomarker Discovery. *J. Am. Coll. Cardiol.* 2016;68: 2819–2830.
7. Arab S, Gramolini AO, Ping P, *et al.* Cardiovascular Proteomics: Tools to Develop Novel Biomarkers and Potential Applications. *J. Am. Coll. Cardiol.* 2006;48:1733–1741.
8. Fert-Bober J, Murray CI, Parker SJ, *et al.* Precision Profiling of the Cardiovascular Post-Translationally Modified Proteome: Where There Is a Will, There Is a Way. *Circ. Res.* 2018;122:1221–1237.
9. Sharma P, Cosme J, Gramolini AO. Recent Advances in Cardiovascular Proteomics. *J. Proteomics.* 2013;81:3–14.
10. Vondriska TM, Ping P. Functional Proteomics to Study Protection of the Ischaemic Myocardium. *Expert Opin. Ther. Targets.* 2002;6:563–570.
11. Leutert M, Entwisle SW, Villén J. Decoding Post-Translational Modification Crosstalk with Proteomics. *Mol. Cell. Proteomics.* 2021; 20:100129.
12. Oliveira T, Thaysen-Andersen M, Packer NH, *et al.* The Hitchhiker's Guide to Glycoproteomics. *Biochem. Soc. Trans.* 2021;49:1643–1662.
13. Melby JA, Roberts DS, Larson EJ, *et al.* Novel Strategies to Address the Challenges in Top-Down Proteomics. *J. Am. Soc. Mass Spectrom.* 2021;32:1278–1294.
14. Jones LM. Mass Spectrometry-Based Methods for Structural Biology on a Proteome-Wide Scale. *Biochem. Soc. Trans.* 2020;48:945–954.
15. Sinha A, Mann M. A Beginner's Guide to Mass Spectrometry–Based Proteomics. *Biochem.* 2020;42:64–69.
16. Matthiesen R, Bunkenborg J. Introduction to Mass Spectrometry-Based Proteomics. *Methods Mol. Biol. Clifton NJ.* 2020;2051:1–58.
17. Aebersold R, Mann M. Mass Spectrometry-Based Proteomics. *Nature.* 2003;422:198–207.
18. Murray KK, Boyd RK, Eberlin MN, *et al.* Definitions of Terms Relating to Mass Spectrometry (IUPAC Recommendations 2013). *Pure Appl. Chem.* 2013;85:1515–1609.

19. Kline KG, Frewen B, Bristow MR, *et al.* High Quality Catalog of Proteotypic Peptides from Human Heart. *J. Proteome Res.* 2008;7: 5055–5061.
20. Westbrook JA, Wheeler JX, Wait R, *et al.* The Human Heart Proteome: Two-Dimensional Maps Using Narrow-Range Immobilised pH Gradients. *Electrophoresis.* 2006;27:1547–1555.
21. Tucholski T, Cai W, Gregorich ZR, *et al.* Distinct Hypertrophic Cardiomyopathy Genotypes Result in Convergent Sarcomeric Proteoform Profiles Revealed by Top-Down Proteomics. *Proc. Natl. Acad. Sci.* 2020;117:24691–24700.
22. Papadaki M, Holewinski RJ, Previs SB, *et al.* Diabetes with Heart Failure Increases Methylglyoxal Modifications in the Sarcomere, Which Inhibit Function. *JCI Insight.* 2018;3:121264.
23. DeLeon-Pennell KY, Tian Y, Zhang B, *et al.* CD36 Is a Matrix Metalloproteinase-9 Substrate that Stimulates Neutrophil Apoptosis and Removal During Cardiac Remodeling. *Circ. Cardiovasc. Genet.* 2016;9:14–25.
24. Kirk JA, Chakir K, Lee KH, *et al.* Pacemaker-Induced Transient Asynchrony Suppresses Heart Failure Progression. *Sci. Transl. Med.* 2015;7:319ra207.
25. Joshi A, Schmidt LE, Burnap SA, *et al.* Neutrophil-Derived Protein S100A8/A9 Alters the Platelet Proteome in Acute Myocardial Infarction and Is Associated with Changes in Platelet Reactivity. *Arterioscler. Thromb. Vasc. Biol.* 2022;42:49–62.
26. Xiang F, Shi Z, Guo X, *et al.* Proteomic Analysis of Myocardial Tissue from the Border Zone During Early Stage Post-Infarct Remodelling in Rats. *Eur. J. Heart Fail.* 2011;13:254–263.
27. Liu T, Chen L, Kim E, *et al.* Mitochondrial Proteome Remodeling in Ischemic Heart Failure. *Life Sci.* 2014;101:27–36.
28. Nishtala K, Phong TQ, Steil L, *et al.* Virus-Induced Dilated Cardiomyopathy Is Characterized by Increased Levels of Fibrotic Extracellular Matrix Proteins and Reduced Amounts of Energy-Producing Enzymes. 2011;11:4310–4320.
29. Lu D, Xia Y, Chen Z, *et al.* Cardiac Proteome Profiling in Ischemic and Dilated Cardiomyopathy Mouse Models. *Front. Physiol.* 2019; 10:750.

30. Aye TT, Scholten A, Taouatas N, *et al.* Proteome-Wide Protein Concentrations in the Human Heart. *Mol. Biosyst.* 2010;6:1917.
31. Lau E, Cao Q, Ng DCM, *et al.* A Large Dataset of Protein Dynamics in the Mammalian Heart Proteome. *Sci. Data.* 2016;3:160015.
32. Doll S, Dreßen M, Geyer PE, *et al.* Region and Cell-Type Resolved Quantitative Proteomic Map of the Human Heart. *Nat. Commun.* 2017;8:1469.
33. Linscheid N, Poulsen PC, Pedersen ID, *et al.* Quantitative Proteomics of Human Heart Samples Collected In Vivo Reveal the Remodeled Protein Landscape of Dilated Left Atrium Without Atrial Fibrillation. *Mol. Cell. Proteomics.* 2020;19:1132–1144.
34. Wojtkiewicz M, Luecke LB, Castro C, *et al.* Bottom-Up Proteomic Analysis of Human Adult Cardiac Tissue and Isolated Cardiomyocytes. *J. Mol. Cell. Cardiol.* 2022;162:20–31.
35. Lambert J. Statistics in Brief: How to Assess Bias in Clinical Studies? *Clin. Orthop.* 2011;469:1794–1796.
36. Pannucci CJ, Wilkins EG. Identifying and Avoiding Bias in Research. *Plast. Reconstr. Surg.* 2010;126:619–625.
37. Wald C, Wu C. Of Mice and Women: The Bias in Animal Models. *Science.* 2010;327:1571–1572.
38. Kilpatrick LE, Kilpatrick EL. Optimizing High-Resolution Mass Spectrometry for the Identification of Low-Abundance Post-Translational Modifications of Intact Proteins. *J. Proteome Res.* 2017;16:3255–3265.
39. Zacharias AO, Fang Z, Rahman A, *et al.* Affinity and Chemical Enrichment Strategies for Mapping Low-Abundance Protein Modifications and Protein-Interaction Networks. *J. Sep. Sci.* 2021;44: 310–322.
40. Gutierrez-Reyes CD, Jiang P, Atashi M, *et al.* Advances in Mass Spectrometry-Based Glycoproteomics: An Update Covering the Period 2017–2021. *Electrophoresis.* 2022;43:370–387.
41. Riley NM, Bertozzi CR, Pitteri SJ. A Pragmatic Guide to Enrichment Strategies for Mass Spectrometry-Based Glycoproteomics. *Mol. Cell. Proteomics.* 2021;20:100029.

42. Swaminathan SK, Olin MR, Forster CL, *et al.* Identification of a Novel Monoclonal Antibody Recognizing CD133. *J. Immunol. Methods.* 2010;361:110–115.
43. Mikesh LM, Ueberheide B, Chi A, *et al.* The Utility of ETD Mass Spectrometry in Proteomic Analysis. *Biochim. Biophys. Acta.* 2006; 1764:1811–1822.
44. Oberg AL, Vitek O. Statistical Design of Quantitative Mass Spectrometry-Based Proteomic Experiments. *J. Proteome Res.* 2009;8:2144–2156.
45. Loos G, Van Schepdael A, Cabooter D. Quantitative Mass Spectrometry Methods for Pharmaceutical Analysis. *Philos. Transact. A Math. Phys. Eng. Sci.* 2016;374:20150366.
46. Meyer JG, Schilling B. Clinical Applications of Quantitative Proteomics Using Targeted and Untargeted Data-Independent Acquisition Techniques. *Expert Rev. Proteomics.* 2017;14:419–429.
47. Mangum K, Gallagher K, Davis FM. The Role of Epigenetic Modifications in Abdominal Aortic Aneurysm Pathogenesis. *Biomolecules.* 2022;12:172.
48. Gao P, You M, Li L, *et al.* Salt-Induced Hepatic Inflammatory Memory Contributes to Cardiovascular Damage Through Epigenetic Modulation of SIRT3. *Circulation.* 2022;145:375–391.
49. Manz XD, Szulcek R, Pan X, *et al.* Epigenetic Modification of the VWF Promotor Drives Platelet Aggregation on the Pulmonary Endothelium in Chronic Thromboembolic Pulmonary Hypertension. *Am. J. Respir. Crit. Care Med.* 2022;205:806–818. doi: 10.1164/rccm.202109-2075OC.
50. Eng JK, McCormack AL, Yates JR. An Approach to Correlate Tandem Mass Spectral Data of Peptides with Amino Acid Sequences in a Protein Database. *J. Am. Soc. Mass Spectrom.* 1994;5:976–989.
51. Kong AT, Leprevost FV, Avtonomov DM, *et al.* MSFragger: Ultrafast and Comprehensive Peptide Identification in Mass Spectrometry–Based Proteomics. *Nat. Methods.* 2017;14:513–520.
52. Käll L, Canterbury JD, Weston J, *et al.* Semi-Supervised Learning for Peptide Identification from Shotgun Proteomics Datasets. *Nat. Methods.* 2007;4:923–925.

53. Perkins DN, Pappin DJC, Creasy DM, *et al.* Probability-Based Protein Identification by Searching Sequence Databases Using Mass Spectrometry Data. *Electrophoresis.* 1999;20:3551–3567. Wiley-VCH Verlag.
54. Zhang J, Xin L, Shan B, *et al.* PEAKS DB: de novo Sequencing Assisted Database Search for Sensitive and Accurate Peptide Identification. *Mol. Cell. Proteomics.* 2012;11:M111.010587.
55. Shilov IV, Seymour SL, Patel AA, *et al.* The Paragon Algorithm, a Next Generation Search Engine that Uses Sequence Temperature Values and Feature Probabilities to Identify Peptides from Tandem Mass Spectra. *Mol. Cell. Proteomics.* 2007;6:1638–1655.
56. MacLean B, Tomazela DM, Shulman N, *et al.* Skyline: An Open Source Document Editor for Creating and Analyzing Targeted Proteomics Experiments. *Bioinformatics.* 2010;26:966–968.
57. Cox J, Mann M. MaxQuant Enables High Peptide Identification Rates, Individualized p.p.b.-Range Mass Accuracies and Proteome-Wide Protein Quantification. *Nat. Biotechnol.* 2008;26:1367–1372.
58. Yuan Z-F, Lin S, Molden RC, *et al.* Evaluation of Proteomic Search Engines for the Analysis of Histone Modifications. *J. Proteome Res.* 2014;13:4470–4478.
59. Parker CE, Mocanu V, Mocanu M, *et al.* Mass Spectrometry for Post-Translational Modifications. In *Neuroproteomics.* Boca Raton, FL: CRC Press/Taylor & Francis, 2010.
60. Bagwan N, El Ali HH, Lundby A. Proteome-Wide Profiling and Mapping of Post Translational Modifications in Human Hearts. *Sci. Rep.* 2021;11:2184.
61. Macklin A, Khan S, Kislinger T. Recent Advances in Mass Spectrometry Based Clinical Proteomics: Applications to Cancer Research. *Clin. Proteomics.* 2020;17:17.
62. Han X, Aslanian A, Yates JR. Mass Spectrometry for Proteomics. *Curr. Opin. Chem. Biol.* 2008;12:483–490.
63. Chait BT. Mass Spectrometry: Bottom-Up or Top-Down? *Science.* 2006;314:65.
64. Fehrenbach DJ, Abais-Battad JM, Dasinger JH, *et al.* Sexual Dimorphic Role of CD14 (Cluster of Differentiation 14) in Salt-Sensitive Hypertension and Renal Injury. *Hypertension* 2021;77:228–240.

65. Lau ES, Binek A, Parker SJ, *et al.* Sexual Dimorphism in Cardiovascular Biomarkers: Clinical and Research Implications. *Circ. Res.* 2022; 130:578–592.
66. Cediel G, Codina P, Spitaleri G, *et al.* Gender-Related Differences in Heart Failure Biomarkers. *Front. Cardiovasc. Med.* 2021;7:617705.
67. Beale AL, Meyer P, Marwick TH, *et al.* Sex Differences in Cardiovascular Pathophysiology. *Circulation*. 2018;138:198–205.
68. Volgman AS, Merz CNB, Aggarwal NT, *et al.* Sex Differences in Cardiovascular Disease and Cognitive Impairment: Another Health Disparity for Women? *J. Am. Heart Assoc.* 2019;8:e013154.
69. Suthahar N, Lau ES, Blaha MJ, *et al.* Sex-Specific Associations of Cardiovascular Risk Factors and Biomarkers with Incident Heart Failure. *J. Am. Coll. Cardiol.* 2020;76:1455–1465.
70. Lindsey ML, LeBlanc AJ, Ripplinger CM, *et al.* Reinforcing Rigor and Reproducibility Expectations for Use of Sex and Gender in Cardiovascular Research. *Am. J. Physiol. Heart Circ. Physiol.* 2021; 321. https://journals.physiology.org/doi/full/10.1152/ajpheart.00418.2021.
71. Meilhac SM, Lescroart F, Blanpain C, *et al.* Cardiac Cell Lineages that Form the Heart. *Cold Spring Harb. Perspect. Med.* 2014;4:a013888.
72. Ericsson C, Franzén B, Nistér M. Frozen Tissue Biobanks: Tissue Handling, Cryopreservation, Extraction, and Use for Proteomic Analysis. *Acta Oncol.* 2006;45:643–661.
73. Schoenhoff FS, Fu Q, Van Eyk JE. Cardiovascular Proteomics — Implications for Clinical Applications. *Clin. Lab. Med.* 2009;29: 87–99.
74. Bruegel M, Planert M, Baumann S, *et al.* Standardized Peptidome Profiling of Human Cerebrospinal Fluid by Magnetic Bead Separation and Matrix-Assisted Laser Desorption/Ionization Time-of-Flight Mass Spectrometry. *J. Proteomics.* 2009;72:608–615.
75. Chen Y, Shi J, Xia TC, *et al.* Preservation Solutions for Kidney Transplantation: History, Advances and Mechanisms. *Cell Transplant.* 2019;28:1472–1489.
76. Fiegle DJ, Volk T, Seidel T. Isolation of Human Ventricular Cardiomyocytes from Vibratome-Cut Myocardial Slices. *J. Vis. Exp.* 2020:61167. doi: 10.3791/61167.

77. Guo G, Chen L, Rao M, *et al.* A Modified Method for Isolation of Human Cardiomyocytes to Model Cardiac Diseases. *J. Transl. Med.* 2018;16:288.
78. Voigt N, Zhou X-B, Dobrev D. Isolation of Human Atrial Myocytes for Simultaneous Measurements of Ca^{2+} Transients and Membrane Currents. *J. Vis. Exp.* 2013:e50235. doi: 10.3791/50235.
79. Coppini R, Ferrantini C, Aiazzi A, *et al.* Isolation and Functional Characterization of Human Ventricular Cardiomyocytes from Fresh Surgical Samples. *J. Vis. Exp.* 2014:51116. doi: 10.3791/51116.
80. Neuß M, Regitz-Zagrosek V, Hildebrandt A, *et al.* Isolation and Characterisation of Human Cardiac Fibroblasts from Explanted Adult Hearts. *Cell Tissue Res.* 1996;286:145–153.
81. Garbutt TA, Zhou Y, Keepers B, *et al.* An Optimized Protocol for Human Direct Cardiac Reprogramming. *STAR Protoc.* 2020;1:100010.
82. Eghbali-Webb M, Agocha AE. A Simple Method for Preparation of Cultured Cardiac Fibroblasts from Adult Human Ventricular Tissue. In *Novel Methods in Molecular and Cellular Biochemistry of Muscle*; Pierce GN, Claycomb WC, Eds. Springer US, 1997:195–198. doi: 10.1007/978-1-4615-6353-2_19.
83. Hsieh S-Y., Chen R-K., Pan Y-H, *et al.* Systematical Evaluation of the Effects of Sample Collection Procedures on Low-Molecular-Weight Serum/Plasma Proteome Profiling. *Proteomics.* 2006;6:3189–3198.
84. Boyanton BL, Blick KE. Stability Studies of Twenty-Four Analytes in Human Plasma and Serum. *Clin. Chem.* 2002;48:2242–2247.
85. Bonilauri B, Santos MDM, Camillo-Andrade AC, *et al.* The Impact of Blood-Processing Time on the Proteome of Human Peripheral Blood Mononuclear Cells. *Biochim. Biophys. Acta BBA — Proteins Proteomics.* 2021;1869:140581.
86. Halvey P, Farutin V, Koppes L, *et al.* Variable Blood Processing Procedures Contribute to Plasma Proteomic Variability. *Clin. Proteomics.* 2021;18:5.
87. Ashworth M, Small B, Oldfield L, *et al.* The Holding Temperature of Blood During a Delay to Processing Can Affect Serum and Plasma Protein Measurements. *Sci. Rep.* 2021;11:6487.
88. Becker K-F. Using Tissue Samples for Proteomic Studies — Critical Considerations. *Proteomics — Clin. Appl.* 2015;9:257–267.

89. Guo M-L, Sun M-X, Lan J-Z, *et al.* Proteomic Analysis of the Effects of Cell Culture Density on the Metastasis of Breast Cancer Cells. *Cell Biochem. Funct.* 2019;37:72–83.
90. Trajkovic K, Valdez C, Ysselstein D, *et al.* Fluctuations in Cell Density Alter Protein Markers of Multiple Cellular Compartments, Confounding Experimental Outcomes. *PLoS One.* 2019;14:e0211727.
91. Brennan K, Martin K, FitzGerald SP, *et al.* A Comparison of Methods for the Isolation and Separation of Extracellular Vesicles from Protein and Lipid Particles in Human Serum. *Sci. Rep.* 2020;10:1039.
92. Yuana Y, Levels J, Grootemaat A, *et al.* Co-Isolation of Extracellular Vesicles and High-Density Lipoproteins Using Density Gradient Ultracentrifugation. *J. Extracell. Vesicles.* 2014;3:10.3402/jev. v3.23262.
93. van den Broek I, Fu Q, Kushon S, *et al.* Application of Volumetric Absorptive Microsampling for Robust, High-Throughput Mass Spectrometric Quantification of Circulating Protein Biomarkers. *Clin. Mass Spectrom.* 2017;4–5:25–33.
94. Molloy MP, Hill C, O'Rourke MB, *et al.* Proteomic Analysis of Whole Blood Using Volumetric Absorptive Microsampling for Precision Medicine Biomarker Studies. *J. Proteome Res.* 2022;21:1196–1203. doi: 10.1021/acs.jproteome.1c00971.
95. Petelski AA, Emmott E, Leduc A, *et al.* Multiplexed Single-Cell Proteomics Using SCoPE2. *Nat. Protoc.* 2021;16:5398–5425.
96. Cong Y, Motamedchaboki K, Misal SA, *et al.* Ultrasensitive Single-Cell Proteomics Workflow Identifies >1000 Protein Groups Per Mammalian Cell. *Chem. Sci.* 2020;12:1001–1006.
97. Li H, Liu C, Bao M, *et al.* Optimized Langendorff Perfusion System for Cardiomyocyte Isolation in Adult Mouse Heart. *J. Cell. Mol. Med.* 2020;24:14619–14625.
98. Omatsu-Kanbe M, Fukunaga R, Mi X, *et al.* An Antegrade Perfusion Method for Cardiomyocyte Isolation from Mice. *J. Vis. Exp.* 2021. doi: 10.3791/61866.
99. Yadav A, Marini F, Cuomo A, *et al.* Software Options for the Analysis of MS-Proteomic Data. *Methods Mol. Biol. Clifton NJ.* 2021; 2361:35–59.

100. Sinitcyn P, Rudolph JD, Cox J. Computational Methods for Understanding Mass Spectrometry–Based Shotgun Proteomics Data. *Annu. Rev. Biomed. Data Sci.* 2018;1:207–234.
101. Zhang Y, Fonslow BR, Shan B, *et al.* Protein Analysis by Shotgun/Bottom-Up Proteomics. *Chem. Rev.* 2013;113:2343–2394.
102. Nesvizhskii AI, Aebersold R. Interpretation of Shotgun Proteomic Data. *Mol. Cell. Proteomics.* 2005;4:1419–1440.
103. Dupree EJ, Jayathirtha M, Yorkey H, *et al.* A Critical Review of Bottom-Up Proteomics: The Good, the Bad, and the Future of This Field. *Proteomes.* 2020;8:14.
104. Catherman AD, Skinner OS, Kelleher NL. Top Down Proteomics: Facts and Perspectives. *Biochem. Biophys. Res. Commun.* 2014; 445:683–693.
105. Gregorich ZR, Chang Y-H, Ge Y. Proteomics in Heart Failure: Top-Down or Bottom-Up? *Pflugers Arch.* 2014;466:1199–1209.
106. White MY, Cordwell SJ, McCarron HCK, *et al.* Proteomics of Ischemia/Reperfusion Injury in Rabbit Myocardium Reveals Alterations to Proteins of Essential Functional Systems. *Proteomics.* 2005;5:1395–1410.
107. Li DK, Smith LE, Rookyard AW, *et al.* Multi-Omics of a Pre-Clinical Model of Diabetic Cardiomyopathy Reveals Increased Fatty Acid Supply Impacts Mitochondrial Metabolic Selectivity. *J. Mol. Cell. Cardiol.* 2022;164:92–109.
108. Herrington DM, Mao C, Parker SJ, *et al.* Proteomic Architecture of Human Coronary and Aortic Atherosclerosis. *Circulation.* 2018;137:2741–2756.
109. St-Germain J, Tehrani PS, Wong C, *et al.* Variability in Streptavidin-Sepharose Matrix Quality Can Significantly Affect Proximity-Dependent Biotinylation (BioID) Data. *J. Proteome Res.* 2020;19:3554–3561. doi: 10.1021/acs.jproteome.0c00117.
110. Berg Luecke L, Gundry RL. Assessment of Streptavidin Bead Binding Capacity to Improve Quality of Streptavidin-Based Enrichment Studies. *J. Proteome Res.* 2021;20:1153–1164.
111. Zauber H, Schüler V, Schulze W. Systematic Evaluation of Reference Protein Normalization in Proteomic Experiments. *Front. Plant Sci.* 2013;4.

112. Li H, Han J, Pan J, *et al.* Current Trends in Quantitative Proteomics — an Update. *J. Mass Spectrom.* 2017;52:319–341.
113. Calderón-Celis F, Encinar JR, Sanz-Medel A. Standardization Approaches in Absolute Quantitative Proteomics with Mass Spectrometry. *Mass Spectrom. Rev.* 2018;37:715–737.
114. Wojtkiewicz M, Berg Luecke L, Kelly MI, *et al.* Facile Preparation of Peptides for Mass Spectrometry Analysis in Bottom-Up Proteomics Workflows. *Curr. Protoc.* 2021;1:e85.
115. Fu Q, Murray CI, Karpov OA, *et al.* Automated Proteomic Sample Preparation: The Key Component for High Throughput and Quantitative Mass Spectrometry Analysis. *Mass Spectrom. Rev.* 2021:e21750. doi: 10.1002/mas.21750.
116. Krieger JR, Wybenga-Groot LE, Tong J, *et al.* Evosep One Enables Robust Deep Proteome Coverage Using Tandem Mass Tags While Significantly Reducing Instrument Time. *J. Proteome Res.* 2019;18: 2346–2353.
117. Ludwig KR, Schroll MM, Hummon AB. Comparison of In-Solution, FASP, and S-Trap Based Digestion Methods for Bottom-Up Proteomic Studies. *J. Proteome Res.* 2018;17:2480–2490.
118. Ding H, Fazelinia H, Spruce LA, *et al.* Urine Proteomics: Evaluation of Different Sample Preparation Workflows for Quantitative, Reproducible, and Improved Depth of Analysis. *J. Proteome Res.* 2020;19:1857–1862.
119. Mikšík I. Coupling of CE-MS for Protein and Peptide Analysis. *J. Sep. Sci.* 2019;42:385–397.
120. Morris CB, Poland JC, May JC, *et al.* Fundamentals of Ion Mobility-Mass Spectrometry for the Analysis of Biomolecules. *Methods Mol. Biol. Clifton NJ.* 2020;2084:1–31.
121. Li C, Chu S, Tan S, *et al.* Towards Higher Sensitivity of Mass Spectrometry: A Perspective from the Mass Analyzers. *Front. Chem.* 2021;9:813359.
122. Riley NM, Malaker SA, Driessen MD, *et al.* Optimal Dissociation Methods Differ for N- and O-Glycopeptides. *J. Proteome Res.* 2020; 19:3286–3301.
123. Burger B, Vaudel M, Barsnes H. Importance of Block Randomization When Designing Proteomics Experiments. *J. Proteome Res.* 2021;20: 122–128.

124. Verheggen K, Raeder H, Berven FS, *et al.* Anatomy and Evolution of Database Search Engines-a Central Component of Mass Spectrometry Based Proteomic Workflows. *Mass Spectrom. Rev.* 2020;39:292–306.
125. Zhang F, Ge W, Ruan G, *et al.* Data-Independent Acquisition Mass Spectrometry-Based Proteomics and Software Tools: A Glimpse in 2020. *Proteomics.* 2020;20:e1900276.
126. Tsiamis V, Ienasescu H-I, Gabrielaitis D, *et al.* One Thousand and One Software for Proteomics: Tales of the Toolmakers of Science. *J. Proteome Res.* 2019;18:3580–3585.
127. Knudsen GM, Chalkley RJ. The Effect of Using an Inappropriate Protein Database for Proteomic Data Analysis. *PLoS One.* 2011; 6:e20873.
128. Vizcaíno JA, Foster JM, Martens L. Proteomics Data Repositories: Providing a Safe Haven for Your Data and Acting as a Springboard for Further Research. *J. Proteomics.* 2010;73:2136–2146.
129. Kushner JS, Liu G, Eisert RJ, *et al.* Detecting Cardiovascular Protein-Protein Interactions by Proximity Proteomics. *Circ. Res.* 2022; 130:273–287.
130. Gök C, Robertson AD, Fuller W. Insulin-Induced Palmitoylation Regulates the Cardiac Na+/Ca2+ Exchanger NCX1. *Cell Calcium.* 2022;104:102567.
131. Zhou T, Li J, Zhao P, *et al.* Palmitoyl Acyltransferase Aph2 in Cardiac Function and the Development of Cardiomyopathy. *Proc. Natl. Acad. Sci. USA.* 2015;112:15666–15671.
132. Marin EP, Jozsef L, Di Lorenzo A, *et al.* The Protein Acyl Transferase ZDHHC21 Modulates α1 Adrenergic Receptor Function and Regulates Hemodynamics. *Arterioscler. Thromb. Vasc. Biol.* 2016;36: 370–379.
133. Ji Y, Leymarie N, Haeussler DJ, *et al.* Direct Detection of S-Palmitoylation by Mass Spectrometry. *Anal. Chem.* 2013;85: 11952–11959.
134. Martin BR, Cravatt BF. Large-Scale Profiling of Protein Palmitoylation in Mammalian Cells. *Nat. Methods.* 2009;6:135–138.
135. Berg Luecke L, Waas M, Littrell J, *et al.*, Surfaceome mapping of primary human heart cells with CellSurfer uncovers cardiomyocyte surface protein LSMEM2 and proteome dynamics in failing hearts. *Nat Cardiovasc Res.* 2023; 2: 76–95.

https://doi.org/10.1142/9789811283376_0010

Chapter 10

Amyloidosis

Diana Canetti

Centre for Amyloidosis, Division of Medicine, University College London, Rowland Hill Street, London NW3 2PF, UK

Contents

Abstract

Amyloidosis is caused by the presence of the amyloid aggregate in the extracellular matrix of tissues throughout the body. Over 30 proteins are known to cause the disease. The molecular characterization of the amyloid deposits consistently showed the presence of apolipoprotein A-IV, apolipoprotein E, and serum amyloid P-component components along with the amyloidogenic protein. The precise amyloid typing is crucial for the correct diagnosis of amyloidosis, and proteomics plays an important role in providing information complementary to immunohistochemistry, genetics, and clinical data. Proteomics offers a rapid and reliable method to identify amyloid proteins, post-translationally modified proteins, truncated proteins, and new amyloid proteins. As with all approaches, it is essential to understand the limitations of the methodology.

1. Amyloidosis

1.1. *Description and classification of amyloidosis*

Amyloidosis represents a group of rare and heterogeneous diseases occurring when proteins in a misfolded state accumulate as insoluble material, called amyloid, in the extracellular space of various tissues throughout the body.

The amyloid deposit is structurally constituted by fibrils, which are composed of twisted protofilaments. "An amyloid protofilament is a stack of protein layers in β-sheet structure, which when twisted about identical stacks, forms an amyloid fibril."[1]

The presence of amyloid deposit in the human body impairs the structure and function of the affected organs, and depending on the amyloid distribution, the disease is classified as localized or systemic. In the former, the fibrils are deposited in a particular organ or tissue, in the latter, multiple organs are involved. Localized and systemic amyloidosis can be also distinguished as hereditary or acquired. Hereditary amyloidosis is caused by genetic mutations generating variant proteins, which are characterized by abnormal structures giving rise to amyloid fibril deposits. Lysozyme (ALys), acquired transthyretin (ATTR), and fibrinogen Aα (FibA) amyloidosis belong to this category.[2–4] The mutations in hereditary amyloidosis are listed at http://amyloidosismutations.com/.[5] Acquired amyloidosis is the result of another previous disease that can generate an inherently amyloidogenic abnormal protein or increase the concentration of a potentially amyloidogenic normal protein.[6] Common examples are monoclonal immunoglobulin kappa and lambda light chains amyloidosis (AL)[7] and serum amyloid A (AA) amyloidosis.[8]

There are some forms of amyloidosis that can be either acquired or hereditary. For example, ATTR amyloidosis is an aging-associated disease caused by wild-type TTR protein. About 25% of people over 75 years old can develop cardiac wild-type ATTR.[9] Hereditary ATTR is caused by destabilized TTR variants, such as V30M, T60A, and V122I. The onset of the hereditary disease usually occurs earlier in comparison with acquired amyloidosis.[10]

From a molecular point of view, amyloid fibril formation is essentially a two-phase process: an initial nucleation phase, where despite the presence of potentially amyloidogenic species, the fibril formation is thermodynamically unfavorable. This represents the rate-limiting step of the amyloidogenic process called lag phase.[11,12] Then, once the nucleus is formed, the equilibrium shifts favoring a rapid exponential growth that leads to the formation and the elongation of amyloid fibrils[12,13] (elongation phase). Some factors can cooperate *in vivo* favoring amyloid formation, such as interactions with extracellular matrix components, local pH, temperature, and oxidation status.[12]

Currently 36 amyloidogenic proteins and peptides, which are listed in Table 1, are known to cause amyloidosis.[1]

1.2. *Characterization of amyloid deposits*

The most common method for amyloid detection in tissues is Congo Red (CR) staining. Amyloid fibrils have a high affinity to the CR dye and show birefringence under polarized light after staining. The amyloid birefringence is usually defined as apple green, but sometimes in CR-stained tissues, the green color may appear feeble and difficult to visualize. Indeed, the amyloid deposits can also be observed as red or yellow material under polarized light, and this depends on how the amyloidotic tissue is dissected with respect to the orientation of the fibrils.

The molecular characterization of the *ex vivo* amyloid deposits showed that the core of the aggregate consists of a fibril protein together with other ubiquitous components including apolipoprotein E (ApoE), apolipoprotein A-IV (ApoA-IV), and serum

Table 1. List of known amyloid fibril proteins and precursors, type of amyloidosis, and amyloid deposition target.

Fibril protein	Precursor protein	Systemic (S) and/or localized (L)	Acquired (A) or hereditary (H)	Target organs
AL	Immunoglobulin light chain	S, L	A, H	All organs, usually except CNS
AH	Immunoglobulin heavy chain	S, L	A	All organs except CNS
AA	(Apo) serum amyloid A	S	A	All organs except CNS
ATTR	Transthyretin, wild type	S	A	Heart mainly in males, lungs, ligaments, tenosynovium
	Transthyretin, variants	S	H	PNS, ANS, heart, eye, leptomeninges
Aβ2M	β2-microglobulin, wild type	S	A	Musculoskeletal system
	β2-microglobulin, variants	S	H	ANS
AApoAI	Apolipoprotein A I, variants	S	H	Heart, liver, kidney, PNS, testis, larynx (C-terminal variants), skin (C-terminal variants)
AApoAII	Apolipoprotein A II, variants	S	H	Kidney
AApoAIV	Apolipoprotein A IV, wild type	S	A	Kidney medulla and systemic
AApoCII	Apolipoprotein C II, variants	S	H	Kidney
AApoCIII	Apolipoprotein C III, variants	S	H	Kidney
AGel	Gelsolin, variants	S	H	Kidney, PNS, cornea
ALys	Lysozyme, variants	S	H	Kidney

ALECT2	Leukocyte chemotactic factor-2	S	A	Kidney, primarily
AFib	Fibrinogen α, variants	S	H	Kidney, primarily
ACys	Cystatin C, variants	S	H	CNS, PNS, skin
ABri	ABriPP, variants	S	H	CNS
Adan	ADanPP, variants	L	H	CNS
Aβ	Aβ protein precursor, wild type	L	A	CNS
	Aβ protein precursor, variant	L	H	CNS
AαSyn	α-Synuclein	L	A	CNS
ATau	Tau	L	A	CNS
APrP	Prion protein, wild type	L	A	CJD, fatal insomnia CJD, GSS
	Prion protein variants	L	H	syndrome, fatal insomnia
	Prion protein variant	S	H	PNS
ACal	(Pro)calcitonin	L	A	C-cell thyroid tumors Kidney
		S	A	
AIAPP	Islet amyloid polypeptide	L	A	Islets of Langerhans, insulinomas
AANF	Atrial natriuretic factor	L	A	Cardiac atria
APro	Prolactin	L	A	Pituitary prolactinomas, aging pituitary
AIns	Insulin	L	A	Iatrogenic, local injection
ASPC	Lung surfactant protein	L	A	Lung
ACor	Corneodesmosin	L	A	Cornified epithelia, hair follicles

(*Continued*)

Table 1. (*Continued*)

Fibril protein	Precursor protein	Systemic (S) and/ or localized (L)	Acquired (A) or hereditary (H)	Target organs
AMed	Lactadherin	L	A	Senile aortic, media
AKer	Kerato-epithelin	L	A	Cornea, hereditary
ALac	Lactoferrin	L	A	Cornea
AOAAP	Odontogenic ameloblast-associated protein	L	A	Odontogenic tumors
ASem1	**Semenogelin 1**	L	A	**Vesicula seminalis**
AEnf	**Enfuvirtide**	L	A	**Iatrogenic**
ACatK	Cathepsin K	L	A	Tumor associated
AEFEMP1	EGF-containing fibulin-like extracellular matrix protein 1 (EFEMP1)	L	A	Portal veins Aging associated

Acronyms: ANS: autonomic nervous system, CNS: central nervous system, PNS: peripheral nervous system, CJD: Creutzfeldt–Jacob disease, GSS: Gerstmann–Straussler–Scheinker syndrome.

Source: Benson *et al.*, *Amyloid*, 2020.

amyloid P-component (SAP); these three proteins are now recognized as amyloid signature proteins when typing amyloid using proteomics.[14]

SAP is a ubiquitous component of amyloid deposits. It is a glycoprotein belonging to the pentraxin family. SAP reversibly binds, in a calcium-dependent manner, to amyloid fibrils, through a coat, which prevents their proteolysis.[15] Its characteristic of reversibly binding to the fibrils allows the use of radio-labeled SAP as a diagnostic imaging tool for amyloid deposition.[16] SAP scanning is only performed in two European centers: University Medical Center Groningen (Netherlands) and National Amyloidosis Centre in London (United Kingdom). A treatment for amyloidosis has recently been reported based on SAP depletion coupled with an antibody-mediated removal of the deposits.[17]

ApoE is a common component of amyloid deposits, but its role in amyloidogenesis is still unknown.[18] It has been reported that ApoE has a key role in Alzheimer's disease (AD) through the interaction with amyloid β protein.[19]

ApoA-IV is an amyloid-associated protein but also an amyloid fibril protein due to its intrinsically amyloidogenic nature. ApoA-IV amyloid deposits were initially detected in the cardiac tissue of a patient with senile transthyretin systemic amyloidosis[20] confirming the deposition of two different types of amyloid in the same tissue.[21] The dual characteristic of ApoA-IV, being a fibril protein and a common component of amyloid deposit, causes some difficulty in identifying ApoA-IV amyloidosis (AApoA-IV). The two largest centers of amyloidosis in the world, the Mayo Clinic in the United States and the National Amyloidosis Centre (NAC) in the United Kingdom adopt different criteria to diagnose AApoA-IV. At the Mayo Clinic, AApoA-IV is diagnosed primarily based on the unusual clinical presentation together with the identification of a high number of ApoA-IV peptides by proteomics in the absence of other amyloidogenic proteins.[22,23] At the NAC, further criteria have been developed for AApoA-IV diagnosis by proteomics. The identification of Apo-IV protein with a Mascot probability score above 2000 or as a top-scoring protein without any other amyloidogenic proteins, and the detection of specific ApoA-IV should generate a Mascot probability score above

2000 or as a top-scoring protein without any other amyloidogenic proteins, and the detection of specific semi-tryptic ApoA-IV signal-containing peptides (p. 18–43, p. 19–43, p. 20–43) increases the confidence of ApoA-IV amyloid identification.[24]

There are additional common components of amyloid deposits including vitronectin[25] and glycosaminoglycans, which act as a scaffold for nucleation and fibril elongation process, and also as a coat for preventing fibril degradation. It has been reported that in AD patients there are higher levels of heparan sulfate proteoglycan and that it is associated with amyloid deposition.[26]

2. Proteomics in amyloidosis

In the last decade, mass spectrometry (MS)-based proteomics has been recognized as the gold standard for amyloid typing[14,27,28] and is now employed for the diagnosis of amyloidosis in state-of-the-art clinical centers around the world. Particularly, the use of laser capture microdissection (LCD), which allows the precise selection of the amyloid material, coupled with MS analysis (LCD-MS), has significantly improved the way to type amyloid deposits.

Amyloid fibril proteins are commonly identified using immunohistochemical (IHC) approaches.[29] Despite some of the disadvantages, such as limited specificity and sensitivity of the commercially available antibodies,[30–32] the approach is still being used successfully in many laboratories.

Several studies have been performed to compare the results obtained by antibody-based techniques and proteomics. For example, a study carried out in 2015 by Mayo Clinic and NAC showed that in 82% of cases, there was a complete agreement between IHC and proteomic results, but in the rest of the 18% of cases the amyloid typing was obtained only by proteomics.[33] The advantages of the proteomic approach have also been shown in other amyloid-related studies.[28,30] Nevertheless, there are some cases where proteomics fails to identify the amyloid protein. This mainly occurs with the tiny amount of tissue sample, insufficient to get acceptable results, and in cases where proteomics provides ambiguous results. In the latter case, the optimization of the protocols for sample preparation would enhance sensitivity and improve amyloid typing.[28]

Considering the limitations associated with IHC techniques, it has been suggested that proteomics will supplant the immunological techniques for amyloid typing in the near future. However, due to the purchase and running costs of a proteomic facility and the requirement for specialized analysts to manipulate samples and interpret MS data, it is likely that only specialized centers will be able to offer this service; others without such resources will probably carry on with IHC.

The need to standardizing proteomic methods for amyloidosis diagnosis and improving knowledge on disease mechanisms led to establishing of the European Proteomics Amyloid Network (EPAN) in 2017 in London as a scientific consortium gathering researchers and clinicians in the field (www.ucl.ac.uk/amyloidosis/national-amyloidosis-centre/uk-amyloid-proteomics-facility/ukampro-research/european-proteomics). During EPAN meetings, members discuss best practices, research updates on amyloid analysis–related challenges, strategies to improve amyloid typing, sample preparation, preanalytical issues, quantification, protocol standardization, MS data sharing, samples exchange, and amyloid protein databases. Further activities include EPAN webinars focusing on the proteomics of different organs with the aim to have a general overview of the protein content and alterations in amyloid-affected tissues.

3. Proteomic protocols for amyloid analysis

One of the EPAN activities is focused on the optimization and standardization of proteomic protocols for amyloid analysis. This work is very important to improve amyloid typing and get reliable, unambiguous results from proteomic analysis, and scientists in the field are putting a lot of effort to achieve this aim.

There are different procedures for the extraction of amyloid proteins depending on the nature of the starting material. Clinical tissue biopsies are classically formalin-fixed paraffin-embedded (FFPE) before any analysis. This represents a common method for storing and preserving tissue samples.

The original protocol developed for amyloidotic FFPE tissues by Professor Ahmet Dogan and colleagues[34] consists in using 10 mM Tris/1 mM EDTA/0.002% Zwittergent 3–16 (TEZ) buffer. Tissue areas containing amyloid are microdissected and collected in the lid of a tube. Proteins are treated at 99°C for 90 min, sonicated in a water bath for 60 min, and digested overnight at 37°C with trypsin. The peptide mixture is treated with dithiothreitol for 5 min at 99°C, and then the buffer is removed by SpeedVac. This procedure is currently employed at NAC and applied to either unfixed fat aspirates or FFPE tissues.[28]

Another procedure for extracting proteins from fresh specimens, such as subcutaneous abdominal fat tissues, requires buffer containing 7 M urea, 2 M thiourea, 4% CHAPS, and 65 mM dithiothreitol. Tissues are crushed with a pestle and sonicated. Lipids and cell debris pellet are removed by centrifugation, and the aqueous layer containing proteins is recovered to be treated with trypsin.[27,35]

Following protein digestion, peptide mixtures are solubilized in acidic solvents, such as 0.1% trifluoroacetic acid or 0.2% formic acid in water, and analyzed by nano-liquid chromatography tandem mass spectrometry (MS/MS).

3.1. *Amyloid protein identification*

LC-MS/MS analyses are commonly performed on Orbitrap mass spectrometers, such as LTQ Orbitrap, Q-Exactive, Q-Exactive Plus, and Exploris.

Acquired MS data can be processed for protein identification by a wide range of bioinformatics platforms, including Mascot, Proteome Discoverer, Peaks, Protein Pilot, and MaxQuant. The Mascot search engine (www.matrixscience.com) is currently used for protein identification at NAC.[28] Oxidation is set as variable modification, and trypsin is set as the *in-silico* search enzyme. Semi-trypsin is used as a secondary search parameter in cases where small proteins, like insulin, are suspected of being the amyloid protein or for identifying ApoA-IV signal-containing peptides to increase the confidence of AApoA-IV diagnosis. Proteomic results are linked to the NAC clinical database, and thanks

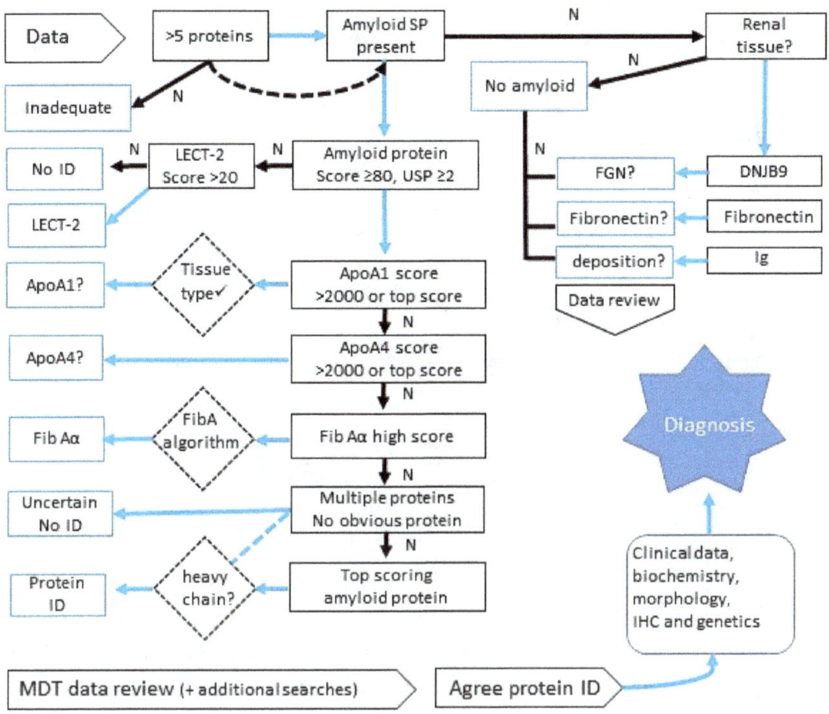

Figure 1. Proteomic algorithm developed at National Amyloidosis Centre (NAC) in the United Kingdom.

Source: Canetti *et al.*, *Clin. Chem. Lab. Med.*, 2020.

to a simple algorithm, the most probable amyloidogenic protein is highlighted[28] (Figure 1). The total number of proteins identified in a sample represents the first step of the algorithm. The minimum number of proteins is set at five with a minimum Mascot score of 80 and at least two unique significant peptides. Contaminants such as keratins and hemoglobin are excluded. If there are less than five proteins, the sample is considered "inadequate." The only exception occurs when there are two signature proteins and the amyloidogenic protein is unambiguously identified or when the renal amyloid protein leukocyte cell–derived chemotaxin-2 (LECT-2) has been detected.[36] LECT-2 amyloid protein is often identified with a low Mascot score and poor protein coverage. In this case, a lower score threshold of 20

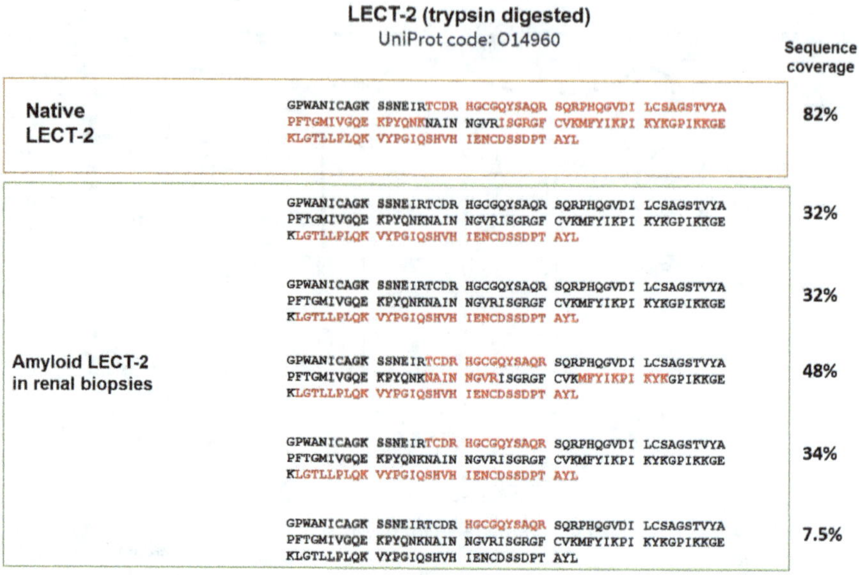

Figure 2. Comparison of LECT-2 protein coverage between native recombinant protein and fibrillary protein detected in clinical samples. The lower LECT-2 sequence coverage in the amyloidotic samples suggests that there is a lower susceptibility to enzymatic digestion.

is selected. Figure 2 shows the comparison of LECT-2 sequence coverage of the native and fibrillary protein. As the native LECT-2 shows a good sequence coverage, the poor protein recovery from clinical samples may be associated with the nature of the fibrillary structure, which limits the enzymatic digestion.[28]

For some amyloid proteins such as ApoA-I and ApoA-IV, a higher minimum score threshold (>2000) is fixed. Both proteins can form amyloid deposits but they are also ubiquitously present in several tissues. Liver, kidneys, and heart can be used as a tissue type discriminator for ApoA-I amyloid identification, as they are the most commonly involved organs.[37] ApoA-IV amyloid is mainly found in the heart and kidneys.[21,23] The identification of ApoA-IV signal-containing peptides also confirms the presence of ApoA-IV amyloid.

An additional algorithm is used for FibA amyloid identification. During the sample collection, contamination with blood easily occurs making it difficult to discriminate the amyloid fibril protein from background contaminants by using the Mascot score alone. FibA amyloid algorithm allows to distinguish amyloid protein from contaminants present in the blood by considering the scores of fibrinogen Bβ and Gγ chains, and the identification of an amyloidogenic FibA variant included in the NAC protein variants database.[38]

The NAC algorithm also highlights potential non-amyloid deposits.[28] For example, the identification of DnaJ homolog subfamily B member 9 (DNAJB9) in CR-negative renal biopsies suggests a fibrillary glomerulonephritis,[39] and the detection of immunoglobulin light and heavy chains without any amyloid signatures may suggest a deposition disease.[40]

Another diagnostic algorithm has been developed by Brambilla and colleagues to accurately identify the amyloid type by proteomics.[27] The α-value algorithm allows us to estimate the abundance of amyloid proteins through the comparison of the protein profile of amyloid affected versus non-amyloid-affected samples by eliminating the effect of contaminants from the blood. The amyloid culprit protein is defined as having the highest α-value, which is calculated by normalizing the patient over control ratio of each amyloid protein's spectral count. The α-value algorithm was initially designed for the identification of four amyloid proteins: immunoglobulin light chain κ and λ, TTR, and SAA in adipose tissue.[27] Then, the algorithm has been updated and extended to the analysis of other amyloid proteins and tissue types by subtracting the baseline value for each amyloid protein from non-specific-subtype affected samples.[41]

4. Challenges in diagnostic amyloid proteomics

4.1. *Detection of multiple proteins in a single biopsy*

One of the major challenges in amyloid proteomics is the identification of more than one potential amyloid protein in a single biopsy.

The detection of multiple amyloid proteins in one sample is usually more likely to be a contamination effect occurring during sampling. In a few cases, it has been reported that the amyloid is constituted by two different amyloid precursor proteins, such as immunoglobulin light chains and wild-type TTR,[7,42] and it is now generally accepted that wild-type TTR amyloid will be present in cardiac tissues collected from elderly patients, due to age-related deposition of wild-type TTR.

In most cases, the amyloid fibril protein can be discriminated from contaminants based on the Mascot score alone although there is a small percentage of samples where blood contamination affects proteomic data interpretation. In the case of FibA amyloid, the NAC algorithm allows differentiating the amyloid protein from contaminants using the FibA algorithm mentioned in the previous section.

An approach to remove multiple amyloid protein ambiguities is decellularization. The strategy was initially applied to an amyloidotic mouse model[43] and has subsequently been adapted and extended to the analysis of unfixed human cardiac tissues and fat aspirates.[44] The key player of the protocol is the detergent sodium deoxycholate (SDC). Serial washes with 4% SDC allow removing soluble proteins leaving intact the amyloid deposits in the extracellular matrix. This approach has not been tested on FFPE tissues.

4.2. *Misidentification of protein variants*

Several factors can lead to the misidentification of a protein variant by proteomics. During MS raw data bioinformatics processing, search engines often generate mismatches putting together peptides from a wild-type protein with a misidentified peptide from another protein.

This issue has been found for renal FibA amyloid, where a mismatch has been reported between proteomics and genetic data in some cases: two variants (R554L and G519E FibA variants) were identified by proteomics although only in one patient it was confirmed by genetic analysis (R554L variant).[28] A further example has been reported for some TTR variants such as p.H110D (H90D) identified in tissue biopsies with high Mascot scores in patients having wild-type TTR confirmed by gene sequencing.[28] A similar issue of misidentification occurred for the p.V142I (V122I) TTR variant. The p.V142I TTR variant was found in more than 60% of FFPE samples from ATTR patients, while the gene typing confirmed its presence in a minority of cases. Following a manual examination of the MS raw data, and the comparison with authentic recombinant p.V142I TTR protein, it was possible to understand that the variant was misidentified in all FFPE samples except for the real p.V142I patients. The problem was caused by formalin-mediated lysine *N*-methylation of wild-type TTR producing the p.146K_{Me} 125–147 peptide, which is isobaric with the p.142I 125–147 variant peptide, and results in almost identical MS/MS spectrum (Figure 3, from Ref. 45). Formalin-mediated lysine *N*-methylation generates a similar result on the lambda light chain, where the K_{Me}-peptide (50K_{Me} 43–59) was initially

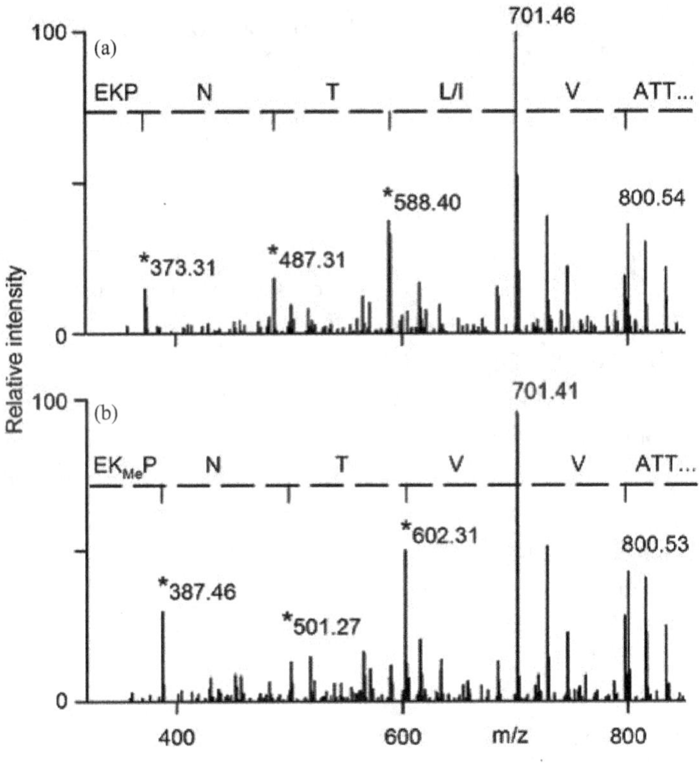

Figure 3. (a) MS/MS spectrum of p.142I 125–147 variant peptide from a patient carrying the p.V142I mutation. (b) MS/MS spectrum of the p.146K_{Me} 125–147 peptide from a non-p.V142I patient. The fragment ions that distinguish the two isobaric peptides are starred.

Source: Canetti *et al*, *Amyloid*, 2017.

misidentified as deriving from an uncommon lambda variant V49I from a myeloma cryoimmunoglobulin.[45,46]

Despite some of these issues, proteomics offers a rapid and reliable method to identify amyloid proteins. Unlike IHC, there is no preselection requirement for protein identification: if the protein is present at sufficient concentration, it will be identified. Proteomics also offers the ability to identify post-translationally modified proteins, truncated proteins, and, indeed, new amyloid proteins. As with all techniques, it is essential to understand the limitations of the method. That is where experience and an ability to critically examine the raw data along with the search engine output will pay dividends when applying proteomics to study important biomedical and clinical problems.

References

1. Benson MD, Buxbaum JN, Eisenberg DS, *et al.* Amyloid Nomenclature 2020: Update and Recommendations by the International Society of Amyloidosis (ISA) Nomenclature Committee. *Amyloid.* 2020 Dec;27(4):217–222. doi: 10.1080/13506129.2020.1835263. Epub 2020 Oct 26. PMID: 33100054.
2. Pepys MB, Hawkins PN, Booth DR, *et al.* Human lysozyme Gene Mutations Cause Hereditary Systemic Amyloidosis. *Nature.* 1993 Apr 8;362(6420):553–557. doi: 10.1038/362553a0. PMID: 8464497.
3. Saraiva MJ, Birken S, Costa PP, *et al.* Amyloid Fibril Protein in Familial Amyloidotic Polyneuropathy, Portuguese Type. Definition of Molecular Abnormality in Transthyretin (Prealbumin). *J. Clin. Invest.* 1984 Jul;74(1):104–119. doi: 10.1172/JCI111390. PMID: 6736244; PMCID: PMC425190.
4. Benson MD, Liepnieks J, Uemichi T, *et al.* Hereditary Renal Amyloidosis Associated with a Mutant Fibrinogen Alpha-Chain. *Nat. Genet.* 1993 Mar;3(3):252–255. doi: 10.1038/ng0393-252. PMID: 8097946.
5. Rowczenio DM, Noor I, Gillmore JD, *et al.* Online Registry for Mutations in Hereditary Amyloidosis Including Nomenclature Recommendations. *Hum. Mutat.* 2014 Sep;35(9):E2403–E2412. doi: 10.1002/humu.22619. Epub 2014 Aug 4. PMID: 25044787.

6. Pepys MB. A Molecular Correlate of Clinicopathology in Transthyretin Amyloidosis. *J. Pathol.* 2009 Jan;217(1):1–3. doi: 10.1002/path.2467. PMID: 19061244.
7. Mahmood S, Gilbertson JA, Rendell N, *et al.* Two Types of Amyloid in a Single Heart. *Blood.* 2014 Nov 6;124(19):3025–3027. doi: 10.1182/blood-2014-06-580720. PMID: 25377563; PMCID: PMC4224198.
8. Simons JP, Al-Shawi R, Ellmerich S, *et al.* Pathogenetic Mechanisms of Amyloid A Amyloidosis. *Proc. Natl. Acad. Sci. USA.* 2013 Oct 1;110(40):16115–16120. doi: 10.1073/pnas.1306621110. Epub 2013 Aug 19. PMID: 23959890; PMCID: PMC3791773.
9. Grogan M, Scott CG, Kyle RA, *et al.* Natural History of Wild-Type Transthyretin Cardiac Amyloidosis and Risk Stratification Using a Novel Staging System. *J. Am. Coll. Cardiol.* 2016 Sep 6;68(10):1014–1020. doi: 10.1016/j.jacc.2016.06.033. Erratum in: *J. Am. Coll. Cardiol.* 2017 Jun 13;69(23):2882. PMID: 27585505.
10. Faravelli G, Mondani V, Mangione PP, *et al.* Amyloid Formation by Globular Proteins: The Need to Narrow the Gap Between *in Vitro* and *in Vivo* Mechanisms. *Front. Mol. Biosci.* 2022 Feb 14;9:830006. doi: 10.3389/fmolb.2022.830006. PMID: 35237660; PMCID: PMC8883118.
11. Naiki H, Gejyo F, Nakakuki K. Concentration-Dependent Inhibitory Effects of Apolipoprotein E on Alzheimer's Beta-Amyloid Fibril Formation *in Vitro*. *Biochemistry.* 1997 May 20;36(20):6243–6250. doi: 10.1021/bi9624705. PMID: 9166797.
12. Merlini G, Bellotti V. Molecular Mechanisms of Amyloidosis. *N. Engl. J. Med.* 2003 Aug 7;349(6):583–596. doi: 10.1056/NEJMra023144. PMID: 12904524.
13. Chiti F, Dobson CM. Protein Misfolding, Amyloid Formation, and Human Disease: A Summary of Progress over the Last Decade. *Annu. Rev. Biochem.* 2017 Jun 20;86:27–68. doi: 10.1146/annurev-biochem-061516-045115. Epub 2017 May 12. PMID: 28498720.
14. Vrana JA, Gamez JD, Madden BJ, *et al.* Classification of Amyloidosis by Laser Microdissection and Mass Spectrometry-Based Proteomic Analysis in Clinical Biopsy Specimens. *Blood.* 2009 Dec 3;114(24):4957–4959. doi: 10.1182/blood-2009-07-230722. Epub 2009 Oct 1. PMID: 19797517.

15. Tennent GA, Lovat LB, Pepys MB. Serum Amyloid P Component Prevents Proteolysis of the Amyloid Fibrils of Alzheimer Disease and Systemic Amyloidosis. *Proc. Natl. Acad. Sci. USA.* 1995 May 9; 92(10):4299–4303. doi: 10.1073/pnas.92.10.4299. PMID: 7753801; PMCID: PMC41931.
16. Pepys MB, Booth DR, Hutchinson WL, *et al.* Amyloid P Component: A Critical Review. *Amyloid.* 1997;4:274–295.
17. Richards DB, Cookson LM, Berges AC, *et al.* Therapeutic Clearance of Amyloid by Antibodies to Serum Amyloid P Component. *N. Engl. J. Med.* 2015 Sep 17;373(12):1106–1114. doi: 10.1056/NEJMoa1504942. Epub 2015 Jul 15. PMID: 26176329.
18. Gallo G, Wisniewski T, Choi-Miura NH, *et al.* Potential Role of Apolipoprotein-E in Fibrillogenesis. *Am. J. Pathol.* 1994 Sep;145(3):526–530. PMID: 8080036; PMCID: PMC1890322.
19. Wisniewski T, Drummond E. APOE-Amyloid Interaction: Therapeutic Targets. *Neurobiol. Dis.* 2020 May;138:104784. doi: 10.1016/j.nbd.2020.104784. Epub 2020 Feb 4. PMID: 32027932; PMCID: PMC7118587.
20. Bergström J, Murphy C, Eulitz M, *et al.* Codeposition of Apolipoprotein A-IV and Transthyretin in Senile Systemic (ATTR) Amyloidosis. *Biochem. Biophys. Res. Commun.* 2001 Jul 27;285(4):903–908. doi: 10.1006/bbrc.2001.5260. PMID: 11467836.
21. Bergström J, Murphy CL, Weiss DT, *et al.* Two Different Types of Amyloid Deposits — Apolipoprotein A-IV and Transthyretin — in a Patient with Systemic Amyloidosis. *Lab. Invest.* 2004 Aug;84(8): 981–988. doi: 10.1038/labinvest.3700124. PMID: 15146166.
22. Dasari S, Amin MS, Kurtin PJ, *et al.* Clinical, Biopsy, and Mass Spectrometry Characteristics of Renal Apolipoprotein A-IV Amyloidosis. *Kidney Int.* 2016 Sep;90(3):658–664. doi: 10.1016/j.kint.2016.04.003. Epub 2016 Jun 2. PMID: 27262366.
23. Bois MC, Dasari S, Mills JR, *et al.* Apolipoprotein A-IV-Associated Cardiac Amyloidosis. *J. Am. Coll. Cardiol.* 2017 May 2;69(17):2248–2249. doi: 10.1016/j.jacc.2017.02.047. PMID: 28449784.
24. Canetti D, Nocerino P, Rendell NB, *et al.* Clinical ApoA-IV Amyloid is Associated with Fibrillogenic Signal Sequence. *J. Pathol.* 2021

Nov;255(3):311–318. doi: 10.1002/path.5770. Epub 2021 Aug 27. PMID: 34331462; PMCID: PMC9291309.

25. Winter M, Tholey A, Krüger S, *et al.* MALDI-Mass Spectrometry Imaging Identifies Vitronectin as a Common Constituent of Amyloid Deposits. *J. Histochem. Cytochem.* 2015 Oct;63(10):772–779. doi: 10.1369/0022155415595264. Epub 2015 Jun 22. PMID: 26101327; PMCID: PMC4823803.
26. Lepelletier FX, Mann DM, Robinson AC, *et al.* Early Changes in Extracellular Matrix in Alzheimer's Disease. *Neuropathol. Appl. Neurobiol.* 2017 Feb;43(2):167–182. doi: 10.1111/nan.12295. Epub 2015 Nov 30. PMID: 26544797.
27. Brambilla F, Lavatelli F, Di Silvestre D, *et al.* Reliable Typing of Systemic Amyloidoses Through Proteomic Analysis of Subcutaneous Adipose Tissue. *Blood.* 2012 Feb 23;119(8):1844–1847. doi: 10.1182/blood-2011-07-365510. Epub 2011 Sep 13. PMID: 21917755.
28. Canetti D, Rendell NB, Gilbertson JA, *et al.* Diagnostic Amyloid Proteomics: Experience of the UK National Amyloidosis Centre. *Clin. Chem. Lab. Med.* 2020 Jun 25;58(6):948–957. doi: 10.1515/cclm-2019-1007. PMID: 32069225.
29. Schönland SO, Hegenbart U, Bochtler T, *et al.* Immunohistochemistry in the Classification of Systemic Forms of Amyloidosis: A Systematic Investigation of 117 Patients. *Blood.* 2012 Jan 12;119(2):488–493. doi: 10.1182/blood-2011-06-358507. Epub 2011 Nov 21. PMID: 22106346.
30. Rezk T, Gilbertson JA, Mangione PP, *et al.* The Complementary Role of Histology and Proteomics for Diagnosis and Typing of Systemic Amyloidosis. *J. Pathol. Clin. Res.* 2019 Jul;5(3):145–153. doi: 10.1002/cjp2.126. Epub 2019 Apr 2. PMID: 30740936; PMCID: PMC6648380.
31. Linke RP. On Typing Amyloidosis Using Immunohistochemistry. Detailled Illustrations, Review and a Note on Mass Spectrometry. *Prog. Histochem. Cytochem.* 2012 Aug;47(2):61–132. doi: 10.1016/j.proghi.2012.03.001. Epub 2012 Jul 20. PMID: 22819384.
32. Gilbertson JA, Hunt T, Hawkins PN. Amyloid Typing: Experience from a Large Referral Centre. In *Amyloid and Related Disorders*; Picken M, Dogan A, Herrera G, Eds. 2012:231–238.

33. Gilbertson JA, Theis JD, Vrana JA, *et al.* A Comparison of Immunohistochemistry and Mass Spectrometry for Determining the Amyloid Fibril Protein from Formalin-Fixed Biopsy Tissue. *J. Clin. Pathol.* 2015;68:314–317.
34. Rodriguez FJ, Gamez JD, Vrana JA, *et al.* Immunoglobulin Derived Depositions in the Nervous System: Novel Mass Spectrometry Application for Protein Characterization in Formalin-Fixed Tissues. *Lab. Invest.* 2008 Oct;88(10):1024–1037. doi: 10.1038/labinvest.2008.72. Epub 2008 Aug 18. PMID: 18711355.
35. Lavatelli F, Perlman DH, Spencer B, *et al.* Amyloidogenic and Associated Proteins in Systemic Amyloidosis Proteome of Adipose Tissue. *Mol. Cell. Proteomics.* 2008 Aug;7(8):1570–1583. doi: 10.1074/mcp.M700545-MCP200. Epub 2008 May 12. PMID: 18474516; PMCID: PMC2494907.
36. Benson MD, James S, Scott K, *et al.* Leukocyte Chemotactic Factor 2: A Novel Renal Amyloid Protein. *Kidney Int.* 2008 Jul;74(2):218–222. doi: 10.1038/ki.2008.152. Epub 2008 Apr 30. PMID: 18449172.
37. Soutar AK, Hawkins PN, Vigushin DM, *et al.* Apolipoprotein AI Mutation Arg-60 Causes Autosomal Dominant Amyloidosis. *Proc. Natl. Acad. Sci. USA.* 1992 Aug 15;89(16):7389–7393. doi: 10.1073/pnas.89.16.7389. PMID: 1502149; PMCID: PMC49715.
38. Taylor GW, Gilbertson JA, Sayed R, *et al.* Proteomic Analysis for the Diagnosis of Fibrinogen Aα-Chain Amyloidosis. *Kidney Int. Rep.* 2019 Apr 15;4(7):977–986. doi: 10.1016/j.ekir.2019.04.007. PMID: 31317119; PMCID: PMC6612008.
39. Nasr SH, Vrana JA, Dasari S, *et al.* DNAJB9 Is a Specific Immunohistochemical Marker for Fibrillary Glomerulonephritis. *Kidney Int. Rep.* 2017 Aug 8;3(1):56–64. doi: 10.1016/j.ekir.2017.07.017. PMID: 29340314; PMCID: PMC5762944.
40. Gallo G, Picken M, Frangione B, *et al.* Nonamyloidotic Monoclonal Immunoglobulin Deposits Lack Amyloid P Component. *Mod. Pathol.* 1988 Nov;1(6):453–456. PMID: 3146753.
41. Canetti D, Brambilla F, Rendell NB, *et al.* Clinical Amyloid Typing by Proteomics: Performance Evaluation and Data Sharing Between Two Centres. *Molecules.* 2021 Mar 29;26(7):1913. doi: 10.3390/molecules26071913. PMID: 33805439; PMCID: PMC8037974.

42. Sidiqi MH, McPhail ED, Theis JD, *et al.* Two Types of Amyloidosis Presenting in a Single Patient: A Case Series. *Blood Cancer J.* 2019 Mar 5;9(3):30. doi: 10.1038/s41408-019-0193-9. PMID: 30837451; PMCID: PMC6401104.
43. Mazza G, Simons JP, Al-Shawi R, *et al.* Amyloid Persistence in Decellularized Liver: Biochemical and Histopathological Characterization. *Amyloid.* 2016;23(1):1–7. doi: 10.3109/13506129.2015.1110518. Epub 2015 Dec 8. PMID: 26646718; PMCID: PMC4819572.
44. Mangione PP, Mazza G, Gilbertson JA, *et al.* Increasing the Accuracy of Proteomic Typing by Decellularisation of Amyloid Tissue Biopsies. *J. Proteomics.* 2017 Aug 8;165:113–118. doi: 10.1016/j.jprot.2017.06.016. Epub 2017 Jun 21. PMID: 28647518; PMCID: PMC5571436.
45. Canetti D, Rendell NB, Di Vagno L, *et al.* Misidentification of Transthyretin and Immunoglobulin Variants by Proteomics Due to Methyl Lysine Formation in Formalin-Fixed Paraffin-Embedded Amyloid Tissue. *Amyloid.* 2017 Dec;24(4):233–241. doi: 10.1080/13506129.2017.1385452. Epub 2017 Oct 10. PMID: 29016222.
46. Lopez de Castro JA, Chiu YY, Poljak RJ. Amino Acid Sequence of the Variable Region of the Light (Lambda) Chain from Human Myeloma Cryoimmunoglobulin IgG Hil. *Biochemistry.* 1978 May 2;17(9):1718–1723. doi: 10.1021/bi00602a021. PMID: 418804.

https://doi.org/10.1142/9789811283376_0011

Chapter 11

Mass Spectrometry–Based Immunopeptidomics for the Cancer Vaccine Science

Sachin Kote

Clinical Peptidomics Group, International Centre for Cancer Vaccine Science, University of Gdansk, Gdansk, Poland

Contents

Abstract

Recent technological advances in chromatography and mass spectrometry have facilitated the identification of endogenous mutated cancer proteins processed and presented peptides with the major histocompatibility complex class (MHC) molecules. Targeting such cancer-specific antigens (neoantigens) would enable immune cells to distinguish between normal and cancerous cells, allowing us to develop personalized cancer vaccine therapies. Mass spectrometry is the only unbiased methodology to comprehensively interrogate the naturally presented MHC peptides. While the isolation of MHC-associated peptides typically uses an antibody-based approach (immunoprecipitation), an antibody-free approach (mild acid elution) exists regarding their respective analytical merits. This chapter compares these protocols for accurate mass spectrometry–based immunopeptidomics, enabling the direct identification of immunopeptidome from cellular models and clinical samples.

1. Introduction

Personalized immunotherapies and patient-specific methods emerged as attractive alternatives to conventional methods. They targeted therapies available to fight cancer.[1,2] Immune checkpoint inhibitors are used to modulate systems of T-cell regulation, and some of them are now the standard of care for melanoma and a growing set of cancers exhibiting a high mutational burden. The recent remarkable outcome of immune checkpoint blocking therapies has motivated the development of T-cell-based killing immunotherapy.[3]

The major histocompatibility complex class I, II (MHC-I, II) is critical to activate the adaptive and humoral immune response and counteract the harmful effects of tumor cells. Therefore, understanding antigen presentation is fundamental to target tumor cell eradication and can potentially improve cancer immunotherapy. Intracellular proteins are degraded in the cytosol, and antigenic precursor peptides are transported to the endoplasmic reticulum, where they are further trimmed to short peptides (8–14 amino acids) that can be assembled with MHC-I and be recognized by $CD8^+$ T cells for foreign antigens that differ from the pool of peptides in a healthy cellular environment. Exogenous antigens can be taken up by professional antigen-presenting cells (APCs) and degraded in the endosomal compartments of the cell. The resultant peptides (10–25 aa) can be subsequently sampled by MHC-II molecules and presented on the surface of these cells. Recognition by $CD4^+$ T cells leads to the T-cell activation and generation of a humoral immune response.

Currently, mass spectrometry (MS)–based immunopeptidomics is the only unbiased analytical methodology to interrogate the repertoire of naturally presented MHC peptides in not only successful in identifying neoantigens on cancer cell lines[4–6] but also in clinical samples.[7] Although modern immunopeptidomics is a cutting-edge application,[8] the technical development in terms of sample preparation, detection, and advanced computational algorithms for identifying the peptides is still in development. Above all, low amounts of neoantigens available per sample require improved processing protocols and more efficacy than the current liquid chromatography–mass spectrometry (LC–MS) methods.

This chapter describes a protocol typically used in two established methods: antibody-based (immunoprecipitation, IP) and antibody-free (mild acid elution, MAE). The MAE is the simplest, yet least specific, approach that involves acid stripping the surfaces of cells in culture, using an isotonic acidic buffer.[9–11] The antibody-free approach is used explicitly to eliminate the antigenicity of MHC-I molecules. The interesting observation of the antibody-free approach is that neither MHC-II antigens nor other non-MHC antigens get removed from the cell surface.[12] Moreover, β2-microglobulin (β2M) is the component of the MHC-I molecule present on the cell's surface. On the other hand, the MAE method facilitates the dissociation of the extracellular non-covalently-bound β2M from the MHC-I complexes on the cell's surface.[10] Antibody-based approach (IP) methods enable the sequential immunoaffinity purification of MHC-I- and MHC-II-binding peptides directly from cellular and clinical samples like tissue and biopsies. The successful identification of these peptides relies on developing appropriate isolation protocols that can efficiently recover immunopeptidome for subsequent MS analyses. Details regarding the analyses of these peptidome samples by MS are included; the downstream computational pipeline setting can be implemented using a new module in PEAKS,[13] MaxQuant,[14] and ProteinPilot.

2. Materials

Note: Prepare all solutions using LC–MS grade water and use MS-grade solvents, reagents, ultraclean glassware, and plasticware. We recommend using high-grade plasticware and always checking tubes and other materials employed for extraction by LC–MS before using them for immunopeptidome sample preparation.

2.1. *MAE buffer*

Citrate-phosphate buffer at pH 3.3 (0.131 M citric acid/0.066 M disodium phosphate [Na_2HPO_4], sodium chloride [NaCl] 150 mM) containing aprotinin and iodoacetamide (1:100) was added to each flask.

2.2. *Affinity chromatography*

1. Five milliliter of growth medium from HB95 hybridoma cells secreting W6-32 anti-pan MHC-I and from HB145 hybridoma cells secreting IVA12 anti-pan-MHC-II,
2. Protein A-Sepharose resin,
3. Dimethyl pimelimidate (DMP),

4. Phosphate-buffered saline (PBS), 137 mM NaCl, 2.7 mM KCl, 10 mM Na_2HPO_4, and 2 mM KH_2PO_4, pH 7.4. The sterile-filtered 10× stock solution is stable at room temperature (RT) for 3–6 months,
5. Termination buffer, 0.2 M Tris, pH 8.0, and
6. Stripping buffer, 0.1 M citrate, pH 3.

2.3. *Sample preparation*

1. Tumor tissues (flash-frozen stored at –80°C),
2. Lysis buffer: 0.25% sodium deoxycholate, 0.2 mM iodoacetamide, 1 mM EDTA, 1:200 protease inhibitors cocktail (we use the cocktail reagent from Sigma-Aldrich), 1 mM phenylmethylsulfonyl fluoride, and 1% octyl-beta-D glucopyranoside in PBS[15] Or lysis buffer: 0.5% (vol/vol) NP-40, 50 mM Tris, pH 8.0, 150 mM NaCl, and protease inhibitor cocktail,[16]
3. Wash buffer 1, 50 mM Tris, pH 8.0, and 150 mM NaCl,
4. Wash buffer 2, 50 mM Tris, pH 8.0, and 450 mM NaCl,
5. Wash buffer 3, 50 mM Tris, pH 8.0,
6. Elution buffer 10% (vol/vol) acetic acid, and
7. ULTRA-TURRAX homogenizer (IKA or comparable equipment).

2.4. *Mass spectrometry*

1. The instrument from SCIEX and Thermo Scientific mass spectrometers (Q-Exactive and Q-Exactive HF, Thermo Fusion, and Fusion Lumos) can be used. It is recommended that a MS coupled online with a nanospray setup,
2. LC–MS compatible trap and analytical column are recommended for instance, 8-µm tip opening and a diameter of 25 or 50 cm × 75 µm and ReproSil-Pur C18 1.9-µm particles of 120 Å pore size resin,
3. Computational platform (32-core processors and 64–128 GB of RAM are necessary for reliable operation, MaxQuant or ProteinPilot or PEAKS software).

3. Methods

The protocol described below provides instructions on purifying and cross-linking antibodies sufficient for the immunoaffinity purification of MHC-I and MHC-II complexes from cellular models and tissue samples (of 1–2 cm^3 each). The procedures can be scaled up or down according to the amount of available tissues. It is also recommended to perform a quality control step before immunoaffinity purification of MHC-I and MHC-II complexes, especially from the cellular model needed to confirm the cell surface expressions of MHC-I or MHC-II molecules on flow cytometry.

3.1. *Isolation of MHC-I peptides by antibody-free approach*

The antibody-free approach is primarily applicable to cellular models. The selected cells can be grown in the recommended cellular growth condition. For instance, melanoma A375 cells were grown in DMEM (Dulbecco's Modified Eagle's Medium; Invitrogen) supplemented with 10% heat-inactivated fetal bovine serum, 1% penicillin/streptomycin (Invitrogen), and incubated at 37°C with 5% CO_2. The wide range of cell numbers from biological replicates of 2–100 × 10^6 cells with treatment or positive/negative control can be used. This method is based on acid elution to purify MHC-I-bound peptides on the cell surface. MHC-I-associated peptides were released by acidic buffer citrate, pH 3.3 buffer using 2–3 mL of 2–100 × 10^6 cell samples. Cell suspensions were then pelleted by centrifugation (1,000*g*, 10 min at 4°C), and the resultant supernatants were considered as MHC-I peptide samples for further purifications. Followingly, the peptide sample extracts were desalted using Oasis HLB cartridges (30 mg; water). Finally, the bound material was eluted with 1 mL of H_2O/80% methanol (MeOH)/0.2% formic acid (FA) vol/vol and diluted to H_2O/40% MeOH/0.2% FA vol/vol. Peptides were loaded on ultrafiltration devices (Amicon Ultra-2 Centrifugal Filters, Ultracel-3k, 2 mL; Millipore) to isolate peptides.

3.2. *Isolation of MHC-I, II peptides by antibody-based approach*

The antibody-based approach is also called immunoprecipitation (IP). The commonly used two antibodies for the purification of MHC-I and MHC-II molecules from the samples are W6-32 and IVA12, respectively. These anti-MHC antibodies can be yielded from the hybridoma growth medium. Hybridoma growth medium of 5–6 mL yields about 12–15 mg of antibodies. These purified antibodies are ready for the first step, that is, the generation of the MHC immunoaffinity column.

a) Exchange the buffer protein A Sepharose supplies a 50% (wt/vol) slurry in 20% (vol/vol) ethanol. The storage buffer exchange with PBS by centrifugation at 3,000–4,000*g*, RT.
b) It is highly recommended to check the quality of antibodies that will be used for IP preparation. It can be done in two ways: either the binding characteristics of the respective MHC of choice using flow cytometry and efficiency of the cross-linking reaction by SDS-PAGE. Therefore, the antibody of choice at 0.5–1 mg/mL in PBS mix with protein A Sepharose resin and rotated gently at RT for 20–30 min to allow the antibody to bind.
c) The next step is to avoid the primary amines' interference with the cross-linking reaction. To remove the residual-free amines, transfer the protein A Sepharose resin and antibody back to the column and allow the antibody to flow through by gravity. Wash the antibody-bound resin with 10-column volumes of borate wash buffer and 10-column volumes of freshly prepared 0.2 M triethanolamine, pH 8.2, at RT.
d) Cross-linking between the protein A Sepharose resin and antibody is crucial to determine the yield of MHC extraction from the samples. For cross-linking the flow, 5-column volumes of freshly prepared DMP are used as a cross-linking agent through the column at RT. Seal the bottom of the column and allow the reaction to proceed at RT for 1 h.

e) For the termination of cross-linking, add 10-column volumes of ice-cold termination buffer, 0.2 M Tris, pH 8.0.
f) Stripping of unbound antibodies. Remove unbound antibodies by washing with 5- to 10-column volumes of 0.1 M citrate, pH 3.
g) The last step is the equilibration of columns. It can be done with the flow of 10-column volumes of PBS, pH 7.4, until the pH of the flow-through is >7. With all these stepwise procedures, the MHC immunoaffinity column is ready for further choice of MHC immunoaffinity purifications.

3.2.1. *Generation of cell lysate*

Cell lysate can be prepared for immunoaffinity purification from cell line models and clinical samples like tissues or biopsies. The cells (1×10^7–1×10^9) are washed with PBS and harvested by centrifugation (2,000*g*, 10 min at 4°C). In the case of tissue, a range of 10–200 mg can be processed. Small cell pellets can be homogenized in lysis buffer by pipetting up and down with or without the sonication step (10–20 min at RT). It also recommends grinding cells under liquid nitrogen or using a bead beater for the lysis of large cell pellets or tissue samples. For cell grinding, place the snap-frozen pellets straight into a Mixer Mill MM 400, and after grinding, scrape the powder into the lysis buffer. Alternatively, tissue samples in lysis buffer can be processed using a tissue homogenizer (ULTRA-TURRAX). Place the homogenizer inside a laminar flow to protect the user from aerosols that may be generated during the homogenization of the tissues or by using Precellys 24 Bead-Beater (five 15-s cycles, speed setting 6,500).

3.2.2. *MHC immunoaffinity purification*

After the cell homogenization, the cell or tissue lysate can be further processed with centrifugation (10 min at 2,000*g*, 4°C) to remove the cell debris and nuclei. Next, take the supernatant and spin it for 75 min in an ultracentrifuge (100,000*g*) at 4°C. Finally, collect the clear supernatant free from other lipid-containing layers. The next step is the pre-clearing of the lysate. For tissue samples, a depletion step of endogenous antibodies is required. Therefore, using gravity flow or a peristaltic pump in a cold room, load the cell lysate onto a protein A Sepharose pre-column that has been pre-equilibrated in 10-column volumes of wash buffer 0.005% (vol/vol) Igepal CA-630, 50 μM Tris, pH 8.0, 150 mM NaCl, 5 mM EDTA, 100 μM PMSF, and 1 μg/mL pepstatin A. In this way, the pre-cleared cell lysate is ready to load with cross-linked mAb column(s) (MHC IP step) and rotate slowly end-over-end at 4°C for 1–2 h. The last two steps include washing and elution of the MHC molecule. It is recommended to prepare freshly all these buffers. Wash the column(s) with 10-column volumes of each wash buffer in the following order: wash buffer 0.005% (vol/vol) Igepal CA-630, 50 mM Tris, pH 8.0, 150 mM NaCl, 5 mM EDTA, 100 μM PMSF, and 1 μg/mL pepstatin A, wash buffer 50 mM Tris, pH 8.0, and 150 mM NaCl (to remove detergent), wash buffer 50 mM Tris, pH 8.0, 450 mM NaCl (to remove non-specifically bound material), wash buffer 50 mM Tris, pH 8.0 (to remove salt to prevent crystal formation). Finally, elute MHC molecules in 5-column volumes of elution buffer 10% (vol/vol) acetic acid. We recommend that you perform RP-HPLC fractionation of the eluate immediately.

3.3. *Purification of MHC-binding peptides*

The eluate from the previous step contains MHC-bound peptides, MHC heavy chain and β2m, leached antibodies, and a contaminating detergent such as Igepal CA-630 polymers, which severely hamper MS analysis. Therefore, the MHC-bound peptides are enriched using a C18 reversed-phase column running on a mobile phase buffer A of 0.1% TFA and buffer B of 80% acetonitrile (ACN)/0.1% TFA. We routinely use a 4.6-mm internal diameter × 50-mm long reversed-phase C18 end-capped HPLC column (Chromolith Speed Rod) on an ÄKTAmicro HPLC system. Separate based on a rapid gradient of buffer A to B, which results in 10–30 peptide-containing fractions (e.g., 2%–40% B for 4 min, 40%–45% for 4 min, and a rapid 2-min increase to 100% B). Using this approach, a small number of early fractions contain >95% of the peptides, whereas the later fractions contain other interfering components. Collect the fractions (500 μL) into LoBind Eppendorf tubes. At this point, fractions can be frozen at -80°C for up to 6–12 months. Next, vacuum-concentrate the peptide-containing fractions (generally, those fractions eluted in up to 42% buffer B) to reduce the concentration of ACN. Typically, dry to 10 μL and dilute to 15–25 μL in 0.1% FA. Peptides could be frozen and stored at –80°C until analysis on an MS. However, it is preferable to proceed directly to the next step.

3.4. *Mass spectrometry analyses*

For the acquisition of purified MHC-associated peptide preparations, it is recommended to use high-end state-of-the-art LC–MS/MS systems. To achieve a comprehensive view of the immunopeptidome, peptides are first loaded onto a trap column using a microflow-loading pump at 5–10 μL/min with 0.1% (vol/vol) FA or TFA in water or 2% (vol/vol) ACN. Nano-LC is always preferred for eluting peptides from a trapping column coupled with an analytical column.

Place 15–20 μL of each MHC peptide samples in the autosampler. Inject the first 5–6 μL of the first sample. Apply a linear gradient of 2%–30% of 80% ACN and 0.1% FA at a flow rate of 250–300 nL/min over 90 min. Based on the current literature, recommended settings for the MS are the following: MS scan range can be set to 300–1,650 m/z with a resolution of 60,000 (200 m/z) and a target value of 3e6 ions. Data may be acquired with the data-dependent "top10/20" method, which isolates the 10 most intense ions and fragments by higher energy collisional dissociation (HCD) with an applied normalized collision energy of 27% and with an AGC target value of 1e5 with a maximum injection time of 120 ms and MS/MS resolution of 15,000 (200 m/z). For MHC-I peptides, no fragmentation should be performed in case of unassigned precursor ion charge states or charge states of four and above. For MHC-II peptides, in case of unassigned precursor ion charge states or charge states of one or of five and above, no fragmentation should be performed. Dynamic exclusion of fragmented m/z values from the further selection can be set for 20 s. Inject 5–6 μL of the first sample (technical duplicate).

Wash the system with two standard wash runs before loading the following biological sample. Mass spectrometers with high-mass accuracy analyzers are preferred for MS and MS/MS. High mass accuracy can improve the reliability of peptide sequence identification; it recommends using high targets and long accumulation times for MS2 spectra for improved spectral quality and spectral yields of up to 30%–40%. For instance, one of the common setup of LC–MS/MS is an Ultimate 3000 nanoLC system coupled online with an Orbitrap Exploris 480 mass spectrometer (both from Thermo Fisher Scientific). However, other comparable instruments could be used.

3.5. *Data analysis*

3.5.1. *Peptide identification and label-free quantification*

The MHC peptides database search differs from proteomics workflow in several ways. First, search the raw data with your favorite search engine. Immunopeptidomics data routinely search with the MaxQuant package (version 1.6.12.0)[14] using the built-in Andromeda search engine, ProteinPilot (v.5) or PEAKS (v.8.5)[13] using the unique parameter setting for MHC-I and MHC-II peptides. Mass tolerances for precursor and product ions were set to 20 ppm and 0.02 Da, respectively. Searches were performed without enzyme specificity and variable modifications for deamidation (N, Q) and oxidation (M). Possible sequence matches were restricted to 7–25 amino acids (aa), with a maximum peptide mass of 4,600 Da. The initial allowed mass deviation of the precursor ion was 10 ppm, and the maximum fragment mass deviation was up to 10 ppm. Further, the statistical analysis can be performed with Perseus software. Strict bioinformatic criteria need to be applied to ensure that only the best hits are chosen for further analysis. Use a decoy database to control the false-discovery rate (FDR). The FDR level considered 1%–5%, and protein abundance calculated using normalized spectral protein intensity (LFQ intensity; label-free quantification) on MaxQuant software. LFQ intensity values were set as Log2 transformed, and peptide groups quantified in all runs could be considered for further analysis.

The human dataset from the UniPort with the removal of known contaminant peptides (peptides identified from the precolumn and other lab-contaminant peptides cataloged over many experimental controls) from the final datasets to ensure high-quality peptide repertoire data. Further data curation, length, motif analysis, and MHC-binding prediction. Moreover, the prediction tools for the evaluation of the obtained experimental peptide data are available from resources such as the Immune Epitope Database and Analysis Resource (IEDB; http://www.iedb.org/)[17] and NetMHC 4.0, (https://services.healthtech.dtu.dk/services/NetMHC-4.0/). Finally, the plot of the length distribution of MHC-I (predominantly are 9-mers) and class II peptides (10–25-mers) is a frequency distribution to provide further confirmation of successful elution of MHC peptide complexes.

References

1. Zhang X, Qi Y, Zhang Q, *et al.* Application of Mass Spectrometry-Based MHC Immunopeptidome Profiling in Neoantigen Identification for Tumor Immunotherapy. *Biomed. Pharmacother.* Dec. 2019;120:109542. doi: 10.1016/j.biopha.2019.109542.
2. Sahin U, Türeci Ö. Personalized Vaccines for Cancer Immunotherapy. *Science.* Mar. 2018;359(6382):1355–1360. doi: 10.1126/science.aar7112.
3. Schadendorf D, Hodi FS, Robert C, *et al.* Pooled Analysis of Long-Term Survival Data from Phase II and Phase III Trials of Ipilimumab in Unresectable or Metastatic Melanoma. *J. Clin. Oncol.* Jun. 2015; 33(17):1889–1894. doi: 10.1200/JCO.2014.56.2736.
4. Bassani-Sternberg M, Bräunlein E, Klar R, *et al.* Direct Identification of Clinically Relevant Neoepitopes Presented on Native Human Melanoma Tissue by Mass Spectrometry. *Nat. Commun.* Nov. 2016;7(1), Art. no. 1. doi: 10.1038/ncomms13404.
5. Carreno BM, Magrini V, Becker-Hapak M, *et al.* A Dendritic Cell Vaccine Increases the Breadth and Diversity of Melanoma Neoantigen-Specific T Cells. *Science.* May 2015;348(6236):803–808. doi: 10.1126/science.aaa3828.
6. Gubin MM, Zhang X, Schuster H, *et al.* Checkpoint Blockade Cancer Immunotherapy Targets Tumour-Specific Mutant Antigens. *Nature.* Nov. 2014;515(7528):577–581. doi: 10.1038/nature13988.

7. Bassani-Sternberg M, Coukos G. Mass Spectrometry-Based Antigen Discovery for Cancer Immunotherapy. *Curr. Opin. Immunol.* Aug. 2016;41:9–17. doi: 10.1016/j.coi.2016.04.005.
8. Yates JR. Recent Technical Advances in Proteomics. *F1000Res.* Mar. 2019;8:F1000 Faculty Rev-351. doi: 10.12688/f1000research.16987.1.
9. Castelli C, Storkus WJ, Maeurer MJ, *et al.* Mass Spectrometric Identification of a Naturally Processed Melanoma Peptide Recognized by $CD8^+$ Cytotoxic T Lymphocytes. *J. Exp. Med.* Jan. 1995;181(1): 363–368. doi: 10.1084/jem.181.1.363.
10. Storkus WJ, Zeh HJ, Salter RD, *et al.* Identification of T-Cell Epitopes: Rapid Isolation of Class I-Presented Peptides from Viable Cells by Mild Acid Elution. *J. Immunother. Emphasis. Tumor. Immunol.* Aug. 1993;14(2):94–103.
11. Storkus WJ, Zeh HJ 3rd, Maeurer MJ, *et al.* Identification of Human Melanoma Peptides Recognized by Class I Restricted Tumor Infiltrating T Lymphocytes. *J. Immunol.* Oct. 1993;151(7):3719–3727. doi: 10.4049/jimmunol.151.7.3719.
12. Sugawara S, Abo T, Kumagai K. A Simple Method to Eliminate the Antigenicity of Surface Class I MHC Molecules from the Membrane of Viable Cells by Acid Treatment at pH 3. *J. Immunol. Methods.* Jun. 1987;100(1):83–90. doi: 10.1016/0022-1759(87)90175-X.
13. Zhang J, Xin L, Shan B, *et al.* PEAKS DB: De novo Sequencing Assisted Database Search for Sensitive and Accurate Peptide Identification. *Mol. Cell. Proteomics.* Apr. 2012;11(4):M111.010587. doi: 10.1074/mcp.M111.010587.
14. Cox J, Mann M. MaxQuant Enables High Peptide Identification Rates, Individualized p.p.b.-Range Mass Accuracies and Proteome-Wide Protein Quantification. *Nat. Biotechnol.* Dec. 2008;26(12), Art. no. 12. doi: 10.1038/nbt.1511.
15. Chong C, Marino F, Pak H, *et al.* High-Throughput and Sensitive Immunopeptidomics Platform Reveals Profound Interferonγ-Mediated Remodeling of the Human Leukocyte Antigen (HLA) Ligandome. *Mol. Cell. Proteomics.* Mar. 2018;17(3):533–548. doi: 10.1074/mcp.TIR117.000383.
16. Purcell AW, Ramarathinam SH, Ternette N. Mass Spectrometry–Based Identification of MHC-Bound Peptides for Immunopeptidomics.

Nat. Protoc. Jun. 2019;14(6):1687–1707. doi: 10.1038/s41596-019-0133-y.

17. Nielsen M, Lundegaard C, Worning P, *et al.* Reliable Prediction of T-Cell Epitopes Using Neural Networks with Novel Sequence Representations. *Protein Sci.* May 2003;12(5):1007–1017. doi: 10.1110/ps.0239403.

https://doi.org/10.1142/9789811283376_0012

Chapter 12

Proteomic Approaches to Viral Infections

Pawel Ciborowski[1,*] and Michael Belshan[2]

[1]*Department of Pharmacology and Experimental Neuroscience, University of Nebraska Medical Center, Omaha, NE*
[2]*Department of Medical Microbiology and Immunology, School of Medicine, Creighton University, Omaha, NE*

Contents

*Corresponding author.
Pawel Ciborowski, Ph.D. Professor Emeritus, Department of Pharmacology and Experimental Neuroscience, University of Nebraska Medical Center, Omaha, NE, USA, E-mail: pciborowski@unmc.edu.

Abstract

Rapid development of mass spectrometry and associated analytical methodology applicable to sample preparation significantly advanced the viral proteomic field. Because viruses are obligatory parasites whose genome code very few proteins, viral proteomics is focused in a major part on changes resulting from viral takeover of the targeted cell's metabolism. In this chapter, we provide several examples of how proteomic approaches were successfully applied to understanding the molecular mechanisms of infections. This knowledge is also critical for new antiviral drug development.

1. Introduction

Viruses are obligatory parasites that have both intracellular and extracellular components to their life cycles. They multiply only within a host cell, releasing infectious particles into the surrounding environment on completion of a replication cycle. If the environment is unaccommodating, viral particles may be inactivated or destroyed. For example, some viruses cannot survive for extended periods when dried. For viruses to proliferate inside cells, they need to take over host metabolic machinery to their advantage to replicate their genome and produce progeny particles. However, it needs to be pointed out that this process is multifaceted and typically quite rapid. Virus replication in a host usually induces a pathological response. The combination of virus replication and the host response results in clinical manifestations of infection. Since the host response is delayed, viral infections typically have a period of incubation before the first symptoms occur. The compounding of these facts profoundly affects how we approach the study of viral infections using any experimental platform, including high-throughput studies, such as proteomics and genomics. In this chapter, we present the major challenges of proteomic studies of viral infections at the cellular and viral levels. Viruses differ in size, shape, and complexity of the viral life cycle. For more information, we recommend other literature sources, that is, Taylor, M.W. (2014). What Is a Virus? In: Viruses and Man: A History of Interactions. Springer, Cham. https://doi.org/10.1007/978-3-319-07758-1_2.

2. Viruses, retroviruses, and bacteriophages — viral life cycle

A virus enters a target cell, takes over the host's metabolic machinery to replicate its genome and synthesize its proteins, eventually producing new viral particles that are released into the surrounding environment by exocytosis, lysing of the cell, or by budding out of the cytoplasmic membrane. Important to note, viruses do not have the ability to manufacture proteins; therefore, they must rely on the host's metabolic machinery. Usually, viral proteins are classified according to their functions such as structural, non-structural, regulatory, and accessory proteins. Other proteins are the host's origin.

Thus, there are two major methods to combat viral infection. One is to prevent viral particles from entering the host cell; the second is to disrupt the viral replication cycle inside the cell and prevent new particle production. Proteomic approaches can be crucial for understanding the viral-directed response of the host cell, thereby providing insights that otherwise would take many years of reductionistic studies.

3. Major objectives of viral proteomics

As the discipline of proteomics emerged, the application of proteomic approaches has been applied to study viral infections.[1–4] Viral infection triggers multiple molecular events after entering the host cell. Many of these events are associated with signaling and regulatory processes such as phosphorylation and ubiquitination. There are also observed alterations in metabolic pathways, that is, methylation, acetylation, and S- palmitoylation. Changes in glycosylation help the virus in the production of the necessary glycoproteins to complete the replication cycle, that is, HIV-1 (human immunodeficiency virus type 1).

Despite unquestionable progress in viral proteomics, there are several outstanding questions related to the proteomic investigations of viral proteins. One, specifically for human and animal viruses causing infections, is to search for viral proteins circulating in body fluids and determine if these proteins can be used as biomarkers. Moreover, can a putative biomarker be used to monitor the progress of pathogenesis (clinical infection), including the progress of clearing infection? Alternatively, can proteomics identify or quantify targets for potential treatments, and so on? How can the sequences and structures of viral proteins be exploited for monitoring epidemiological changes in spreading infections? Can proteomics of viral proteins facilitate the development of new vaccines and/or treatment strategies? All questions presented here are not fully resolved and research still progresses in these areas.

3.1. *Major considerations*

Proteomics, as well as other -omics experimental approaches, of viral infections present a set of challenges. For example, viruses do not infect all target cells at the same time in a host *in vivo,* nor in most model systems. Thus, it should be recognized that if a sample collected for investigation contains a pool of infected and uninfected cells, the changes induced by infection will be "diluted" by the contribution of non-infected cells and/or those that have activated defensive mechanisms, such as apoptosis. This is true in *in vitro* cell culture, infected tissues of patients, and infected persons. Another major challenge in examining infected cells is the inherent variability of responses within a population of human subjects (donors of cells, i.e., macrophages)[5] in contrast to model systems that employ a homogeneous population of transformed cells from clonally expanded cell lines. A third challenge, although not specific to viral infection, is the dynamic range of protein concentrations within a sample. This challenge has, in part, been addressed by quantitative proteomics.[6,7]

A fourth challenge is *post hoc* data analysis — normalization of samples, statistical and bioinformatic analyses, and validation of results.[6]

Significant challenges remain in the quest to identify proteins that consistently change in one direction or that remain totally unchanged in the broader population of human subjects (cells and tissues) and that could serve as normalization factors. We propose to refer to these proteins as "canonical proteins." For example, there might be a small number of proteins, whose directionality of change in viral infection, that is, HIV-1-infected macrophages, allows them to be used as indicators of the status of a high-throughput experiment.

4. Host factors in viral infections

Irrespective of their nucleic acid type, viral particles can be divided into two groups: enveloped and non-enveloped. Enveloped viruses are encased in a host cell–derived membrane during assembly, which usually results in the envelopment/acquisition of host components. Due to the presence of these additional host components, the proteomic analysis of enveloped viruses is more complicated than those of non-enveloped viruses.

It must be noted that enveloped virus–host interactions are the most complex processes due to the multifaceted nature of such infection. Moreover, initial viral infection is highly dynamic (acute phase) in time regardless of the following potential persistence and length of the latent infection. Lack of full understanding of virus–host interactions and lack of adequate *in vitro/in vivo* experimental models creates limitations to ultimately finding a cure for viral infection. The premise of high-throughput studies, both genomics and proteomics, was to gain more insight into the complexity of latent infection.

Kattenhorn *et al.*[8] used high-throughput proteomics to identify proteins associated with murine cytomegalovirus (CMV) particles. The authors identified 58 viral proteins, 20 of which were uncharacterized. They also identified numerous host factors including actin, annexin, cofilin, histone H2A, elongation factor 1α, glyceraldehyde-3-phosphate dehydrogenase, and rho GDP dissociation inhibitor originating from host cells.[8] Interestingly, histone H2A was not found in a separate study of human CMV–associated proteins.

Virions of small non-enveloped viruses such as polyoma and papilloma display a simpler composition as they contain just one major and one minor capsid protein.[9, 10] As with all viral infections, the major question, which may lead to the discovery of antiviral targets, is how the virus is redirecting the metabolism of the host cell. It has been proposed that such a mechanism is mediated by extensive post-translational modifications (PTMs) of histones. This mechanism is not fully understood; however, Schaffhausen *et al.* showed that in the polyomaviruses, virions associated with histones are highly acetylated.[11] This indicates that epigenetic regulation mechanisms can be somehow transmitted in progeny virions. Further studies by Fang *et al.* supported preceding discoveries indicating the involvement of epigenetic-based mechanisms.[12] Nanoflow LC–MS/MS (liquid chromatography with tandem mass spectrometry) techniques and instrumentation facilitated these discoveries and can play a profound role in further studies of epigenetic modulation and facilitation of the life cycle of, that is, papillomavirus. Additional knowledge of the viral life cycle using -omics approaches will improve the understanding of virus replication and enable the development of other therapeutic strategies.

In the quest for effective treatments and vaccine development, HIV-1 has been one of the most investigated enveloped viruses.[13] More specifically, multiple studies of HIV-1 virus particles were performed and identified Cyclophilin A (CypA) as a cellular factor incorporated into HIV-1 virions. Members of the cyclophilin family are linked to multiple viral infections. CypA acts as a proviral factor in such viruses as hepatitis C virus, coronavirus, and HIV, leading to the hypothesis that inhibitors targeting this protein may inhibit the life cycle of these viruses. CypB was found to enhance HIV-1 infection.[14] DeBoer *et al.* used the mass spectrometry (MS) SWATH-based proteomic approach to investigate changes in T-cell nuclei during HIV-1 infection.[15] They revealed a total of 13 and 38 unique proteins in infected and uninfected cells, respectively. An additional 15 proteins were found to be differentially regulated between infected and controlled nuclei. Some of these proteins, that is, actin were previously reported. Cyclophilins are one example that illustrates the complexity of viral infection understanding, which is crucial for the design of proteomic-based studies.[16]

As much as the development of MS-based proteomics has led to new discoveries, the role and/or functions of newly discovered differentially expressed proteins still require more functional research, employment of various experimental approaches, and orthogonal validation.

5. Viral proteins and antiviral antibodies in body fluids

Most viral diseases are associated with a single virus that produces prominent pathological effects. However, epidemiological studies showed that in viral pneumonia coinfections have been reported for several respiratory viruses, notably with influenza[17] when detected using multiplex respiratory virus assays. Another example is HIV/hepatitis C virus (HCV) coinfection, common in the illicit drug–using populations due to needle sharing. Coinfection of HIV-1 with hepatitis B virus (HBV) or HCV is linked to increased mortality.[18] Most epidemiological studies are based on quantitative polymerase chain reaction (qPCR) or enzyme-linked immunosorbent assay.[19–21] The most common are the antibody-screening tests that detect the presence of antiviral antibodies or antibodies against viral proteins. Such immunoassays are generally accurate in providing a yes/no answer for initial screening.[21]

Van Rijswijck and colleagues[22] reported an experimental approach to screen qualitatively and quantitatively clonal immunoglobulin G1 (IgG1) repertoires in plasma. They purified a pool of IgG, cleaved them into Fc and Fab fragments, and separated them based on molecular weight to profile them by LC–MS. This approach assumed that each clone has a distinct feature not only based on mass but also based on retention time in LC.[23] Quantification was achieved by spiking in recombinant IgG1 monoclonal antibodies as an internal standard. Harvey *et al.*[24] in their review summarized accomplishments on mutations of the SARS-CoV-2 (severe acute respiratory syndrome coronavirus 2) spike protein, the primary antigen, focusing on antigenicity. Such an approach can be used in vaccine development.

While infection detection assays are rapid, sensitive, and specific, they do not provide much information about molecular mechanisms underlying virus–host interactions. Here, proteomic along with genomic approaches emerge for discovery and more in-depth investigations.

6. Interactions of viral proteins with host's proteins, structures — proteomic studies

As we stated before, viruses cannot reproduce outside the host cells and must usurp the host cell's metabolic machinery. To accomplish this task, viral proteins, usually in small numbers, interact with intracellular proteins redirecting the metabolic effort to the virus's advantage. This aspect of viral infections has been investigated not only to better understand the viral life cycle but also to stop and/or cure the infection. Proteomics played and still is playing a profound role in these studies. Here, we point out three examples which of course do not exhaust all knowledge in this area.

6.1. *Epstein–Barr virus*

Epstein–Barr Virus (EBV) is an enveloped virus. Cheerathodi and Meckes, discussing the oncogenic potential of EBV, made a case that the latent membrane protein 1 (LMP1) is essential and enough for the transformation of fibroblasts and primary lymphocytes.[25] Proteomic studies showed that among many interacting proteins LMP1 interacts with galectin-9[26] and the germinal center kinase TNIK.[26] TNIK activates canonical NF-κB and JNK triggering numerous downstream signaling pathways, which will undoubtedly affect the overall function of viral host cells.

Galectins, intracellular and extracellular, play multiple roles in many pathological processes[27] including immune escape. A comprehensive proteomic study of galectins-1, -3, and -9 showed statistically significant differences in the expression of these galectins in human monocyte–derived macrophages (hMDMs) infected with HIV and exposed to methamphetamine (Meth).[27] Although these differences were significant, the differences might be easily missed as measured using other methods such as western blot analysis. Although this study did not explain the exact effect of these galectins on the immune system, the authors showed concurrently the lack of differences at the qPCR level.

6.2. *Human immunodeficiency virus type 1*

HIV-1 infection has been the subject of intense research for several decades. Despite the unquestionable success of developing antiretroviral therapies, how this virus operates inside hosts' cells, that is, in the latency phase, is still unknown. Schweitzer *et al.* investigated the interactions of viral and cellular proteins after HIV-1 entry into a host's cell. The goal of this study was to discover the formation of preintegration complex that moves viral-reversed transcribed DNA to the nucleus.[28] This study reported that out of 1155 identified proteins, 94 were unique to infected cells while 121 proteins were unique to control cells. The authors identified several previously reported factors, such as XRCC6, TFRC, and HSP70. The presence of XRCC6 was validated in infected cells and shown to be associated with HIV-1 DNA by immunoprecipitation–PCR experiments. Overall, the presented study showed additional novel candidate proteins contributing to HIV-1 replication.

A comprehensive proteomic study published by Jager *et al.*[29] investigated a physical interaction between HIV-1 and cellular proteins in two cellular systems. These systems were based on HEK293 and Jurkat cells. The authors used 18 HIV-1 proteins and affinity tagging to pull down interacting proteins using LC–MS/MS. Further analysis revealed 497 cellular and HIV-1 proteins interactions and 435 human proteins. Interestingly, only 40% of these proteins were identified in both cellular models. These results are aligned with our previous statement about the appropriate experimental model employed as well as the variability of responses. It is possible that HIV-1 has multiple ways of affecting cellular metabolism that will be used depending on the local environment as well as the response of targeted cells.

More in-depth studies, while limited, targeted localization of cyclophilins that for quite some time were reported to be involved in support of the HIV-1 life cycle was performed by several groups.[30, 31] CypB, a member of the immunophilin family and intracellular chaperone, predominantly localizes to the endoplasmic reticulum. It also contains a nuclear localization signal as well as is secreted from cells.[32] De Boer *et al.* reported that passive transfer experiments showed that although secreted, CypB did not impact HIV-1 infection; however, it modulates the pathway of HIV-1 nuclear import.[32]

Summarizing, viral proteins, although some viruses have a few of them coded in their genome, work to the virus' advantage in multiple ways by hijacking the host's metabolism. At the same time, we need to note that viral entry is the first early step and is based on specific surface ligand–viral surface protein interaction.

6.3. *Oncoviruses*

Oncoviruses are a diverse group of viruses that can initiate cancer. These viruses are EBV, human T-lymphotropic virus 1(HTLV-1), HBV, human papillomavirus (HPV), HCV, Kaposi's sarcoma-associated herpesvirus (KSHV or HHV-6), and Merkel cell polyomavirus (MCV or MCPV). EBV and KSHV belong to the herpes viruses group while others represent other groups: papillomaviruses (HPV), retroviruses (HTLV-1), flaviviruses (HCV), hepadnaviruses (HPV), and polyomaviruses (MCV). The relationship between viruses and cancer remains under debate, asking for more research from various disciplines, such as genomics and proteomics. This creates an opportunity for those who aim to apply proteomics as well as other high-throughput studies for a better understanding of onco-virology and its impact on human health.[33] Therefore, the rapid development of MS and associated analytical methodology applicable to sample preparation significantly advanced the viral proteomic field. Complexes of viral and cellular proteins can be isolated by affinity purification (AP) and then followed by MS/MS identifications. Quantitative MS will also support the measurement of relative and absolute changes as well as measurements of the stoichiometry of components participating in such complexes.

EBV causes Burkitt, Hodgkin, and post-transplant B-cell lymphomas by remodeling metabolic pathways. The mechanism remains unknown. Wang and co-workers applied proteomics and identified more than 8000 host proteins; the authors quantified 29 viral proteins. Results indicated the contribution of one-carbon metabolism (1C) revealing mitochondrial remodeling and the contribution to nucleotide synthesis, NADPH (nicotinamide adenine dinucleotide phosphate) production, and redox defense.[34]

The use of fluorescent tags was used to enable the location and their interaction(s) of the viral proteins in living cells and correlated with protein interactions at various times postinfection.[35] A further study by White *et al.* identified and compared the interaction partners of a series of E6 proteins from different HPVs using AP–tandem MS.[36] Proteomic investigation of EBV EBNA1 (EBV nuclear antigen 1) using AP–LC–MS/MS methodology revealed interactions in EBV-associated cancers in latent and lytic infections.[37]

7. How proteomics can help to fight COVID-19?

After the worldwide eruption of SARS-CoV-2 [COVID-19 (coronavirus disease 2019)] in 2019 and 2020 as an acute pandemic, now it is accepted that this infection will not be eradicated. New variants now emerge worldwide. With people traveling in high numbers, variants are easily transferred from country to country and continent to continent. Two tests are available, one based on qPCR and one based on protein antigens (nasal swabs). The first effective treatment to prevent SARS-CoV-2 infection was based on vaccination. Later on, the treatment for ongoing infection became available [nirmatrelvir with ritonavir (Paxlovid), remdesivir (Veklury), and molnupiravir (Lagevrio)]. This virus and coping with the resulting disease will stay with us forever; thus, there is quite a broad area for proteomics to contribute to expanding our knowledge.

8. Metabolomics and viral infections

Metabolomic investigations are outside the scope of this chapter; nevertheless, metabolites are products of enzymes inside cells, and expression and/or localization of such enzymes can be altered by invading virus(es). For the most part, to find early fingerprints associated with infection, investigation of metabolites is focused on biofluids.[38] As reviewed by Noto *et al.*, most observed changes were related to the metabolism of nucleotides, carbohydrates, lipids, and amino acids. Thus, metabolomics could be considered as an option to discover new biomarkers; however, the origin or the presence of the primary or secondary effect of altered metabolites will remain unknown. This leads us to the conclusion that the comprehensive application of multiomics approaches including proteomics, metabolomics, genomics, transcriptomics, and cell omics is necessary for the treatment of viral infections. This, of course, opens another challenge of integration and adequate validation of results obtained from such diverse platforms. We face this challenge in the fight against SARS-CoV-2 infection.[39]

9. Examples of successful proteomic studies of viral infections

The number of successful applications of proteomics to understand viral infections and the implications of associated comorbidities is growing. Here, we describe a few examples showing proteomic studies supported by orthogonal validation.

One study, including quantitative proteomics, RNA sequencing (RNAseq), and immunohistochemistry is focused on intracellular and extracellular galectins in HIV-1-infected macrophages with concurrent exposure to Meth. Interestingly, the authors did not observe a significant effect of Meth at the RNAseq level while SWATH-MS analyses showed differences in fold changes of galectin-1, -3, and -9, between experimental conditions.[6] As expected, quantitative LC–MS/MS (multiple reaction monitoring and SWATH–MS) measurements showed significant differences in some conditions with some differences close to the borderline. Validation by immune-cytostaining showed a quite even distribution of galectins throughout the cytoplasm with some overlap with lysosomes stained with LysoTracker. These data suggest that differences in expression might be quantitatively small, yet significant and that some methods are unable to show such small changes as significant. Undoubtfully, further fractionation of cytosolic fraction is required in additional studies. Further bioinformatic investigations show proinflammatory phenotype of hMDM and suggest a possibility of a much broader effect particularly in the local milieu of macrophages residing in tissues such as the brain. Bioinformatic investigations of metabolic pathways combined with various experimental designs might be critical in deciphering the exact role of galectins, an important regulator of the innate immune system, as potential biomarkers of disease progression and efficacy of therapies.

Another example was the search for biomarkers recognized in latently infected cells. Such a discovery is critical in the path leading to the cure of HIV-1 infection. Belshan and colleagues created a panel of Jurkat T-cell lines latently infected with the HIV-1 virus containing fluorescent markers to identify cells infected with actively replicating virus versus latent virus. They used the SWATH–MS approach to compare the membrane proteomes of one of the cell lines to parental Jurkat cells. This study reported several candidates significantly altered expression. Western blot validation narrowed data to three potential biomarkers: LAG-3, CD147, and CD231. The authors also confirmed that phenotypic differences in latently infected cells exist, which creates an additional challenge in data interpretation.[40]

Progress in understanding viral infections from accumulated genomic and proteomic studies leads to the development of gene delivery systems as part of gene therapy. Such systems are based on retro-, adeno-, adeno-associated, herpes, and other viruses and lead replacement of gene products for therapeutic purposes. Many systems such as virosomes containing reconstituted influenza, Sendai, VSV, or EBV fusion proteins have proven to be effective delivery vehicles for DNA, drugs, peptides, and toxins in tissue cultures. Additional genomic and proteomic research is needed that will lead to testing in animal models and eventually as therapeutics in humans.

References

1. Lehmberg E, Traina JA, Chakel JA, *et al.* Reversed-Phase High-Performance Liquid Chromatographic Assay for the Adenovirus Type 5 Proteome. *J. Chromatogr. B Biomed. Sci. Appl.* 1999;732:411–423.
2. Maxwell KL, Frappier L. Viral Proteomics. *Microbiol. Mol. Biol. Rev.* 2007;71:398–411.
3. Viswanathan K, Fruh K. Viral Proteomics: Global Evaluation of Viruses and Their Interaction with the Host. *Expert Rev Proteomics.* 2007;4:815–829.
4. Munday DC, Surtees R, Emmott E, *et al.* Using SILAC and Quantitative Proteomics to Investigate the Interactions Between Viral and Host Proteomes. *Proteomics.* 2012;12:666–672.
5. Grabowska K, Harwood E, Ciborowski P. HIV and Proteomics: What We Have Learned from High Throughput Studies. *Proteomics Clin. Appl.* 2021;15:e2000040.
6. Grabowska K, Macur K, Zieschang S, *et al.* HIV-1 and Methamphetamine Alter Galectins -1, -3, and -9 in Human Monocyte-Derived Macrophages. *J. Neurovirol.* 2022;28:99–112.
7. Macur K, Zieschang S, Lei S, *et al.* SWATH-MS and MRM: Quantification of Ras-Related Proteins in HIV-1 Infected and Methamphetamine-Exposed Human Monocyte-Derived Macrophages (hMDM). *Proteomics.* 2021;21:e2100005.
8. Kattenhorn LM, Mills R, Wagner M, *et al.* Identification of Proteins Associated with Murine Cytomegalovirus Virions. *J. Virol.* 2004;78: 11187–11197.
9. Schowalter RM, Buck CB. The Merkel Cell Polyomavirus Minor Capsid Protein. *PLoS Pathog.* 2013;9:e1003558.
10. Wang JW, Roden RB. L2, the Minor Capsid Protein of Papillomavirus. *Virology.* 2013;445:175–186.
11. Schaffhausen BS, Benjamin TL. Deficiency in Histone Acetylation in Nontransforming Host Range Mutants of Polyoma Virus. *Proc. Natl. Acad. Sci. USA.* 1976;73:1092–1096.

12. Fang CY, Shen CH, Wang M, *et al.* Global Profiling of Histone Modifications in the Polyomavirus BK Virion Minichromosome. *Virology.* 2015;483:1–12.
13. Goodsell DS. Illustrations of the HIV Life Cycle. *Curr. Top Microbiol. Immunol.* 2015;389:243–252.
14. DeBoer J, Madson CJ, Belshan M. Cyclophilin B Enhances HIV-1 Infection. *Virology.* 2016;489:282–291.
15. DeBoer J, Jagadish T, Haverland NA, *et al.* Alterations in the Nuclear Proteome of HIV-1 Infected T-Cells. *Virology.* 2014;468–470:409–420.
16. Zav'yalov VP, Denesyuk AI, Lundell J, *et al.* Some New Aspects of Molecular Mechanisms of Cyclosporin A Effect on Immune Response. *APMIS.* 1995;103:401–415.
17. Crotty MP, Meyers S, Hampton N, *et al.* Epidemiology, Co-Infections, and Outcomes of Viral Pneumonia in Adults: An Observational Cohort Study. *Medicine (Baltimore).* 2015;94:e2332.
18. Parikh N, Nonnemacher MR, Pirrone V, *et al.* Substance Abuse, HIV-1 and Hepatitis. *Curr. HIV Res.* 2012;10:557–571.
19. Yamayoshi S, Sakai-Tagawa Y, Koga M, *et al.* Comparison of Rapid Antigen Tests for COVID-19. *Viruses.* 2020;12:1420.
20. Ong DSY, Fragkou PC, Schweitzer VA, *et al.* How to Interpret and Use COVID-19 Serology and Immunology Tests. *Clin. Microbiol. Infect.* 2021;27:981–986.
21. Mohit E, Rostami Z, Vahidi H. A Comparative Review of Immunoassays for COVID-19 Detection. *Expert Rev. Clin. Immunol.* 2021;17: 573–599.
22. van Rijswijck DMH, Bondt A, Hoek M, *et al.* Discriminating Cross-Reactivity in Polyclonal IgG1 Responses Against SARS-CoV-2 Variants of Concern. *Nat. Commun.* 2022;13:6103.
23. Bondt A, Hoek M, Tamara S, *et al.* Human Plasma IgG1 Repertoires Are Simple, Unique, and Dynamic. *Cell Syst.* 2021;12:1131–1143 e5.
24. Harvey WT, Carabelli AM, Jackson B, *et al.* SARS-CoV-2 Variants, Spike Mutations and Immune Escape. *Nat. Rev. Microbiol.* 2021;19: 409–424.
25. Cheerathodi MR, Meckes DG, Jr. The Epstein-Barr Virus LMP1 Interactome: Biological Implications and Therapeutic Targets. *Future Virol.* 2018;13:863–887.

26. Pioche-Durieu C, Keryer C, Souquere S, *et al.* In Nasopharyngeal Carcinoma Cells, Epstein-Barr Virus LMP1 Interacts with Galectin 9 in Membrane Raft Elements Resistant to Simvastatin. *J. Virol.* 2005;79: 13326–13337.
27. Chen L, Chen K, Hong Y, *et al.* The Landscape of Isoform Switches in Sepsis: A Multicenter Cohort Study. *Sci. Rep.* 2022;12:10276.
28. Schweitzer CJ, Jagadish T, Haverland N, *et al.* Proteomic Analysis of Early HIV-1 Nucleoprotein Complexes. *J. Proteome Res.* 2013;12: 559–572.
29. Jager S, Cimermancic P, Gulbahce N, *et al.* Global Landscape of HIV-Human Protein Complexes. *Nature.* 2011;481:365–370.
30. Han J, Lee MK, Jang Y, *et al.* Repurposing of Cyclophilin A Inhibitors as Broad-Spectrum Antiviral Agents. *Drug Discov. Today.* 2022;27: 1895–1912.
31. Zhong Z, Ning J, Boggs EA, *et al.* Cytoplasmic CPSF6 Regulates HIV-1 Capsid Trafficking and Infection in a Cyclophilin A-Dependent Manner. *mBio.* 2021;12:e03142.
32. Borozan PG, Walker HS, 3rd, Peterson GJ. True Tibial Artery Aneurysms: Case Report and Literature Review. *J. Vasc. Surg.* 1989;10: 457–459.
33. Cao J, Li D. Searching for Human Oncoviruses: Histories, Challenges, and Opportunities. *J. Cell Biochem.* 2018;119:4897–4906.
34. Wang LW, Shen H, Nobre L, *et al.* Epstein-Barr-Virus-Induced One-Carbon Metabolism Drives B Cell Transformation. *Cell Metab.* 2019;30:539–555 e11.
35. McBride AA. The Promise of Proteomics in the Study of Oncogenic Viruses. *Mol. Cell. Proteomics.* 2017;16:S65–S74.
36. White EA, Kramer RE, Tan MJ, *et al.* Comprehensive Analysis of Host Cellular Interactions with Human Papillomavirus E6 Proteins Identifies New E6 Binding Partners and Reflects Viral Diversity. *J. Virol.* 2012; 86:13174–13186.
37. Malik-Soni N, Frappier L. Proteomic Profiling of EBNA1-Host Protein Interactions in Latent and Lytic Epstein-Barr Virus Infections. *J. Virol.* 2012;86:6999–7002.
38. Noto A, Dessi A, Puddu M, *et al.* Metabolomics Technology and Their Application to the Study of the Viral Infection. *J. Matern. Fetal Neonatal Med.* 2014;27(Suppl. 2):53–57.

39. Ma J, Deng Y, Zhang M, *et al.* The Role of Multi-Omics in the Diagnosis of COVID-19 and the Prediction of New Therapeutic Targets. *Virulence.* 2022;13:1101–1110.
40. Belshan M, Holbrook A, George JW, *et al.* Discovery of Candidate HIV-1 Latency Biomarkers Using an OMICs Approach. *Virology.* 2021;558:86–95.

https://doi.org/10.1142/9789811283376_bmatter

Index

www.ingramcontent.com/pod-product-compliance
Lightning Source LLC
LaVergne TN
LVHW010327220926
848484LV00004BB/854